Crude Britannia

How Oil Shaped a Nation

James Marriott and Terry Macalister

PLUTO PRESS

First published 2021 by Pluto Press
345 Archway Road, London N6 5AA

www.plutobooks.com

British Library Cataloguing in Publication Data
A catalogue record for this book is available from the British Library

ISBN   978 0 7453 4109 5   Hardback
ISBN   978 1 7868 0638 3   PDF
ISBN   978 1 7868 0639 0   EPUB
ISBN   978 1 7868 0640 6   Kindle

This book is printed on paper suitable for recycling and made from fully managed and sustained forest sources. Logging, pulping and manufacturing processes are expected to conform to the environmental standards of the country of origin.

Typeset by Stanford DTP Services, Northampton, England

Simultaneously printed in the United Kingdom and United States of America

For: Doreen Massey

The next generation

'Is it not an irony that those who live on top of wealth should be the poorest people in the nation?'
Sam Badilo Bako, school teacher in Taabaa, Ogoni, Nigeria

'Remember . . . it always happens first on records.'
Bored Stiff fanzine

'The future is already here, it is just that it is not yet evenly distributed.'
William Gibson, novelist

Contents

Maps and Table

Tracks

To hear these tracks please visit
www.plutobooks.com/crude-britannia-playlist/

Prologue
The Last Living Rose

Goddamn Europeans!
Take me back to beautiful England
And the grey, damp filthiness of ages,
And battered books and
Fog rolling down behind the mountains,
On the graveyards, and dead sea-captains.
Let me walk through the stinking alleys
To the music of drunken beatings,
Past the Thames River, glistening like gold
Hastily sold for nothing.

—PJ Harvey, 'The Last Living Rose', 2011

First we visit the ruins. We follow the Manorway down through Corringham towards the river's edge. This road was once busy with petrol tankers. Those that were laden ground up the hill, those that were empty passed them as they rushed towards the refinery to refill. It was a constant back and forth. Bees coming and going from the hive, day and night, ceaselessly. Now the road is almost empty of traffic.

As we drive from the ridge the estuary is laid out before us; fields, scraps of marshland and beyond, the wide brown Thames merging with the sea. The tide is far out and there are few vessels in the shipping channel. The Kent hills to the south are sleepy grey forms.

Between the 1940s and the 2000s the mouth of the Thames, 50 miles east of London, was Britain's major zone for oil refineries and gas importation docks – similar to the Elbe near Hamburg or the Rhine at Rotterdam. In the late 1960s around 15,000 men and women were employed running three refineries, working on building three more and operating a couple of gas plants. Storage tanks and chimneys dominated sections of the horizon. The flares and lights of the plants punctured the night sky. Oil companies built towns and villages, housing the families who moved in from across the country.

The refineries dominated the life of the region. Workers at Coryton joined the Pegasus Club, its clubhouse an archetypal 1960s building up on the ridge above. The club had 'sections' each devoted to a different activity: there were sections for swimming, Sunday football, bowls, tennis, sailing, golf, cricket, hockey, badminton, drama, film making, camping, karting, rifle shooting, horticulture, angling and carnivals.

The shipping channel was busy with tugs and pilot boats nipping between tankers carrying crude from the Middle East and smaller bunker barges delivering petrol or diesel to depots up river or up the coast. The sky was hung with plumes of smoke from chimneys and power stations. Now, after two generations, this industry has almost gone. The last refinery closed in 2012.

We park some distance from the gates of the old Coryton Refinery, hoping that there will be a delay before the inevitable visit from the site security guards. To our left is a wide plain of pale brown concrete. Acres of foundations for storage tanks and cracking towers, pipework and buildings. In the middle distance there's a set of hills formed of rubble and dust, on its ridgeback a JCB digger lumbers along. Soaring above it all is a solitary chimney, a sandy-coloured tube rising 300 feet into the air topped by a ring of black paint and a tiny handrail. The view of the estuary from up there must be extraordinary. Near its base the chimney has gaping holes where the steel flues that fed it with fumes have been ripped away. This machine has been dismembered.

Apart from the distant rumble of lorries and the 'beep beep' of reversing trucks involved in demolition work, it is surprisingly quiet. A robin sings powerfully from the bare branches of a cherry tree that must have been planted to bring a touch of colour to the avenue that led into this great industrial facility. It seems as though the nature of the marshes is reasserting itself after a century-long interlude.

Above a gatehouse at the entrance to the site flaps a shredded Union Jack. There's clearly someone in the building but they don't bother to come out, so we walk across the parking lot towards an office block with 'Thames Enterprise Park' emblazoned on its side. It is relatively new but there's a rusted heavy chain and padlock around the door handles. We peer into a hallway littered with the detritus of clothing and indecipherable corporate leaflets.

Eventually a man approaches us in a hi-viz waistcoat emblazoned with a company logo. He's thickset and in his fifties with black stubble on his chin and black work boots. Contrary to expectations he's not really bothered by our presence. We stand beside him for a while, gazing over the car park and the men and machines who are painstakingly destroying one of the old administrative buildings.

'I've been here 15 years. A security guard for BP, for Petroplus, for Vopak, for Thames Oil, for Thames Enterprise, and now back to Thames Oil', he says. The refinery went through many changes of ownership in its 60-year life. It was built by Socony-Vacuum which evolved into Mobil, who later sold the plant to BP. We survey the almost empty car park.

'This place used to be full with people coming on and off shift. There were 800 staff here when Petroplus closed it in 2012, and loads of contractors', he tells us. Among the vehicles on the tarmac are a number of minibuses. 'Bringing guys who are working for Brown & Mason, the demolition company. Mainly they are blokes from the North of England … They left the big chimney as a historic monument and blew up the other three.' There's no sense of bitterness in his voice, just resignation. The scale of the deconstruction of this place is dizzying. The plant closed over five years ago. Since then hundreds have laboured to level the site and still the work is not completed. It will have taken longer to pull this machine apart than the two years it took to construct it.

* * *

We had known each other for several years before making this visit. Our lives were intertwined by a common interest – Terry, a journalist who had reported on the oil and gas industry for 35 years, latterly as an energy editor; and James, a writer and activist who'd spent the past three decades studying the sector. And we had a shared love of music.

We are children of the oil revolution. We grew up in a country saturated in petrol, plastics and pesticides. The first passenger jet flew from London to New York when Terry was four. The first motorway in Britain opened three years before James' birth. Both of us travelled in cars before we were born. We suckled on synthetic rubber teats and dummies that were the output of petrochemical plants. We played with toy trucks and our teenage daydreams were filled with sports cars. Later music gripped us, carrying with it the promise of sex and the open road. Plastic toothpaste tubes, holidays abroad, nylon clothes, food packaging, detergents, collecting oil company stickers and garage giveaways, we grew up in the abundance of cheap oil that fuelled the optimism of an era.

By the 2010s, in later middle age, we found ourselves in a country dazed by constant shocks – the financial crash and the growing impacts of climate change, the surprises of the Scottish independence vote and the Brexit referendum, and then the virus that brought the oil world to a sudden halt. We had spent our working lives writing on the oil and gas industry and its impacts around the world, but what was its role now in Britain's turbulence?

In the EU referendum on 23 June 2016, the Castle Point district around the former refinery of Coryton voted 72.7 per cent to leave and 27.3 per cent to remain. It was the third strongest leave area in the UK. Why did these men and women vote so strongly to exit the European Union? We noticed that the leave cause had been similarly powerful in other areas where the oil industry had closed plants and laid off workers, for example in Port Talbot, south Wales. Was there any correlation between these things? Was there a link between oil and 'left behind Britain', deindustrialised Britain?

The era of optimism that characterised our childhoods came in the aftermath of World War II, a conflict fuelled by oil and in many respects fought for oil. After the war, Britain's late empire was an oil empire. How does the memory of these events, often shaped by the oil companies, affect our country in our present times?

Both of us are deeply concerned by the human-driven changing of the Earth's climate. It will dominate the rest of our lives, and those of friends and family. The greatest source of carbon dioxide in the atmosphere is the extraction and consumption of fossil fuels. What is the role of the oil companies, whose profitability is built on this process, in shaping the UK's response to climate chaos?

All around us the same oil companies eagerly promote their desire to decarbonise themselves and the UK. Are they doing this in response to civil society pressure about the climate? Or do they see themselves under a greater threat from the rise of the Big Data corporations? A threat so powerfully illustrated by the Covid pandemic that shifted the country from commuting to computers, from cars to Zoom.

With these questions in mind we have travelled Britain for the last five years trying to understand how oil has shaped war, cities, finance, transport, industries, communities, regions, international relations, consumption patterns, popular culture and more. In short, how oil has shaped a nation.

As we've journeyed, three observations in particular have guided our understanding.

Helen Thompson, an academic at Cambridge University, explained to us: 'I think it's too difficult for people to think about oil. It's almost like it's a parallel world, because it's so big. It permeates everything. We can't think about our day-to-day life without oil, even though most people don't consciously understand that, with the implications of actually realising that we are all complicit in lots of those bad things that result from it.'

Paul Stevens was Professor of Petroleum Policy and Economics at the University of Dundee, a chair created by BP. Early in our explorations he observed to us: 'They steered away from publicity. They preferred to be covert rather than overt because oil companies had never been popular. And they didn't really have to lobby too hard because [of their big tax revenues] they were pushing an open door with government.' Stevens helped us to understand the oil sector as a 'submarine industry.'

A senior oil executive told us privately: 'People think the big oil majors have loads of influence but in fact it is really quite small. The big corporations are takers of policy, not makers of policy. Perhaps all they can do is play games at a small scale.'

The UK that we attempt to describe in this book is a land formed by oil. London, shaped by petroleum. Four estuary regions – the Thames, Severn, Mersey and the Forth and Dee – moulded by the industry for almost a century. And most distinct, the North Sea, turned into an offshore colony run largely by a handful of corporations. In all of these places we have talked to people from all walks of life, trying to understand through the example of the local how oil has, and still does, shaped the economics, politics and society of our country.

In this mapping of a nation formed by oil we have also understood just how the nation has begun to shape oil. How the people of our country are changing the oil industry and forming a new era with accelerating speed.

* * *

After a short drive back up to the ridge we're at the Pegasus Club on the outskirts of the village of Corringham. This was built by Mobil and was – and is at the time of our first visit – a social centre for those that worked in Coryton. On the terrace in the hot sunshine we meet Tony Wade, the man who'd arranged this gathering. He's in conversation with another grey-haired man: 'Did you hear about Elmey? He's passed away. Asbestosis.'[1]

Together we look out over the club's football pitches, the Manorway, the marshland and the distant river. 'Yes, it is a nice view. But it was better when it included the sight of a couple of refineries.' Wade's a rumbustious soul in his seventies, full of welcoming bluster. 'They left the one chimney, from the Fluid Catalytic Cracker. They say it's for historic reasons, but it's obvious that it's so that they can get planning permission for three-story warehouses if they want. Have you seen the container ports in China? That's what Dubai Ports want to build here on the refinery site.'

In the distance, at the edge of the brown smudge of the Thames, we can see a row of twelve massive gantry cranes, positioned to lift brightly coloured containers on and off the ocean-going vessels that slide in from the sea. Next door to Coryton there had been a second, larger refinery – Shell Haven – owned by Shell and constructed in the late 1940s, to process Kuwaiti crude. It closed in 1999, over a decade before Coryton. The site was cleared of storage tanks and cracking towers and sold as a shipping and distribution hub to Dubai Ports, owned by the Dubai Sovereign Wealth Fund. There's a poetic twist in the fact that Shell and the Dubai ruling families got rich on the exploitation of oil in the Gulf, and now Dubai is buying up the company's cast-offs. Although not until the UK taxpayer had funded the decontamination of the soil after decades of continuous refining.

'Ah gentleman, you are honoured. Here's Mr John Vetori, one of the former managers at Coryton come to join us. This is a rare treat', says Wade. Out into the sunshine strides a tanned man in his sixties, neatly dressed in golfing clothes. He exudes an assertiveness quite distinct from the others gathered about us. 'I don't want to be outside because I need some shade', announces Vetori, so we all retire to the seating in the bar. Others join us and pretty soon there are 14 men gathered around quietly waiting for the conversation to begin. The day seems auspicious. On this date Prime Minister Theresa May is triggering Article 50, the UK's formal request to leave the European Union.

Stefan Rogocki worked in the Tank Farm at Coryton, among the vast off-white steel cylinders filled with petrol and diesel, asphalt and jet fuel. His face is remarkably pale. He stopped work at the plant prior to its closure because of an industrial illness. His son Mike started at the refinery and had only recently finished his apprenticeship when Petroplus went bust and he was laid off. Mike found a job doing security at the Fawley Refinery, but returned to the area when he got work at the Dubai Ports docks driving one of the cranes. 'He's a "sky dog"', says his father. 'He earns £30,000 a year. Half he would have got if the refinery had stayed open.'[2]

It seems that all the men share the same sense of shock that Petroplus closed the place. 'It was making a profit.' 'They only closed it so as to support their other plants in Germany, Belgium and France.' However, Petroplus itself did not long outlast Coryton. The company folded shortly after the refinery shut down owing $1.5 billion to creditors. The 800 staff at Coryton who were laid off got no redundancy packages. 'I just got a pitiful state redundancy', says Dave Musson. These former employees believe that the people who owned the now defunct company were asset strippers. However, their strongest criticism is not aimed at the head of Petroplus, an American called Thomas O'Malley. Rather, several blame the EU. 'It's supposed to provide a level playing field, but the UK played by the rules whereas the other governments did not. The Germans, French and Belgians defended their plants and we were shut down.' No criticism is made of the British government. It seems that the Prime Minister David Cameron and Chancellor George Osborne, despite being vocal critics of the EU, failed to resist the manoeuvring of a Swiss-based corporation, Petroplus.

With hindsight we can see that the fate of Coryton was probably sealed a decade before its closure. BP had acquired the refinery in 1996 as part of an asset swap with Mobil. However, demand for petrol, diesel and jet fuel was falling in the UK, while at the same time refineries were being constructed in India and China, designed to feed the booming Asian markets. Under BP's chief executive, John Browne, the corporation strove to drive down costs at Coryton in a bid to make the refinery competitive. Two years later, in June 1998, workers at the plant were stunned by the news that Shell was to close its neighbouring refinery and lay off its 290 direct employees. Nearly a decade later, in February 2007, John Browne and the then head of BP Refining, John Manzoni, oversaw the sale of Coryton to Petroplus.

Away from the main group gathered in the Pegasus Club bar, we talk with Eddie Horrigan. He lives just to the north in Basildon, which was largely built in the 1950s, partly to house the families of men employed in the refineries.[3] The oil company helped finance the construction of this new town. 'In the small area around where I lived in Basildon there were 40 to 50 people who worked in Mobil.'

Horrigan explains how in about 2006, when he was nearing retirement, he went with a couple of apprentices from Coryton to visit the school that he'd attended in Basildon. It was careers day and he was there to tell the sixth form about working in the refinery. 'Not one, not one, knew what a refinery was! I was shocked! I told the teacher so.'

In a matter of years these massive plants, whose lights and flare stacks lit up the sky day and night, have been erased from the landscape and the lives of the people in the Thames Estuary. The refineries exist now more in the minds and hearts of the former employees, and in the tissue of their lungs, than in concrete reality out on the marshes.

* * *

On the drive away from Coryton we stop off at a petrol station near Basildon and, after filling up, ask the woman serving at the till about when the lorries come to resupply the underground storage tanks. She's friendly, but all she knows is that it is a BP tanker that comes in the very early mornings. She's wearing a shirt with a BP logo on it, so we ask if she's employed by the company. She replies, 'Oh no, by MFG'.

Back in the car, aided by a chart we've compiled, we put together a trail of how the unleaded petrol got to our engine.

The garage with its BP sign we refilled at is not owned by BP, but by MFG,[4] the largest forecourt operator in the UK, itself owned by a US private equity company. It was supplied by a road tanker, which, although it was marked with a BP logo, was owned by XPO Logistics,[5] a US transport multinational. The tanker driver does a circuit of tens of petrol stations, delivering mostly at night, from Thames Oilport at the old Coryton site. This storage depot is owned by Greenergy, a private equity company from London.[6] The Thames Oil depot also supplies aviation fuel to Stansted, Gatwick, Heathrow and RAF bases such as Honington in Suffolk, via the UK's Oil Pipeline Authority. For 60 years these pipelines were owned by the British state, as a matter of national security; now they are owned by CLH[7] of Spain, whose largest shareholder is CVC, a private equity company registered in Luxembourg. The fuel that is delivered from Thames Oil is supplied to the storage depot by ship. Most likely a tanker working for an oil-trading corporation such as Vitol, a private company based in Switzerland. The tanker would have arrived from anywhere across the globe, but it could have shipped fuel processed to meet EU standards in the refinery at Haldia, in West Bengal, India. Where did the crude that fed that refinery come from? We know that BP has a contract to supply Haldia with oil, but from where it does so is commercially confidential. At this point the trail has so many variables that we can no longer trace it through the thickets of the industrial world.

On the sound system PJ Harvey sings:

Past the Thames River, glistening like gold
Hastily sold for nothing.

We talk it over. So what does all this mean? It means that a decade ago, when these roads were thick with traffic, BP refined fuel at Coryton, their own road tankers delivered to an array of petrol stations and we might just have pulled into one that belonged to the company. But now, although the traffic is just as heavy, this entire system, from a refinery in India to the garage we stopped at, is owned and controlled by a mesmerising array of private equity companies. These firms feast on the ruins left by the closure of BP's and Shell's refineries. They have moved in on Britain as the big corporations have slowly moved out.

But why is this important, if we can still fill up the car? Because these companies are almost all registered abroad, many in tax havens such as Luxembourg or Monaco. And the men who run them are often millionaire tax exiles.

For all the talk of the end of fossil fuels, the boom in electric vehicles and the rise of renewable energy, the role of oil in the UK economy remains vitally important. As things return to 'normal' after the pandemic there are still as many trucks on the road and almost as many planes in the sky, there is still as much plastic in the kitchen and pesticides on the fields. Yet the control of this bloodstream of our society is now in the hands of a very few wealthy men entirely hidden from public view, beyond the reach of politicians, trades unions or civil society groups.

Those that own the private equity companies want from Britain the highest return on their capital and the lowest taxes possible, with the least stringent regulations. The last thing they desire is for limits to be put on the sale of fuel as a way to tackling climate change. Whereas the heads of BP, Shell and other corporations may profess a commitment to decarbonise to appease an alarmed public, the private equity men are hidden from view and pressure.

These owners may know next to nothing about the communities such as those in the Thames Estuary that live next to assets like Thames Oil through which the bloodstream flows. They are certainly not in the business of building homes for their staff like the oil companies once did at Coryton. And those that work in a place like Thames Oil may have little job security.

It is the 'submarine industry' still. Only now the submarine has dived further into the deep, partly in response to the pressure of civil society. Our journey is to follow it, and to find out how this industry helped build Britain's past and still has a part in shaping Britain's future.

We drive on. As we come over the crest of Benfleet Hill on the A130, the Thames Estuary fills the view. Down there, on the flat land around Coryton are the dirty white cylinders that make up the Thames Oil storage facility. Beyond it, moored to a jetty, is the hulk of a tanker, pumping fuel into the bloodstream.

PART I

1940–1979

1

The Whole World Was Aflame

> The oil began to catch alight
> And no land remained unscarred.
> The whole world was aflame
> But oil was our country's name.
>> 'Die Muschel von Margate' by Kurt Weill, part of
>> *Konjunktur (Oil Boom)*, a musical by Léo Lania, 1928[1]

In the summer of 1941 Terry's father, Lennox 'Mac' Macalister, was a surgeon in an RAF mobile field hospital in the midst of the North African desert war. The British forces had been fighting for a year, trying to slow the advance of German army under the command of Rommel. The ultimate goal of the Axis forces was the oil fields of Iran and Iraq.

Mac Macalister's life was utterly determined by oil. It was lack of diesel for tanks that stopped Rommel taking Cairo, and aviation fuel that kept RAF fighters in the air, harassing the German supply lines. In the desert, petrol was as important as water.

In the same months Richard Marriott, James's father, aged ten, lived in a world almost devoid of oil. The Keeper's Cottage near Leeds was heated by firewood. The trains that took him south to school were run on coal. Toys were made of wood, lead or Bakelite. There was not a shred of plastic in his daily life. Every drop of crude was for the war effort.

North Africa was not the only theatre in which oil was a central war aim. On 22 June 1941 German forces invaded the Soviet Union determined to conquer western Russia and especially the oil fields of the Caucasus. A month later the US imposed an embargo on oil exports to Japan. This was swiftly followed by a similar ban from the British government and the Dutch government-in-exile. In retaliation, on 25

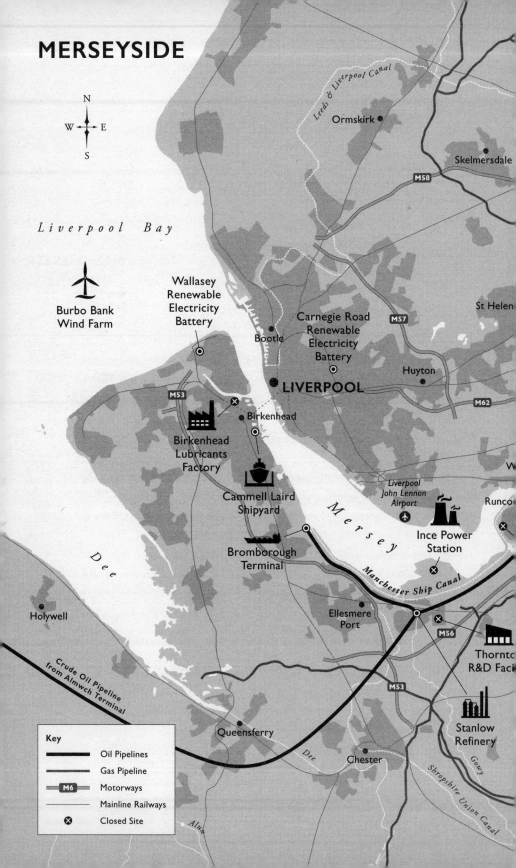

MERSEYSIDE

N
W · E
S

Liverpool Bay

Burbo Bank
Wind Farm

Wallasey
Renewable
Electricity
Battery

Leeds & Liverpool Canal

Ormskirk

Skelmersdale

M58

Bootle

Carnegie Road
Renewable
Electricity
Battery

M57

St Helen

Huyton

LIVERPOOL

M62

M53

Birkenhead

Birkenhead
Lubricants
Factory

Cammell Laird
Shipyard

Liverpool
John Lennon
Airport

Bromborough
Terminal

Mersey

Ince Power
Station

Runco

Dee

Manchester Ship Canal

Holywell

Ellesmere
Port

M56

Thornto
R&D Faci

Crude Oil Pipeline
from Almwch Terminal

M53

Stanlow
Refinery

Queensferry

Dee

Chester

Shropshire Union Canal

Gowy

Alun

Key

▬▬▬	Oil Pipelines
———	Gas Pipeline
M6	Motorways
———	Mainline Railways
⊗	Closed Site

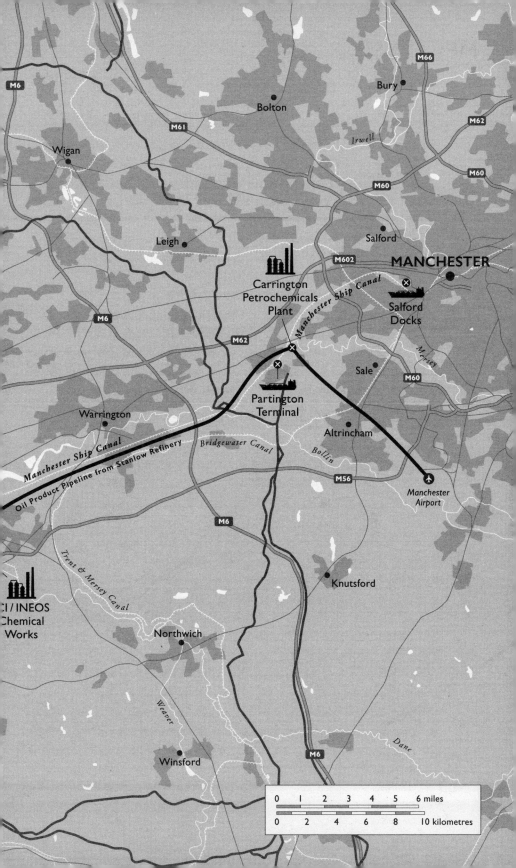

July 1941 Japan invaded southern Indochina heading for the prized oil fields of Borneo, mainly controlled by Shell.[2]

In the summer of 1941, Britain's chances of survival as an independent state looked slim. The US had not yet entered the war, the Japanese forces swept all before them, Rommel was winning in North Africa and the Red Army was retreating before the German advance. Across the globe the extraction, refining and distribution of oil was central to every conflict.

In this moment of global oil wars, Britain took a pivotal switch away from coal. For over 50 years long lines of fuel wagons had lumbered across the nation, running on rails from station to station. Clearly moving petrol in this way was acutely vulnerable to aerial attack.

Every gallon of oil product utilised in Britain had been shipped into the eastern estuaries such as the Forth and Thames. But on the outbreak of World War II in 1939 the east coast was exceptionally vulnerable to the German air force, navy and mines. The mouth of the Thames was a killing ground. Consequently oil, vital to keep Britain at war, was shipped into western ports, such as Bromborough on the Mersey next to the Stanlow Refinery, and Avonmouth on the Severn Estuary near Bristol.

Plans to address this were commissioned from Shell in early 1941 by the UK government's Petroleum Department during the Blitz, the bombardment of London. On 2 April that year drawings for a pipeline running from Avonmouth to Walton were approved by the Petroleum Board, headed by Sir Andrew Agnew. Construction began in May. For a nation at war a vast amount of steel and labour was diverted to this project. By November 1941 the Avonmouth/Thames, or the A/T, pipe was pumping petrol for cars, diesel for tractors and kerosene for lamps[3] across the country. Within a couple of years the pipeline had been extended to the Shell Haven terminal and the Isle of Grain, both in the Thames Estuary.

The North/South, or N/S, pipeline from Avonmouth to Bromborough, next to Stanlow, was soon added. From Stanlow a pipe was laid through the Midlands to Misterton in Nottinghamshire and Sandy in Bedfordshire. Another ran from Old Kilpatrick, on the River Clyde not far from Glasgow, across Scotland to Grangemouth, on the Firth of Forth near Edinburgh. The next ran from Thames Haven in the

Thames Estuary up to Norwich and Saffron Walden, supplying the bomber airfields with aviation fuel. Seventy-five thousand tons of steel pipe, financed by the state, linked the estuaries of Britain. These were the subterranean trunk roads of the country, as fundamental to its functioning as the highways, canals and railways above ground. A thousand miles of oil became the bloodstream of the nation. Hidden, not only from enemy aircraft, but also from prying eyes and the public's imagination, owned by the state and effectively managed by two private corporations.

It's a good mile through the suburban streets of Walton-on-Thames as we walk from the station to the river. At the foot of the bridge that links the Surrey and Middlesex banks, we turn on to the towpath and follow it downstream.

On the journey to understand the background to the maelstrom of our times, we need to return to some point of origin. Of course there are myriad such moments and places, but we need to find one, a thread that we can pick up and follow. We have come to the navel of Britain's oil system.

There's a nest of six off-white storage tanks, set down low behind a grass-covered mound and high wire fences. We walk along its northern perimeter to reach the gateway. There's barely any indication of the compound's function, except a signboard explaining the site to be under the ownership of CLH – Compania Logistica de Hidrocarburos.

*   *   *

World War II was an oil war, both in its experience and in its memory. In the 80 years since, World War II has been distilled into images of Spitfire fighters and Lancaster bombers, tanks in the North African desert and ships in the Atlantic convoys. Behind all these icons lay combustion engines, pipelines, refineries and drilling rigs. The conflict years saw an astounding increase in the global extraction of crude.[4] The war proved to be Britain's great leap into the world of oil. Behind this lay an alliance between the private oil corporations and the nation state.

In accordance with an agreement made between Anglo-Iranian, Shell, three other oil companies and the British government, at the outbreak of war the businesses were to pool their resources and put them under state control for the duration of the fighting. The

Petroleum Board was set up to oversee the delivery of oil from overseas, the operation of terminals, the management of refineries and the retail of fuel to both civilians and the armed forces. Road tankers and petrol stations had their Shell and BP logos removed and, painted in dull camouflage, were rebranded POOL.

The companies' workforce both in the UK and overseas was effectively requisitioned. David Barran, Shell's representative in Egypt, wanted to enlist in the services at the start of the conflict but was told by the War Office[5] to stay in post and coordinate fuel supplies to the 8th Army fighting the Germans in Libya.

Refineries were retooled to prioritise the production of aviation fuel of a new high-octane standard, and Stanlow – guided by the Shell research facility at Thornton – was at the forefront of this effort. A new plant to manufacture high-octane jet fuel was built by Shell at Heysham.[6] The entire operation of the Petroleum Board was run out of the Shell-Mex and BP Ltd offices on the Strand in the heart of London, directed by Sir Andrew Agnew, chairman of Shell Transport and Trading, and a key figure in the Shell group of companies.

At noon on 1 December 1939 Captain A.R. Hicks checked the position of his vessel, Shell's *San Calisto* oil tanker, as it headed into the Thames Estuary.[7] There was a deafening explosion. Leaning over the chart table Hicks had the sensation of the amidships being lifted bodily, and then the ship plunging down by the bow. A few moments later a second mine detonated, and a more violent explosion drove a huge column of water up through the shattered hull and afterdeck. Six men were killed instantly. Another six were wounded. This was the first of Shell's tankers to be sunk.[8]

By the following spring the shipping lanes through the Mediterranean and Suez Canal were closed to Allied tankers due to Axis attacks, so oil imports from Iran stopped almost entirely. Refined products and not crude oil were shipped across the Atlantic from the US, Caribbean and South America.[9] The tankers had been requisitioned by the state, painted with camouflage and coordinated by the Trade Control Committee, once again under the chairmanship of Sir Andrew Agnew.[10] By the end of the war 87 Shell vessels had been lost and a further 50 seriously damaged.[11] Fifty Anglo-Iranian tankers were sunk, claiming the lives of 657 sailors.[12] Almost half of both companies' fleets were destroyed.

It was Anglo-Iranian tanker crews that braved the U-boats of the Atlantic to deliver oil to Avonmouth and Shell workers who laboured under the threat of bombardment at Stanlow. It was Shell-Mex and BP staff who oversaw the supply, by pipeline, rail and road tanker, of fuel to RAF Spitfire crews at airfields across Britain. All of these men – and women – were employed by the companies, but they were under the command of the British state. As the historian of BP writes, 'The machinery of wartime control expanded until it was so complete that the seam between the company, other oil firms and the government seemed scarcely to exist at all.'[13]

* * *

In October 1934 the chairman of Royal Dutch Shell, Sir Henri Deterding,[14] stayed with Hitler for four days at the head of Germany's private retreat, Berchtesgaden,[15] 20 months after the latter had come to power. The meeting was reported in the *New York Times*, as was Deterding's aim to secure a monopoly for Shell over petrol distribution within the Reich for 'a long period of years'. There were more American consular reports the following year stating that Shell were aiming 'to obtain a monopoly in Germany' and that Deterding had offered a major oil loan to the Nazi government to facilitate this agreement.[16,17]

Germany provided the strongest market in Europe for oil products between 1933 and 1939 with the largest growth in aviation fuel, partly on account of the rearmament of the Luftwaffe. BP had 9.6 per cent of the market and Shell 22.1 per cent, with Standard Oil[18] holding 26.1 per cent.[19] Just as the Shell parent group operated in the UK via its subsidiary Shell-Mex and BP Ltd, so in Germany it worked through the company Rhenania-Ossag. Between 1932 and 1938 this Shell arm in the Reich increased in turnover from 180 million Reichsmarks to 500 million Reichsmarks and in staff from almost 6,000 to 10,000 – a more rapid growth than their UK operations experienced.[20]

During the 1930s business in Germany, including both Anglo-Iranian and Shell, was increasingly dominated by Nazi ideology. In places this was embraced. The Shell company Rhenania-Ossag adopted anti-Semitic policies, with Jewish members of the board being forced to resign two and half years before this was demanded by the Nuremberg Laws.[21,22] This would have required the consent of the Shell Group board, upon which sat Sir Andrew Agnew.[23] These changes took place not only at the level of senior executives but all the way through the company at drilling rigs, oil terminals, refineries, distribution depots and petrol stations. The nature of the Nazi regime was well understood by Shell and Anglo-Iranian both in Germany and outside it. This was a time when concentration camps, such as Dachau and Buchenwald, were operational and the rule of the Gestapo absolute. However, as long as trade could flow in and out of Germany the companies continued operating. Shell imported crude from, for example, its fields in Venezuela and refined it in Hamburg, while BP refined crude in Abadan in Iran, and shipped it into the German market.

Fascism not only provided a bulwark against communism but was also a governance structure that promised private ownership of property, stability, state-stimulated economic growth and a constraint on the disruptions that could come from organised labour.[24] Corporations are invariably drawn to creating monopolies – just as Deterding desired in Germany. For control of a market, as opposed to free competition, ensures a greater return on capital at a lower risk. Consequently the discipline of Nazi Germany looked in many respects more appealing, and was more profitable, than the changeability of parliamentary democracies such as Britain. There was the threat that the state would nationalise strategic industries but this, it seemed, could be kept at bay through the close cooperation between corporations and governments. The 1930s illustrates well the point that the oil companies have no intrinsic loyalty to democratic and open societies. Their loyalty is always, first and foremost, to return on capital.

* * *

The likelihood of conflict between the Allies and the Axis powers grew during 1938 and the potential impacts of war became ever greater. This posed a question to British oil executives of which side to back in the event of fighting. The choice for Anglo-Iranian was relatively simple as their assets in Germany were likely to be confiscated as soon as war was declared; for the BP petrol stations[25] and supply systems would then be majority owned by the government of an enemy nation.[26] However, Shell decided to effectively divide itself into an Allied corporation and an Axis corporation.

Rhenania-Ossag, the Shell company in Germany, swung in behind the government as the Nazi state began to invade other countries. Upon the German annexation of Austria in March 1938 and Czechoslovakia in March 1939, the group sanctioned Shell companies in these states to come under the control of Rhenania-Ossag.[27] After the fall of France, in June 1940, Royal Dutch Shell in all the Axis-occupied states became a key part of the German economy, under its head, or *Verwalter*, Eckhardt von Klass. Von Klass was appointed with the approval of the *Reichskommissar* Seyss-Inquart, the highest German civil authority in the occupied Netherlands.[28] A Swastika flag flew outside the Shell head office on Carel van Bylandtlaann in The Hague. The refinery at Pernis near Rotterdam kept operating throughout 1940,[29] and three years later it was recommissioned to process crude from the Axis country of Romania.[30] The Shell laboratory in Amsterdam continued its work, including investigating the production of the pesticide DDT.

As the official company history describes, von Klass 'strove to maintain the integrity of Royal Dutch as a holding company with an eye to keeping the business strong so it could serve Germany after the war. To that end he travelled around to exercise supervision over (Shell) Group companies in occupied Europe, negotiating successfully to acquire companies in Hungary, Yugoslavia and Greece.' When Germany invaded the Soviet Union, von Klass immediately contacted Berlin to resuscitate the group's claim on the oil fields that had been nationalised in 1921.[31] In September 1942 von Klass volunteered 106 Shell staff and equipment to assist the German forces in building a refinery on conquered Soviet territory. A plan was set in place to dismantle the Shell refinery at Petit-Couronne near Rouen in France and move it to the Caucasus. However, the German army's advance to Baku was halted and the plan stalled.

In 1940 the oil fields in Romania delivered 1.4 million tons of oil a year to Germany, climbing to three million tons by autumn 1943. Shell was the biggest oil producer in Romania.[32] Much of this oil for the German war machine came from Shell's refinery at Ploesti, which despite being bombed by the Allies from June 1942 was kept in operation. 'The Germans had driven huge numbers of forced labourers to rebuild the oil refineries around Ploesti',[33] reads the company history.[34]

Shell staff worked together with the *Reichsbahn* to oversee the transport of crude from company-owned oilfields in Romania to the Harburg refinery in Hamburg or the Pernis refinery in Rotterdam. Shell workers assisted the delivery of aviation fuel from those refineries to the airfields of the Luftwaffe in Germany or occupied territories. In spring 1941,[35] as the pipeline was being laid to Walton, a dogfight over the Channel between a Messerschmitt and a Spitfire could have seen both planes powered by Shell fuel.

Shell, divided into two corporations, continued to work in both spheres throughout the war operating profitably under both fascism[36] and democracy. Four years after the war Shell published a four-volume work entitled *Shell War Achievement*, illustrated by the renowned artists Paul Hogarth and Ronald Searle. It was aimed mainly at its own staff, describing the role of the company in support of the Allies during the conflict. Today the company's official history, published in 2007, makes much of von Klass's inefficiencies as a chairman and the passive resistance of staff to the German occupation. There is an embarrassment about Shell's role in the Nazi state.

Shell operated on both Axis and Allied sides. Corporations often try to declare that they are above politics, but as the war illustrated, such is their scale that they cannot help but affect politics. Shell's actions and inactions impacted Britain in the midst of the conflict. The course of the UK in the war would have been different if Shell had dynamited its refineries in the Netherlands and France and destroyed its oil wells in Romania. The oil companies helped shape the war and this shaped the nation.

* * *

There's a gentle north-easterly breeze and the morning sun is bright. Up on the viewing gallery of the dome of St Paul's Cathedral is a good place from which to survey the city, from which to gaze into the past. We have in our minds the iconic image of this dome rising above a sea of smoke and fire, as the bombs of the Luftwaffe scream down in the Blitz. While all about was aflame St Paul's remained miraculously intact.[37] There is another photograph of Winston Churchill with his walking stick and homburg hat,[38] inspecting the rubble of a city devastated by air attack. A metropolis destroyed with the aid of aviation fuel.

By 1945 much of the centre of London, and especially the City, was utterly disfigured by bombing. The docks, which stretched along the Thames from the Estuary to Wandsworth, that since the 1870s had been the main channel of petrol supply into the metropolis, had been devastated. London was not laid to waste like Hamburg, Dresden or Warsaw. At the end of hostilities these cities were oceans of rubble on which shipwrecked citizens attempted to act out a semblance of the life they had known before.[39] London had nothing equivalent to the German *Stunde Null*, but the Blitz was a defining moment. In the roar of bombs over 43,000[40] Londoners were killed. Perhaps the intensity of the shock readied the citizens for radical change. Out of the ruins grew a new London, an oil city, a Petropolis.

In the midst of war, as London was still under the threat of bombing, Patrick Abercrombie, a professor of town planning at University College London and consultant for the London County Council, presented a vision of the future city at the Royal Geographical Society. His proposal was to create a 'centre of the Empire', and in order to do this he felt that six determinants needed to be addressed. The first was the control of road and rail traffic. Road traffic he declared was a 'menace' that 'slaughters so many men, women and particularly, children'.[41] The second determinant was housing; the third, industry; the fourth, the village units of London. The fifth, open space, and the sixth, the subterranean conduits of water, gas, electricity and so on. All these aspects had to be addressed in order to make a new London.

As the president of the Royal Geographical Society said in responding to this description of the planners' work: 'Their dreams may not take practical shape as they have dreamed them, but in what they have done they have started the younger generation on the road that they have to travel if their London is to be a London that is more worth living in than the old slummy, ramshackle London that older men like myself have known and loved.'[42]

Abercrombie's vision was turned into a concrete proposal and published by the London County Council while the war raged.

Ensconced in a café, we spread out a copy of the map that was issued with the subsequent Greater London Plan 1944. We gaze at the blotches of purple, brown and green designating areas for industry, housing and parks and we can see the world we have grown up in. Here are the New Towns such as Harlow. Here are the yellow patches that denote the future airports of Heathrow and Gatwick. And here, arcing around the metropolis, is the broad white line of what became the M25 with a crossing of the Thames at Dartford. It was not to be completed until 1986, 42 years after Abercrombie's team had drawn it up.

To look at the map induces in us a feeling of vertigo as we realise that the city we had both lived in since our early twenties did not just evolve organically, but was in large part designed by a handful of men in the middle of the war. We are the statistics that inhabit their design. A design undertaken in order to create what Abercrombie called a 'balance' between humanity and the environment.[43] This was not a city built to overawe its inhabitants or visitors, such as Hitler and Speer's New Berlin or Lutyens' New Delhi. The bombing would not make way for a *Prachtallee* (Avenue of Splendours) as in Berlin, but something that approximated a citizens' city. At the heart of an empire built on a racialised hierarchy was to be constructed a city with a degree of equity and diversity, bearing out what Peter Ackroyd describes as 'London's democratic and egalitarian instincts'.[44]

However, this new social democratic London had a key foundation: oil. Nineteenth- and early twentieth-century London had evolved out of the coalmines of England and Wales.[45] Anthracite, cheap because capital was not invested in improving working conditions and because miners' wages were so low, had made it possible to build the largest city in history. Cheap coal had fired the kilns that turned clay into innumerable bricks. The heat of cheap coal had fused chalk with mud and created an ocean of cement. Cheap coal had glowed in the narrow grates of millions of poorly insulated rooms. Cheap coal had fuelled the trains that brought labour and its sustenance into the metropolis. Cheap coal had made the town gas that lit buildings and streets. Cheap coal fuelled the power stations that generated the electricity that ran the trams and underground. Without cheap coal there would have been no modern London.

Abercrombie's Plan[46] marked a shift from coal state to oil state, from coal city to oil city.[47] Coal may have kept the home hearths burning but civilian airports blossomed at Heathrow and Gatwick. To feed thousands of planes burning aviation fuel, new spurs ran off the main pipeline system built in the war. The New Towns such as Harlow and Basildon were focused on motorcars, built around petrol. Basildon was part financed by Mobil and Shell. Oil made possible the arterial roads, such as the A40 Western Avenue and the A12 Eastern Avenue. The highroads, and in due course the motorways, were designed not to carry electric-powered trams but diesel-driven buses and petrol-fuelled private cars. The traffic that Abercrombie had described as a 'menace' was not tackled by setting limits or bans upon it but by planning bigger roads. London was emulating the interstate highways of the US[48] and the autobahns of Nazi Germany where mass ownership of cars was seen as the road to national advance, as exemplified by the Volkswagen, the 'people's car'.[49]

* * *

From our café close to St Paul's we cross the Millennium Bridge, gazing at the tower of Tate Modern. For 90 years until its closure in 1981, various power stations upon this site of Bankside had generated electricity at the heart of the city, the vast majority from burning coal. We walk upstream along the Thames Embankment, the river almost empty of traffic. Only the ghosts of the coal-fired tugs and steamers, and the tide-powered lighters and freighters.

The implementation of Abercrombie's vision required both an end of hostilities and a government with the political will to drive it through. That government came in the form of Attlee's Labour Party riding on the back of its surprising 1945 general election victory. Aneurin Bevan was appointed minister of health with a remit that covered housing[50] and immediately set about the task of working with the London County Council to realise the 1944 Greater London Plan.[51] The ability of the state to undertake such a transformation of the capital was enabled by having the nation's war-ravaged finances underpinned by massive loans from the US government and by gaining control of the 'commanding heights' of the British economy.[52]

In 1918 the Labour Party had adopted a new constitution that included the famed Clause IV: 'To secure for the workers by hand or by brain ... the common ownership of the means of production, distribution and exchange.' The groundswell of political support for the 1945 election manifesto enabled Attlee to enact this vision despite the determined opposition of many interested parties. The coalmines were nationalised in 1947, as well as electricity generating and transmission, town gas (manufactured from coal) production and distribution, the railways, water supplies and in Scotland a hydroelectric scheme that stretched across the Highlands.[53]

Much of this infrastructure had effectively been in government hands since the outbreak of war in September 1939. Six years of conflict had demonstrated that the state could run industry efficiently, even in hugely demanding circumstances. Indeed it conclusively proved that private capitalist concerns could be brought under public control and function effectively.

From the vantage point of hindsight, one sector stands out from this group – the oil industry. The war had illustrated just how fundamental oil was to the industrial economy – it had replaced coal as the engine of victory. The entire oil system had been run under state control. Why did this arrangement, so similar to that which pertained in the coal industry, not lead to the state taking the British oil industry into public ownership?[54]

Shell and BP were able to engineer an historic compromise with government similar to that brokered by private banks and other British institutions which might have appeared to be at odds with the new era of social democracy, such as private schools. The compromise involved oil companies remaining private, with the assets of their shareholders left untouched, but becoming 'national champions'[55] supporting the great rebuilding of Britain following the war. Both were keen to emphasise their loyalty to the country after their divided loyalties during the war. The compromise had a profound impact on the UK's development in relation to social justice and equality, Britain's foreign policy and war, and the entire society's impact on global ecology.

The nationalisation of the coalmines meant that industry became an arm of the state. But because the oil companies remained privately owned, in some respects this relationship was reversed – the British state became an arm of the oil companies. These corporations helped shape the nation's future.

* * *

After a ten-minute stroll along the Thames we arrive at the South Bank. This is the site of the Festival of Britain which opened in May 1951, and still here is the Royal Festival Hall, the centrepiece of the exhibition. Created by the Labour government and overseen by Herbert Morrison, it was a celebration of peace after war and the hope of New Britain. Mac Macalister and Richard Marriott were among the eight and a half million people that flocked to the exhibition during its five-month life. By the time the new Conservative government came to power the exhibition had closed. Winston Churchill, the new prime minister, was widely assumed to be against the festival, apparently referring to it as 'three-dimensional Socialist propaganda'.[56]

Beneath the futuristic Skylon, events in every art form were supplemented by the Festival of Britain *About Britain Guides* which described each region of the country and encouraged readers to discover their land. It outlined favourable routes to take everywhere from north-east Scotland to south-west England. Routes to be taken by private car, the new country laid open by the combustion engine. Appropriately the Festival of Britain site was overlooked by the head offices of Shell-Mex and BP Ltd, the largest provider of petrol to the nation.

In the pre-war era cars were an elite form of travel, driven by 'bright young things' and advertised as a luxury. Post-war they became mass transport, the people's transport, with publications supporting the idea of 'everyman' travelling the nation by car. The Special Roads Act had passed in 1949 and work was well underway on the planning of Britain's motorway system. The M1 was opened in November 1959 by Ernest Marples MP, the minister of transport who also enacted the dramatic closure of Britain's railways, with their coal-fired trains, as he wielded the 'Beeching Axe'.[57]

* * *

On 16 December 1955[58] Elizabeth II, the young monarch, opened the Queens Building at the new London Airport, later renamed Heathrow. This modernist block was created by Sir Frederick Gibberd, who was also busy with the designs for Shell's research centres at Sittingbourne and Thornton. The Queens Building housed offices for the state-owned British Overseas Airways Corporation (BOAC) and other airlines, and had a viewing platform and rooftop gardens, which became one of the most visited attractions in London as sightseers gazed upon the modern world.

Out on the tarmac runways were gleaming silver Comet jet airliners. Eight BOAC Comets a week would take off for Johannesburg, Tokyo, Singapore, Colombo and many airports en route.[59] The world's first passenger-carrying jet airliner flew from London to Johannesburg in May 1952. In their first year the Comets carried 30,000 passengers.[60] In June 1953 the Queen, the Queen Mother and Princess Margaret graced the Comet in an inaugural Royal flight.[61] On 6 October 1958 a BOAC Comet made the first transatlantic passenger jet flight from London to New York.[62]

This was the new era of jet travel and the world's pioneering aeroplane was made just on the northern perimeter of London. The Comet was developed by De Havilland, financed and directed by the British government at the Hatfield Aerodrome factory, the site of one of Abercrombie's future airports which was never realised. De Havilland had a workforce of 4,000 people. Hatfield was a new aircraft town. BOAC ordered a fleet of Comets to serve flights to distant cities; many, like Nairobi, Tehran and Delhi, parts of the formal or informal empire.

The flight from London to Tokyo in propeller airliners had taken 86.5 hours; now the Comet reduced that time to 36 hours even with nine stops.[63] However, the new jets guzzled fuel compared to the airliners they replaced. Shell-Mex and BP secured an exclusive supply contract for the Comet[64] and 150,000 gallons an hour of aviation spirit ran through the fuelling machinery of Heathrow. By 1956 the Comets were being outclassed by their American rivals, the Boeing 707 and the Douglas DC8, which had greater range and carried more passengers. However, the American planes consumed three times the amount of fuel. Shell-Mex and BP duly built a spur pipeline from the Walton-on-Thames depot, and by 1959 were pumping jet spirit into the tanks of 707s and DC8s through a new refuelling facility.[65] The sales successes at Heathrow were a vindication of the Anglo-Iranian policy on aviation spirit set out in 1947 a decade before by Frederick Morris, director of distribution, to aggressively enter markets 'wherever and whenever possible'.[66]

The scale of oil production in World War II, with an intensity of energy turned against enemy forces, was now turned on civilian markets, as witnessed in the leap in production of aviation fuel at Stanlow and Abadan. Vast sums of capital had been invested to ramp up wartime refining, which at the end of hostilities was soaked up by the rise in civil aviation and private motoring. This was the transformation of everyday British life by oil.

From the viewing platform of the Queens Building visitors could admire the new galleons of modern Britain. These Comets charted the supply routes and command lines of an empire that had moved from the wind, coal and oil of shipping to the jet fuel of air travel. The coal empire was now an oil empire, overseen no longer by Victorian clerks in frock coats or Edwardian generals in khaki, but by Elizabethan businessmen and civil servants in lounge suits. Behind it all were the oil companies determinedly pumping fuel to Heathrow, which became one of the highest energy users in the country, the great bonfire without flame to the west of London.[67]

* * *

Britannic House, in the City of London, was the head office of the Anglo-Iranian Oil Company (AIOC). Designed by Sir Edwin Lutyens, an Edwardian architect of the British Empire who was simultaneously creating the imperial capital of New Delhi. Its front was adorned with sculpted figures in headdresses representing the company's assets in Persia and Iraq.

Within was the boardroom and suite occupied by the chairman Sir William Fraser Lord Strathalmond. The tall and hawk-like Fraser had been born into the company, his father was a key figure in its early days. As chairman he was renowned for being authoritarian and abrasive.[68]

Using capital from both its own resources and the state, AIOC undertook substantial infrastructure projects such as the building of the refinery at the Isle of Grain. Also in the Thames Estuary, Shell built a new research facility at nearby Sittingbourne in Kent; while on the Manchester Ship Canal at Carrington, Shell financed the construction of a new chemicals plant and another research unit at Thornton next to Stanlow.[69]

On Teeside, Wearside, Tyneside, Clydeside and Merseyside, Shell and BP both invested massively in the shipbuilding industry by ordering new tonnage. The companies had lost nearly 140 tankers to enemy action and now they were eager to replace those losses as well as expand their shipping fleets. Between 1945 and 1965 Shell ordered 136 vessels and BP 168 from 25 yards such as Swan Hunter and J.L. Thompson in Sunderland, Lithgows and A. & J. Inglis in Glasgow, as well as Harland & Wolff in Belfast and Cammell Laird on Merseyside.[70]

Lairds was the third largest builder.[71] In all the Birkenhead facility built 18 ships for Shell[72] and nine for BP.[73] Orders from these formed a high portion of the yard's output after military orders. With Carrington upstream, Stanlow, Thornton and Bromborough midstream, and Lairds near the mouth – as well as a vast array of chemical plants such as ICI at Runcorn – the Mersey was one of Britain's great oil rivers.

Jack Rogers was a joiner at Lairds, and in the mid-1960s worked on the last three tankers built there for Shell.[74] 'I preferred merchant work because the standard of the work was far superior. Navy work in those days was pretty plain. The *Opina*, *Opalia* and *Oscilla* were three tankers for Shell and they were a lovely job. They all had wrought iron spiral staircases that led to the wheelhouse and at the back, instead of ordinary panels, they had a big beautiful mural made from eight by four panels which had to be cut and lined before they were fixed. That was one of the nicest jobs', he told author David Roberts.[75]

Alongside constructing Royal Navy ships, oil tankers were the lifeblood of the UK's shipbuilding industry. Indeed during the 1940s, 1950s and 1960s the yards around the country were effectively arms of the defence and oil industries; the fate of so many workers hung on decisions made in the Ministry of Defence and the head offices of Shell and BP.

The Royal Navy and the merchant fleet were icons of British imperial power. Officers in white shorts and peaked caps on the bridges of ships ploughing across the Atlantic or Indian Oceans, or patrolling the Mediterranean, symbolised the resurgent country in the aftermath of victory in 1945. A large number of the UK's merchant vessels were owned or operated by the Anglo-Saxon Co Ltd and the British Tanker Company,[76] the shipping arms of the two oil corporations. Meanwhile the ordering of ever-larger tankers from the shipyards of the UK was key to making shipbuilding a pillar of the British economy. This made the oil industry the largest client of British shipbuilding after the Royal Navy, helping to bind in the support of trades unions for the corporations and build the sense of BP and Shell as national institutions. It was cemented by the name of every BP ship having the prefix 'British' and images of Queen Elizabeth launching tankers in front of Union Jack-waving crowds. On 17 March 1965 the sovereign named the largest tanker to be built in Europe the MV *British Admiral* at the Vickers Armstrong yard in Barrow-in-Furness, declaring 'Ships and oil are two of the most powerful sinews of our economy.' The 100,000 ton supertanker slid down the slipway.[77]

2

روشــنــايي آسـمــانهــــــــــــــا

The Brightness of the Heavens

باتهم!
نــان اي مونس عــــاشقــــان
ا روشـنــايـي آسـمــانهـــــــــــــ
باب آسـمــان چراغ يا!مهتــــــــــ
نان روشـنـي بخش جهــــــــــ
ماهم كـو؟
نزدت چ ه ش بها،با اودرآنجا بـودیـم
فارغ زدنیا ،لـبها بـه لـبها بـودیــــم

—Viguen, 'Moonlight', 1954[1]

Rose Hammond was born in Charlton, on the banks of the Thames, half way between the Isle of Grain and London. As a teenager she worked in the Siemens cable factory before falling for a man in the Royal Army Service Corps stationed in the next-door barracks town of Woolwich. After marriage and the birth of their first daughter they moved back to where Arthur Hammond had come from, the village of Stoke.

Overlooking the River Medway, Stoke was a couple of miles from Grain and the largest settlement in the area, having grown on the back of an industry of digging blue clay from the marshlands at the river's edge. The Hammonds were a family of blacksmiths and Arthur found work as a steel erector constructing oil storage tanks for the Admiralty on the Isle of Grain. In 1914 he was called up to serve on the Western Front. After months in the trenches he was gassed and invalided out with a small War Pension.

By the late 1920s Rose and Arthur had four girls and four boys. Ten people lived in a four-room cottage rented off Jimmy Muggeridge on Middle Street in Stoke. It had no electricity or gas, no indoor toilet nor running water. Both Rose and Arthur worked on the farms around

Stoke and the nearby village of Allhallows. The world moved by cart, the fields were ploughed with horses. There were no buses and few cars. Milk came from Bennets' farm and meat from the village butcher, Goldsmith's Empire. The Hammonds would walk the several miles to the fields and back. When in season, work on Muggeridge's Binney Farm was pulling potatoes by hand. Some of the crop was loaded onto Thames barges in Yantlet Creek, the wide muddy inlet that separated the marshes of Allhallows from the marshes of the Isle of Grain.

We walk up the track towards the ruined farmhouse and rusting barns at Binney's. Skylarks fill the air above the wide land. In the distance is the flat outline of Grain, closer to the continuous grassy bank of the seawall. We can imagine the tan sails of the Thames barges as they wait in the creek. Rose Hammond is among the bent figures in the field, picking potatoes from the pale sandy soil, filling the sacks, heaving them onto the cart in this pre-petroleum world.

Rose's youngest son, Frank, was born in 1927. When he was two the family moved into a new cottage in Stoke. It had electricity and water. In his seventies Frank Hammond recalled: 'I remember the first night my mother lifted me up to turn on the light … we had a bath indoors and taps with water coming out which was fascinating for me. We kept filling the bath up because we couldn't believe it.'[2]

* * *

For the best part of a year between 1947 and 1948, the central figures in Anglo-Iranian Oil Company (AIOC)[3] debated the merits and demerits of where to build two new refineries that they wanted. Should they be located close to the oilfields operated in Iran, Iraq and Kuwait? Or should they be located near the main markets of Britain and Europe where refined products were sold? There was concern around building in the Middle East due to the rising labour costs and labour militancy. And there was the threat of political upheavals that could lead to states, such as Iran, seizing company assets or changing the terms of the agreements under which AIOC extracted oil. On the other hand, the concern about building refineries in Western Europe was about whether there was sufficient market for the petrol, diesel and aviation fuel that AIOC and its competitors were trying to sell.[4]

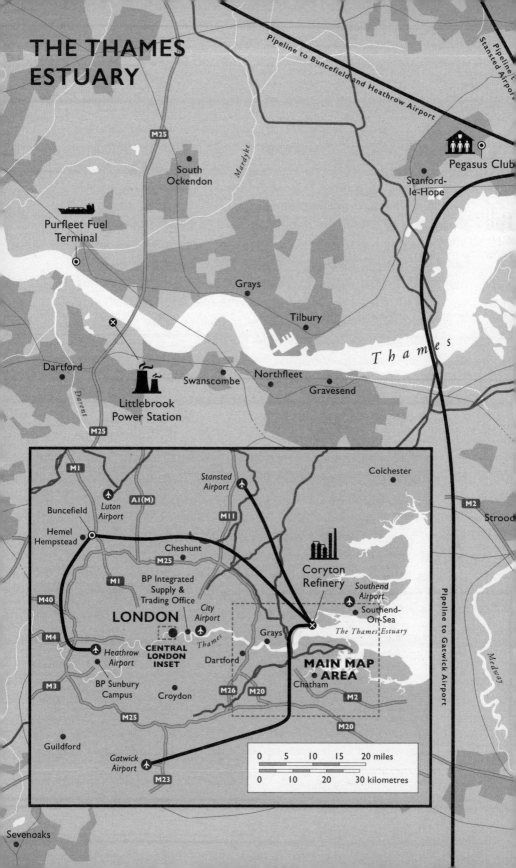

THE THAMES
ESTUARY

Pipeline to Buncefield and Heathrow Airport

Pegasus Club

Stanford-le-Hope

M25

South
Ockendon

Mardyke

Purfleet Fuel
Terminal

Grays

Tilbury

Thames

Dartford

Littlebrook
Power Station

Swanscombe

Northfleet

Gravesend

Darent

M25

Colchester

M1

Stansted
Airport

Luton
Airport

A1(M)

M2

Strood

Buncefield

M11

Hemel
Hempstead

Cheshunt

Coryton
Refinery

BP Integrated
Supply &
Trading Office

M25

M1

Southend
Airport

Southend-
On-Sea

M40

LONDON

City
Airport

M4

CENTRAL
LONDON
INSET

Thames

Grays

The Thames Estuary

Heathrow
Airport

Dartford

MAIN MAP
AREA

BP Sunbury
Campus

Croydon

M3

Chatham

M26

M20

Pipeline to Gatwick Airport

Medway

M2

M25

M20

Guildford

0   5   10   15   20 miles

0   10   20   30 kilometres

Gatwick
Airport

M23

Sevenoaks

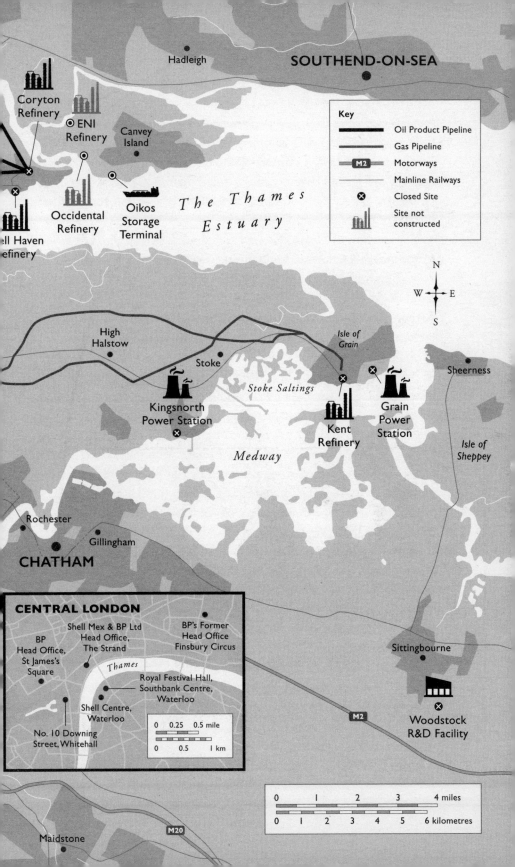

SOUTHEND-ON-SEA

Hadleigh

Coryton
Refinery

ENI
Refinery

Canvey
Island

Occidental
Refinery

Oikos
Storage
Terminal

ell Haven
efinery

*The Thames
Estuary*

Key

	Oil Product Pipeline
	Gas Pipeline
M2	Motorways
	Mainline Railways
⊗	Closed Site
	Site not constructed

N
W   E
S

High
Halstow

Stoke

Kingsnorth
Power Station

Stoke Saltings

*Isle of
Grain*

Kent
Refinery

Grain
Power
Station

Sheerness

*Isle of
Sheppey*

Medway

Rochester

Gillingham

CHATHAM

Sittingbourne

CENTRAL LONDON

BP
Head Office,
St James's
Square

Shell Mex & BP Ltd
Head Office,
The Strand

BP's Former
Head Office
Finsbury Circus

Thames

Royal Festival Hall,
Southbank Centre,
Waterloo

Shell Centre,
Waterloo

No. 10 Downing
Street, Whitehall

| 0 | 0.25 | 0.5 mile |
| 0 | 0.5 | 1 km |

M2

Woodstock
R&D Facility

| 0 | 1 | 2 | 3 | 4 miles |
| 0 | 1 | 2 | 3 | 4 | 5 | 6 kilometres |

M20

Maidstone

The men in their fifties at the centre of these discussions, Sir William Fraser, Harold Snow, George Coxon and Frederick Morris, spent their hectic days in a petroleum-fuelled whirl of cabs and cars in London and propeller aeroplanes that carried passengers from the newly opened London Airport to the cities of Europe, America and the Middle East. Despite being a mere 30 miles east of Britannic House, the head offices of AIOC in the City of London, the world of the Isle of Grain was starkly different.

The location of Grain was well suited for a British refinery. Close to the prime market of London and south-east England. Close to deep shipping channels in the Thames and Medway, and near to the naval docks at Chatham.[5] The Admiralty terminal on Grain, which Arthur Hammond helped to construct, was a key node in the web of oil lines built after 1941. It was a starting point of the PLUTO system of undersea pipes which supplied the Allied forces after the invasion of Normandy in 1944. Fraser, Morris and Snow had all been central to the fuelling of the military throughout the war and knew well the existence of the Grain storage tanks.

However, the scale of the planned refinery was such that it would require far more space than the acreage under Admiralty control. Land needed to be purchased or requisitioned by AIOC. The main farmers on Grain were the Muggeridge family, indeed they had tenancies at a patchwork of farms running all along the Hoo Peninsula. A deal was cut with Richard Muggeridge, the lands of Red House Farm were swallowed up by the refinery, and in 1949 housing was rapidly erected at the Grain hamlet of Wallend to accommodate the construction workers. The same year Arthur Hammond died. His son Frank, who'd spent the previous seven years working in the power plant at the Admiralty terminal while living with his mother in Stoke, got a job building the refinery.

The vast majority of the labour that descended upon Grain was not recruited locally. In Stoke, housing was built for men and their families, some of whom who came from Aberfan in south Wales to work in the construction. In the next-door village of High Halstow the company was soon building further housing for refinery staff. Each of these new homes with electricity and indoor toilets stood out from the rest of the houses as the new, modern world. And the occupants who worked for the oil company often had better wages than their neighbours.

Marion Stopps was in her forties and living four miles west of Grain. She recalled, 'Halstow was a lovely village but of course it changed when BP came. We used to have a lot of pleasure with tennis and the dances, then of course BP came and they built a lot of houses and it's all different now. High Halstow people complained a lot, the BP people brought money to the community and it was different. If you have money, life is different.'[6]

At the time that work commenced on the refinery, the farms in the area had begun to use tractors, a few Fordsons and Fergusons run on diesel, but this was as nothing compared to the army of machines that passed through Stoke on its way to Grain. Lorries and bulldozers, pile drivers and graders, were utilised to drain the marshes, level the land and erect the refinery. The noise of engines and steam hammers seemed constant.

The view from the potato fields at Binney's changed utterly. Grain was no longer a slight rise of grey green fields, but a vast wall of black steel towers. In the darkness it twinkled with a thousand lights like a city of skyscrapers. Through every hour of the day and night the flares at the refinery's top roared in the east wind.

At the end of 1952, after three years in construction, Frank Hammond started shift work at the refinery:

I was an operator. We all started at the bottom. We went to school for about twelve weeks and were taught what a refinery did, not just about oil, but about the power. I didn't know the half of what they were teaching me. I was operating machinery which could handle say one million gallons of crude oil a week, or something like that … They taught me how to handle valves and pumps and what the chemical reaction was. Some of us were brighter than others and they got the top jobs but gradually as we got older, we less intelligent, we took longer to get there but, eventually, we made it. In that sort of job you progress or you don't progress.

The following year his mother, Rose, got a place at the refinery. 'She worked there in the canteen which she enjoyed. The easiest job she ever had and she worked there until she was 76. They had to force her to go out the door and that was the finish of her, she was so popular down there, being an old lady, and they thought a lot of her.'[7]

* * *

After far too long a delay, we are heading to Peter Pickering's flat in east London.[8] The street of Victorian terraces lies almost parallel with the M11 motorway as it cuts through the city. You can hear the roar of one of Abercrombie's designs where Pickering lives. The constant rush of traffic gives the area a battered feel.

At Wadley Road we approach the door wondering if Pickering will be well enough to answer. Through the glass we can see a pale figure moving and then he's there, smaller than we remembered, but with that same energy in his face. It's clear that he lives in his bedroom and the front room. This latter is crammed with books and papers, there's barely space to sit. The walls of the passage that leads to the kitchen are covered with scribbles and graffiti. Pickering's delighted with these drawings by his grandchildren. They are filled with wild abandon.

We settle down to record him. He frequently complains that his memory is going and that Alzheimer's is setting in, but his stories are vivid. As we're leaving after several hours, he gives us a copy of a chronology of his life that he's typed up for family and friends. The entries are often just puzzling fragments but they help us piece together a sense of days lived.

Pickering has been radical since he was 14, influenced by his father who was an early member of the Left Book Club and by reading poets such as Auden, Spender and MacNeice. After serving in Italy during World War II, Pickering was demobbed in April 1947. Two months later he travelled by train with a group of British volunteers to work on the Peoples' Youth Railway project in the new communist Yugoslavia. Young men and women from across Europe were constructing the line that ran from Šamac to Sarajevo. Hundreds were involved in helping this shattered country build itself a new socialist future. The British contingent included several people who later became well known – the left-wing historian E.P. Thompson, the Labour MP, John Stonehouse; the artists Paul Hogarth and Ronald Searle,[9] and the future Thatcherite Alfred Sherman. Pickering spent a month building the railway in the heat of summer. And the nights with Pam, a new lover he'd met at the camp.

In the autumn of 1947, back in London, Pickering joined DATA films and was soon an assistant director and editor travelling around in an Austin van working on a documentary called *All Eyes on Britain*, about the revival of the country after the devastation of war. By this time love with Pam had foundered and Pickering had met Sheila at a family wedding. His diary notes: 'Wild times at her parents maisonette in Hayes. Folding Put-U-Up falls on Frances visiting my mother's house in Beckenham, while Sheila and I in bed upstairs.'[10]

In 1949 DATA was based at 21 Soho Square, in the heart of the West End of London. The company's executive producer, Donald Alexander, and one of the film directors, Jack Chambers, were both members of the Communist Party. As Pickering describes, DATA was run in a democratic way – one person, one vote – and everyone had a vote from the executive producer to the office boy. He made films about the coal mines that had been nationalised on 1 January 1947, and the building of a new plant at the state-owned Steel Company of Wales in Port Talbot. The following year Sheila and Pickering were busy Labour Party activists in the Dulwich constituency of south London. The diary records cryptically 'Sheila's lovers. Names from time. Alan Matejas. Pat Quinn' and then 'On September 4th 1948, I marry Shelia.'

At about the same time DATA got its commission from AIOC to create a documentary on the construction of its refinery on the marshes of the Isle of Grain. The leftist output of the unit evidently didn't bother the corporation. Pickering would direct and John Ingram write the script for the 25-minute film that would become *The Island*.

In the warmth of his room Pickering describes the process of making the film. The man in the oil company that the DATA team liaised with was Ronnie Tritton. 'He was fairly right wing. He was tricky', says Pickering as his eyes twinkle. The contract with AIOC was to shoot 'coverage' of the construction process over four years beginning in spring 1949. He explains that together with John, and at times Ron Bicker and John Gunn who oversaw the photography, he travelled to Rochester, 30 miles south-east of London, and put up in the Rochester Hotel. From there they would head out each day, driving the Austin van along the Hoo Peninsular to the Isle of Grain and attempting to capture some of the building work that was underway. When it rained, or there wasn't enough light, they retired to the nearby Royal Victoria Hotel and played billiards.

The editing of *The Island* was stretched out over the better part of three years. In the darkness of a cutting room in Wardour Street, Pickering stitched together the film with a score by Malcolm Arnold. In between work time was spent with DATA colleagues in Soho pubs such as the Dog and Duck and the Carlisle Arms, and cafes like Victor's on Greek Street. By now Sheila was pregnant and the diary notes: 'I begin to spend time in the evenings in the West End with Jean, the office girl at DATA.'

No sooner was *The Island* released in 1952 than work began on shooting and editing its sequel, *The Tower*. The project rolled out through the following year until at the end of 1953 Pickering wrote: '"*The Tower*" is completed. My platonic relationship with Jean, deepens with evenings in cafes around Fleet Street'. Finally in 1954 came a third film, *The Kent Refinery*, largely as a compilation of footage from the other two. The contract was now concluded. The refinery itself was completed the same year and inaugurated by a visit from the Queen, the Duke of Edinburgh and chairman of BP, Sir William Fraser. On 9 March 1954, Sheila and Peter Pickering's daughter Jo was born.

We read the notes in the booklet that accompanies the British Film Institute reissuing of the work in 2010:

> Very much in the tradition of 1930s rural documentaries, *The Island* concentrates on the tensions between the timeless and the modern – here, between the Isle of Grain's existing community and environment and the incoming oilmen. Predictably, the film concludes that these can be reconciled ... *The Tower* is a conventionally cheerful documentary ... Low angle shots and suspenseful music over key sequences ... attempt to inject a sense of heroism and romance into an essentially process-based story with a technocratic message. The conclusion: the refinery has been a stunning success.[11]

What strikes us is the tone of cynicism in the way the film's described. The author implies that of course this film was biased and of course the optimism of the filmmakers was forced.

But it seems to us that Pickering saw the refinery as one part of the evolution of a whole society. That this half-state-owned oil company was contributing to the building of a New Britain after the war and he too was playing a role in that. Just as he'd contributed to the building of a socialist Yugoslavia. The optimism that shines out of the film reflects a feeling of political purpose intertwined with the exhilaration of a love affair, a marriage, a child and a romance.

* * *

Dylan Thomas, the 37-year-old poet from Swansea, was smitten with remorse as he wrote to his beloved wife Caitlin from Iran.[12] She had just discovered that her husband was having an affair with Pearl Kazin, an executive at the American magazine *Harper's Bazaar*. Over the space of five weeks he wrote four times begging her forgiveness, writing as he journeyed around Iran. Between 8 January and 14 February 1951, Thomas, together with his producer Ralph 'Bunny' Keane, were researching a film for Anglo-Iranian. The oil company had commissioned Thomas to write the script for a piece that would reveal the workings of the industry, and the benefits that it brought to Iran, to be shown to audiences back home and beyond.

Thomas was one of Britain's foremost poets at the time, had made over a hundred broadcasts for BBC radio, and written scripts for wartime propaganda movies. As a man from Swansea he'd lived in the shadow of Anglo-Iranian all his life.[13] The company's oil refinery, in the hills behind the town, had started construction in 1919, when Thomas was five. Taking the surname of the founder of Anglo-Persian, William Knox D'Arcy, and joining it with the Welsh word for church, the plant was given the title Llandarcy. It was part of the British war machine and the minister for munitions, Winston Churchill, approved the siting of the refinery behind a low hill that hid its position from the sea. Opened three years later it was fed by tankers arriving regularly in Swansea's Queens Dock, from where the Iranian crude was pumped up to the works. The oil company was one of the largest concerns in the area, and the refinery had its own model village for its workers. Swansea switched from a coal town to a coal-and-oil town.

'I saw rows and rows of tiny little Persian children suffering from starvation: their eyes were enormous, seeing everything and nothing, their bellies bloated, their matchstick arms hung round with blue, wrinkled flesh', wrote Thomas to Caitlin as he described a visit to a Tehran hospital. 'After that, I had lunch with a man worth 30,000,000 pounds, from the rents of peasants all over Iran, and from a thousand crooked deals. A charming, cultivated man.'[14]

Thomas was clearly shocked by the poverty he witnessed, the 'dirty and wretched clothing' of the mass of people, of women who covered their 'rags' of clothes with *chadhurs*, and 'babies huddled in with the poverty'. He described the contrast of his visit to the Anglo-Iranian refinery town of 'puking Abadan', the 'evergreen, gardened, cypressed, cinema'd, oil-tanked, boulevarded, incense-and-armpit cradle of Persian culture'. He disliked the casual racism of the company staff, 'Scotch engineers running down the Persian wops'.

In the reception room of the refinery Thomas was introduced to a 29-year-old office worker, Ebrahim Golestan, who later became one of Iran's leading filmmakers. Golestan took him to a bar, and over McEwans they talked of Chaucer, Bach and poetry. The Iranian described his country:

Everything is made of clay, wherever you travel here you see the houses are made of clay and mud bricks. We have a narrow strip of jungle by the side of the Caspian Sea – the rest is deserts, mountains and stones. Building with mud is a lot easier than building with stones despite having a lot of stony mountains. From the time of Alexander onwards there are no houses, palaces or temples made of stones here. They are all made of clay – baked or not. We come from the earth and go back to the earth and there is nothing left of us when we go. That's how it is, isn't it?[15]

Thomas never completed the script, the film was never made, and he died within two years. He was not unaware of his role in Anglo-Iranian, 'my job was to pour water on troubled oil', but he can have had little understanding just how intensely political the hiring of such a high-profile poet was for the company.

Four weeks after Thomas ended his journey, the Iranian Parliament, the Majlis, made an historic announcement that it would nationalise all of Anglo-Iranian's assets in the country. The seven giant fields[16] that produced more than 30 million tons of oil a year, the 1,990 miles of pipelines,[17] the pump stations and terminals, the world's largest refinery at Abadan, the export docks, the workers' towns, the petrol stations and offices across the country – this was Britain's biggest overseas asset. More than that, it was 51 per cent owned by the British state.[18] There had been nationalisations before – notably the Soviets had confiscated Shell's oil fields in the Caucasus 30 years previously, and the Mexicans likewise in 1938. However, neither of these had been so central to Shell as Iran was to Anglo-Iranian; the oilfields in the Zagros Mountains and the refinery at the head of the Persian Gulf had been the heart and soul of the company for nearly 50 years.

The events that surrounded the nationalising of Anglo-Iranian's assets were momentous in the history of Iran, Britain, the company and the global oil industry.

In 1933 Anglo-Iranian signed an oil concession with the autocratic Shah Riza on terms that were highly favourable to the company. In 1941 the British, fearful of Iran falling into Axis hands, sent in its army. The Shah abdicated and fled. Over the next decade, under his weaker successor Muhammad Riza Shah, there was an upsurge in political parties and activism. The most potent of these was Dr Muhammad Mossadegh,[19] a 70-year-old parliamentarian who had been imprisoned under the previous Shah. Mossadegh became the figurehead of a broad political movement, the National Front, which called for the nationalisation of Anglo-Iranian. On winning a powerful position in the elections for the Majlis that convened in February 1950, the National Front pushed for state control. Just over a year later a bill adopted that principle, which passed into law on 1 May 1951.

Eric Drake was AIOC general manager in Iran, based at Khorram-shahr. On 10 June the board of the new National Iranian Oil Company arrived at the AIOC offices. There were speeches, the ritual slaughter of a sheep and the raising of the Iranian flag over the main building. Amir Ala'i, governor-general of Khuzistan, told a crowd of about a thousand that oil was now nationalised. Drake was met at the door of his office by a soldier with fixed bayonet. He was informed that he was now an employee of the Iranian government and any refusal to cooperate was illegal.[20]

The British considered military intervention. However the US, frightened of destabilising Iran and thereby tipping it towards the Soviet Union, pushed against this. Consequently the Labour Attlee government began a tortuous process of attempting to negotiate Iran out of nationalisation. The Anglo-Iranian board, under Sir William Fraser, took an exceptionally hard line, causing frustration to both the UK Foreign Office and the US State Department. Mossadegh negotiated with a remarkable tenacity. At various stages he ratcheted up the pressure, such as by forcing the evacuation of all British staff on 3 October 1951.

Anglo-Iranian placed an embargo against any other company purchasing what they claimed to be 'their oil' from the loading terminals at Abadan. The process of negotiating continued as Britain and the company tried to regain the assets and compensation, while the Iranians strove to get an agreement as to how their crude could be sold onto the world market. Significantly the Attlee Labour government was replaced by Churchill's Conservatives in October 1951, and the British and Americans then focused on forcing the replacement of Mossadegh. Churchill berated his predecessors 'who had scuttled and run from Abadan when a splutter of musketry would have ended the matter'. He wrote to the new foreign secretary, Anthony Eden, 'If we had fired the volley you were responsible for at Ismalia at Abadan none of these difficulties would have occurred.'[21]

In August 1953 Mossadegh and the National Front were overthrown in an army coup led by General Zahedi who was backed by the CIA and MI6. Anglo-Iranian, and a consortium of oil companies including Standard Oil and Shell, negotiated a new deal with the new army-backed government. British staff returned to operate the oil fields and refinery, and the US began nearly three decades of financial and military aid to an increasingly autocratic Iran. Mossadegh was sentenced to three years' solitary confinement and died in 1967. He was denied a public burial by the Shah for fear of stirring up popular unrest.[22]

* * *

One of the staff who went to work with the assets reclaimed by Anglo-Iranian was John Browne, posted to the oil field at Masjid-i-Suleiman. A year later, in 1958, he was joined by his wife and only son, also John Browne, and future CEO of BP. In his memoirs Browne junior described his experiences as a nine year old. The journeys from boarding school in England were 'flights back and forth from Iran on the latest aeroplanes such as the Comet 4',[23] wrote Browne, adding, 'In many ways, the organisation and functioning of expatriates in Iran still resembled colonial societies, with a prevalent superior mindset and no integration into local communities.'[24] He continued:

> Certainly the male white British bias had been made explicit in the Anglo-Iranian staff manual. This directed that recruitment should be restricted to British subjects and those of European origin or descent, except in foreign countries where there was a legal or conces-sional obligation to employ local nationals. It was not the company's policy 'to employ women to work which is normally regarded as men's work' and female staff who married were expected to resign and 'no exceptions to this rule can be contemplated'.[25]

Browne's description is understandably a child's perspective – filtered through the autobiography of a man in his sixties eager to embody progressive views – but it gives a sense of the world of the expat oilmen and their families. A world located in Iran, but so clearly set apart from it: in fact detached from the political realities that the actions of the company had helped shape. The Shah's rule, as Britain and the US had expected, became infamously autocratic, with an appalling human rights record. Much of the torture in Iran's prisons was carried out by the Shah's secret police, some of whose officers had been trained in Britain in 1957 when the force was established.[26]

* * *

In the field at Binney's we gaze across to the Isle of Grain and the tangle of wire mesh fencing and buddleia where the Kent Refinery once stood. With a deeper understanding of the intense struggle in Iran, Peter Pickering's film now seems naïve and the cynical tone of the review of *The Tower* seems to miss the point. The refinery could be seen as a weapon in a war of Britain against Iran, a war of the late empire, the social democratic empire. It is intriguing how the lengthy process of Pickering making three films of Grain, from first commission in autumn 1948 to final release in spring 1954, stretched across the full length of the war against Iran. Two worlds of 'Grain and Soho' and 'Abadan and Whitehall' evolved in parallel, apparently unconnected.

During the negotiations over nationalisation the Iranians declared that all crude extracted in the country should be refined in the country, a move that invariably increases the revenue of a producer state and certainly helps build industrial employment. The AIOC's decision not to site a further refinery in Iran and to build at Grain meant that the company now had three plants in Britain – Llandarcy, Grangemouth and Kent. Consequently there was a greater number of refining jobs in Britain and fewer in Iran, and the income to the Exchequer in London grew, at the expense of Tehran.

Building on Grain was a key step in reducing AIOC's dependence on Abadan, and consequently made the company able to ride out any future loss of overseas refining capacity. The construction at Grain made it easier for Fraser to obstinately face down Mossadegh and refuse to cut a deal. It made it possible for the company to demand, and enforce, an embargo on Iranian crude and refined products and thus put immense pressure on the Iranian state. Grain was a weapon in the economic war which underpinned the covert war and helped fell the democratic government of the largest state in the Middle East.

However, siting a refinery in Kent only made sense if the British economy could consume sufficient oil, so it also became a weapon in a second battle that radically increased the UK market for petroleum products. The flow of the pipeline built in World War II from Walton to Grain was reversed. Aviation spirit was pumped west via Walton to Heathrow and into the tanks of Comet 4 jets that carried the likes of the young John Browne on his flights to Abadan.[27]

From the moment that Britain realised it did not have the strength to undertake military action on its own, the struggle between AIOC and Iran illustrated that London's power in the region was dependent upon Washington. The Suez Crisis of 1956 is usually seen as a marker of Britain's declining imperial strength and an affirmation of US hegemony, but this new reality had been highlighted five years earlier, in May 1951, when the US blocked the British plan to decisively act against the 'theft of their assets'.

By the end of the 1950s the dominance of the US over its foremost ally in World War II could be seen not only in geopolitics but also in Britain's domestic economy. BOAC began to supplement its fleet not with Comet 4s built in Hatfield, but with Boeing 707s from Seattle. The Kent Refinery was opened by the Queen in April 1955, but eleven months earlier her mother had opened the rival refinery at Coryton built by the Socony-Vacuum corporation of New York.[28]

In January 1953 Coryton received its first load of oil. A tanker had made fast to the import jetty and crude from beneath the Kuwaiti desert was pumped ashore. This liquid geology had been extracted by the AIOC and the US corporation Gulf Oil from the subsurface of Kuwait. The British Foreign and Imperial Office had enabled the creation of the state of Kuwait, carved from the Ottoman Caliphate in 1899. It was held as a British Protectorate until 1961 and was a keystone in the post-war British Empire.[29]

At Coryton the crude was driven into a Thermofor Catalytic Cracking plant. This machine, the most advanced of its kind in the UK, had been built in Greenwich and installed at the heart of the refinery. Subjected to intense heat, the compound structure of the oil of Kuwait was split on the marshlands of Essex. At the centre of the process was the catalyst silica. In this monumental chemistry set 30,000 barrels of crude a day were broken down into diesel and petrol for trucks and cars, radically altering the economy of southern England. Kuwait City became a little part of America built on the shores of the Gulf, just as Abadan and the oil fields that supplied it had become part of Britain in Iran. The movement of traffic on the roads of the Thames Valley, and of planes in the sky above, was chained to the transfer of oil from the Middle East.

Socony-Vacuum had two problems: how to supply refined product to their lubricant factories at Wandsworth in London and at Birken-head on Merseyside, and how to deal with the challenges in the international transfer of dollars after the birth of the Bretton Woods system. To answer these they built Coryton. Like the Marshall Plan, the refinery played its role in US post-war geopolitics akin to the Ford factory at Dagenham and the Exxon refinery at Fawley. It took two years to build Coryton and construction was led by a company team known as the 'Texas Taskforce'.

After 15 months of successful refining the plant was officially launched on 27 May 1954. Queen Elizabeth the Queen Mother journeyed by Port of London launch down the Thames from St Katharine's Dock in the City to open Coryton. She was accompanied by Winthrop Aldrich, the US ambassador, and Bernard Brewster-Jennings, the president of Socony, over from New York. The management staff formed a line, all wearing new bowler hats.[30] At the entrance to the plant stood Geoffrey Lloyd MP, the minister of fuel and power, beneath a Union Jack and a Stars and Stripes that fluttered in the breeze. That same summer in Memphis, Elvis cut his first record, 'That's Alright'. Within two years he was high in the British charts. America was coming, and with it rock and roll carried on vinyl.

In April 1959 TV viewers gazed in amazement at the gyrating pelvis of Vince Taylor. Dressed in black leather, Britain's answer to Elvis punched out one of the country's first rock & roll songs:[31]

> Well, my baby drove off in a brand new Cadillac
> Ooh, my baby drove off in a brand new Cadillac
> Well, she looked at me, 'daddy, I ain't never comin' back'
> I said baby-baby-baby won't you listen to me?
> Come on sugar, come on hear my plea
> Well she looked at my Ford, we'll never agree
> Cadillac car, oh yeah.

This explosion of cultural energy not only knelt before America, it also worshipped the car. The automobile was rock and roll, sex and youth, all rolled into one. What was pumped out of Kent and Coryton refineries was the fuel for wild rebellion. The refineries that had provided the energy for tanks and bombers in the war now also fed the sleek Comet 4s and the dreamed-of Cadillacs.

Around Abadan the world moved by cart and foot, but that world in Grain – the world of Rose Hammond – had been transformed into the oil world; transformed into a voracious market for the products of the refinery and the oil fields.

3
Baby, You Can Drive My Car

Asked a girl what she wanted to be
She said baby, can't you see
I want to be famous, a star of the screen
But you can do something in between
Baby, you can drive my car
Yes, I'm gonna be a star
Baby, you can drive my car
And maybe I'll love you
Beep beep'm beep beep yeah.

—The Beatles, 'Drive My Car', 1965

Bang on time Lazarus Tamana arrives at our appointed meeting on Level 5 of the Royal Festival Hall. It is a quiet place for an interview, with thick carpets, easy chairs and a wide view over the Thames. Lazarus, as ever, has a broad smile, but there is such sad weariness in his heavy lidded eyes. Impeccably dressed, he folds his tall frame into a sofa.

Tamana was born in the British Empire – just as we were. His mother was a farmer and his father a primary school teacher in the small town of Bodo in what was then the Owerri District of the colony of Nigeria. It was ultimately administered by the British secretary of state for the colonies responsible to the secretary of state for foreign affairs. In the year of Tamana's birth, 1957, these were respectively the Right Honourable Selwyn Lloyd, Conservative member for the Wirral,[1] and the Right Honourable Alan Lennox-Boyd, Conservative member for mid-Bedfordshire. Both men worked at the Foreign and Colonial Office in Whitehall, ten minutes walk from where we are sitting. At the time of Tamana's birth the Foreign Office had much of its work absorbed by the aftermath of the Suez Crisis.

THE SEVERN ESTUARY

Teifi

W   A   I

Tywi

Carmarthen

St. Bride's Bay

Haverfordwest

Milford Haven

Llandarcy Refinery

Pembroke

Llanelli

Angle

Carmarthen Bay

M4

Nea

Milford Haven Refinery

Angle Bay Terminal

SWANSEA

Port Talbot

Swansea Bay Tidal Power Scheme

Baglan Petrochemicals Plant

Bristol Ch

Lundy Island

Ilfracombe

N

W   E

S

Barnstaple Bay

Barnstaple

Bideford

Torridge

Taw

Tamar

Key

	Crude Oil Pipeline
	Gas Pipeline
M5	Motorways
	Mainline Railways
→	Imported Crude Oil
⊗	Closed Site

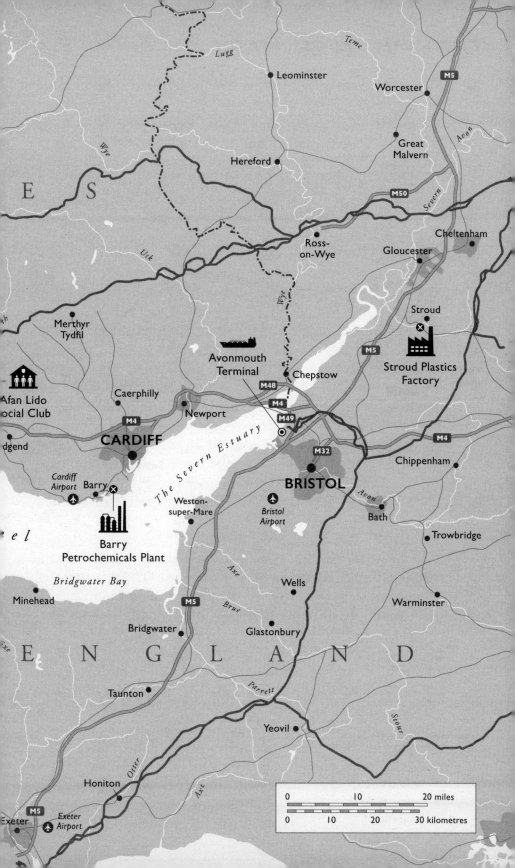

Leominster

Worcester

M5

Teme

Lugg

Great
Malvern

Wye

Hereford

Avon

M50

Severn

Cheltenham

M5

Ross-
on-Wye

Gloucester

Usk

Stroud

Merthyr
Tydfil

Stroud Plastics
Factory

Avonmouth
Terminal

Wye

Chepstow

M5

ESS

Afan Lido
ocial Club

Caerphilly

M48

M4

M49

M4

CARDIFF

Newport

Chippenham

The Severn Estuary

BRISTOL

M32

dgend

Cardiff
Airport

Barry

Avon

Barry
Petrochemicals
Plant

Weston-
super-Mare

Bath

*Bristol
Airport*

Trowbridge

el

Bridgwater Bay

Axe

Wells

Warminster

Minehead

M5

Brue

Glastonbury

E   N   G   L   A   N   D

Taunton

Parrett

Stour

Yeovil

Otter

Axe

Honiton

M5

Exeter

*Exeter
Airport*

0		10		20 miles

| 0 | 10 | 20 | | 30 kilometres |

Bodo is a town in Ogoniland, or Ogoni as it is also known, a country of 385 square miles and a population of 850,000.[2] After prolonged conflict it had been occupied by the British Army in 1901 and absorbed into the colony of Nigeria. Ogoni lies on the eastern edge of the massive Delta, an area the size of Belgium which is in essence a rainforest watered by the River Niger and the storms that blow in from the Gulf of Guinea. Ogoni, with its entirely distinct language, was essentially a self-sufficient farming and fishing culture, harvesting yams and vegetables in the alluvial soil and netting carp from the myriad creeks. Boats were powered by paddle – except for the vessels that passed up the river to Port Harcourt.

Tamana remembers his childhood. 'When I was in primary school we would go to fish in the nearby waters. My house was just 200 yards away from the creek. We would finish school, drop our bags, run into the creek and use our bare hands to catch fish and mudskippers and crab and all those things.'[3]

The construction of Port Harcourt in the 1910s, as a railhead and harbour just beyond the western edge of Ogoni, accelerated the integration of these lands into the imperial economy as farmers began to grow palm oil which was sold to British merchant companies such as the Royal Niger Company,[4] exported to ports such as Liverpool and used to lubricate engines or to be processed into soap.

However, the truly seismic shift in the life of the Ogoni came as a consequence of a commercial deal signed in Whitehall in 1936[5] by Sir Henri Deterding of Shell, Sir John Cadman of Anglo-Iranian and the secretary of state for the colonies, Right Honourable William Ormsby-Gore. The agreement gave the Shell D'Arcy Exploration, a joint venture between the two companies, a concession to drill for and extract oil anywhere in Nigeria's 370,000 square miles. It was, and remains, one of the largest concession deals ever signed.[6]

It triggered the start of oil exploration, especially in the Niger Delta whose geology had been known to harbour oil deposits for three decades. Shell-BP seismic teams prospected in the creeks and mangrove forests on and off for more than 14 years until in 1956 in Ijawland, just to the west of Ogoni, they finally found oil in sufficient quantities to be profitable.[7] In February 1958 Eric Drake, then BP chairman of supply and development, who had formerly been AIOC's key person in Iran, flew from Heathrow to Lagos airport, then travelled by road to Port Harcourt for the ceremonial announcement of the first shipment of Nigerian crude – known as Bonny Light.[8] Discoveries within Ogoniland itself quickly followed and the life of the land and the history of its people began its terrible descent.

The impact of diesel-fuelled ships and petrol-fuelled vehicles, building landing jetties and access roads, cutting down forests for seismic tests and drilling rigs, constructing workers accommodation and compound fences, and laying pipelines and telegraph wires came as an acute shock to Ogoni. Here was the sudden intrusion of the petroleum world into a people who had next to no use for oil except as fuel for kerosene lamps.

As Tamana describes:

I remember very, very vividly when the tractors were rolling in some part of Bodo. Those heavy equipments. Shell were rolling in and laying pipes and digging the ground. We used to go and see them, look at them because for us kids it was exciting. After some time we started seeing some black stuff appearing on the surface of the mangrove forests, on the creek because we'd normally go and bathe in those waters before we'd go to school in the morning.

He continues:

They had money to throw around. They were spending it on themselves and all that, so they were very, very affluent. They can easily buy something in the market. ...There were a lot of whites because of the heavy equipment they were using and it was whites who were driving them, who were directing these things. The menial jobs like digging were done by the Nigerians.

But only once oil extraction began were the real impacts felt. The sinking of a drill deep into the rocks below the Delta was accompanied by spills. The soil around the drill sites became contaminated by crude and in a world of water it seeped, or was dumped, into the creeks. A thick sheen covered the rivers and killed the fish and plants. As with almost any oil well, the crude within the rock was found together with natural gas. This was vented to the surface, up a pipe rising 60 feet above ground, and set on fire. The resulting gas flares burned day and night. The flames illuminating the forest, the roar drowning out the birdsong.

* * *

Opposition to this industrial invasion began to mount as Tamana recalls: 'The people were not happy about what was going on. Fishermen who used to fish heavily were directly affected and then they started complaining to the district officers, to other people there, but nothing much was done about it. Some complaints were made directly to Shell.'

In 1965 Sam Badilo Bako, an Ogoni school teacher from Taabaa nearby, wrote an open letter to Shell-BP:

> Today we need education, we need employment, we want to eat, we want to live … We do not want our children to remain out of school … We do not want to lack money because Shell has removed land, which used to be our main source of revenue. In short, we do not want to live in a vicious circle in which we shall see Shell-BP as the authors of our misfortune and our oppressors … Is it not an irony that those who live on top of wealth should be the poorest people in the nation?[9]

But these protests were almost entirely ignored by the Shell-BP company in Port Harcourt,[10] by the headquarters of Shell and BP in London and The Hague, and by the government of Nigeria in Lagos. For on 1 October 1960, five years previously, Nigeria had achieved independence from Britain. The Nigerian government was technically free to run its own affairs as an independent country within the Commonwealth, but that freedom did not extend to the control of its mineral rights, to the realm of oil. Shell and BP were in command of the extraction and export of Nigeria's key source of foreign exchange. Their power echoed that of the AIOC in Iran a decade previously. The two multinationals were guided by their need to ensure return on capital. Others as well as them benefited from the oil extracted, but those benefits were unevenly distributed, as shown by the contrast between shareholders in London and villagers in Bodo.

* * *

In the autumn of 1965, not long after Sam Badilo Bako wrote his letter, crude travelled from deep beneath the Delta soil up to a wellhead in Bodo. The liquid was then pushed through the pipeline that ran from Bodo to the Bonny Oil Terminal. Compressors at several pump stations along the 20-mile route[11] drove the liquid over farmland and across the creeks. The line juddered as the crude pulsed through the steel tubes. Farmers and children had to step over the pipe that ran directly across paths from the villages to the fields and waterside. This pipe was a giant syringe drawing wealth out of Ogoni, out of Nigeria.

At Bonny the crude was gathered in massive white storage tanks awaiting the arrival of a ship from Europe or the US. The MV *British Admiral* moored at the quayside was on her second voyage. She was BP's first Very Large Crude Carrier (VLCC), with a deadweight of over 100,000 tonnes and capable of carrying a cargo of half a million barrels of oil. The massive vessel had been built at the Vickers Armstrong shipyard in Barrow and launched in March 1965 with full ceremony by the Queen, with Maurice Bridgeman, chairman of BP, in attendance.[12]

In the heat of the Gulf of Guinea the crew, in their white shorts and knee-length socks, oversaw the *British Admiral* slip her moorings and set course for the Atlantic horizon. The coast of West Africa shrunk to a thin purple line. Days later the tanker rounded Cap Vert and headed north, bound for Britain. It was a calm crossing of the sea roads east of the Azores and after a two weeks the crew sighted land at Scilly. Passing between those islands and Land's End, she entered the Celtic Sea and made for Angle Bay Ocean Terminal at the mouth of Milford Haven.[13,14]

In Pembrokeshire, to the west of Llandarcy, lay Milford Haven whose qualities as a deepwater harbour were renowned. In these long inlets parts of the Allied navy had gathered before the invasion of Normandy. In 1959 the directors of BP sanctioned the financing of the Angle Bay Ocean Terminal.

Into the waters on the southern shore of Milford Haven a concrete jetty was constructed, leading to a pumping station on the fields of the village of Angle. From there a trench was dug running 65 miles to Llandarcy,[15] passing under the estuaries of the Taf and the Tywi. Into the trench was laid a steel tube a foot and half in diameter. In contrast to Ogoni, the oil in Wales passed entirely underground, out of view and out of mind of farmers and schoolchildren. This pipe was a syringe injecting a drug into the bloodstream of Britain.

The amount of crude arriving at the gates of Llandarcy increased from a million tons a year to nearly eight million tons at the start of the 1960s. The refinery was expanded to turn this raw material into its standard products, but such was the scale of the throughput that it could also provide naphtha feedstock, the lifeblood of petrochemicals plants. So from Llandarcy a further pipeline was built to run two miles under the River Neath and pump naphtha to the heart of the BP factory that had just been built on Baglan Burrows.[16]

Baglan Burrows, or Baglan Moors, was an extensive area of sand dunes that curved around the northern edge of Swansea Bay, stretching for four miles between the Avon and Neath Rivers. This plain of sand and marram grass ran inland up to the foot of the hills, the mynydd of south Wales. On the hillside clustered the homes of coal miners in the village of Baglan and on the slopes behind were farms such as Cwmclais. The sand dunes were occasional grazing for sheep, a place to collect dewberries in summer and a wonderful playground for the children of Port Talbot. In 1960[17] it was chosen as the site for a revolution, as transformatory for Wales as the drilling rig in Bodo was for Ogoni.

The previous year the National Coal Board had announced the closure of three mines in Port Talbot – Glenhafod, Aberbaiden and Pentre. As pits were 'rationalised' in the West Glamorgan Coalfield, it was another indication that employment in the south Wales mines was declining fast.[18] Some found work in the Port Talbot Steel Works, others moved away, like the families from Aberfan who migrated to Kent and built the refinery on Grain.[19]

The age of coal was clearly waning and that of oil waxing.[20] In this new era a field of rapid growth was petrochemicals. Prior to the 1950s chemicals from oil had been almost entirely manufactured in the US, but by 1962 Britain was the largest petrochemicals producer in Europe making plastics, synthetic rubbers, manmade fibre textiles, detergents and more.[21]

The board of BP under Sir Neville Gass recognised that the world of petrochemicals offered the prospect of a whole new market for oil products, a whole new territory of profit. And if BP did not stake its claim on this land, then its competitors would beat them to it. The essential weapon in this conquest was the cracker. Similar to the catalytic cracker in a refinery such as Coryton, this was a machine that took naphtha and broke it down, into ethylene, propylene, olefins and butadiene. These chemicals themselves became the raw materials for products such as polystyrene, synthetic rubber, PVC and so on. They were the building blocks for the industrial manufacture of food, housing, clothing, transport, household goods, pesticides and much, much more. Out of a cracker came a whole new world.

In 1960 BP already had three crackers in operation at its Grange-mouth plant outside Edinburgh. By building another it would become the largest ethylene producer in the UK, outstripping its rivals Shell and ICI.[22] Ethylene was the most flexible and valuable chemical, or 'intermediary', to come out of a cracker. The board debated carefully as to whether a fourth should be constructed on the Grangemouth site, but concluded that there was not sufficient market for chemicals in Scotland to absorb further output. So Baglan was decided upon and BP built a bridgehead into a new territory for plastics primarily in southern England. The company's aim was to make Baglan as important as Grangemouth and the new cracker in south Wales was to add 10 per cent to the UK's ethylene-manufacturing capacity.[23]

On 31 January 1961[24] BP made public its decision to open the chemical plant in Baglan[25] at a cost of £13 million, and thereby created 600 new jobs. The company[26] requested that Port Talbot Council build 100 houses for key workers,[27] adding them to the Sandfields estate which had been under construction on the southern end of Baglan Moors for a decade.

Two and a half years later, on 16 October 1963, the massive new plant was built,[28] up and running and ready for its official opening. The minister for housing and local government – incorporating Welsh affairs – represented the Conservative Macmillan cabinet, standing alongside John Morris, Labour MP for Aberavon. Ironically, given his future reputation as a Thatcherite opposed to state industries, the minister was Sir Keith Joseph. The *South-West Evening Post* announced that this project, which had obliterated the sand dunes with its dewberries and grazing, had meant that Baglan Moors was a 'Wasteland turned into a wonderland'.[29]

The previous day, 15 October, Harold Macmillan, labouring under the weight of the Profumo affair sex scandal and lying in hospital after a prostate operation, informed the cabinet that he would be resigning as prime minister. He felt that he had been killed off by a gang of backbench MPs, but in truth he already looked like a relic of the 1950s. Sir Keith Joseph was whisked away from the ceremony back to London to deal with the cabinet emergency.[30] He travelled by navy helicopter, because the M4 motorway, intended to stretch from London to south Wales, was still under construction. By this date it had reached no further west than Maidenhead in Berkshire.

The Sunday before the opening, on 13 October, the Beatles performed their hit 'From Me to You' live on the ITV programme *Sunday Night at the London Palladium*.

Through this programme 'Beatlemania' was officially born in Britain. Sex scandals, pop music, plastics[31] – the Swinging Sixties were well underway.

* * *

The ethylene plant was the heart of the new Baglan factory.[32] Each year 60,000 tons of the chemical was produced from the naphtha feedstock. Combined with chlorine imported to Baglan by rail, 1,000 tons a week[33] from the Distillers factory at Sandbach in Cheshire,[34] it produced ethylene dichloride, or EDC. This was then exported to other factories, particularly those within a hundred-mile radius such as Barry, Newport[35] and Stroud.[36] Several times a week a train of tank cars would travel south from Baglan to the Distillers factory at Barry. There the EDC was turned into vinyl chloride monomer and then polyvinyl chloride, or PVC, the wonder material of the plastics age.[37] Rapidly a Severn Estuary oil province was evolving, stretching from Angle Bay to Stroud near Gloucester, with Baglan at its heart and BP firmly in command. Wales was transformed through the building of infrastructure that could ingest a massively increased crude load and then process it into petrol, plastics and pesticides.

Some of Barry's PVC output was loaded onto road tankers and rumbled up the A48 and the new M4 motorway to the EMI record plant in Hayes, west London. The Beatles recorded their new album *Rubber Soul* at EMI's Abbey Road studios over 13 sessions between 12 October and 15 November 1965. The tracks were cut into lacquers on lathes at Abbey Road and these were used to create the metal master at Hayes. The PVC was poured into the moulds and, with all the machines running, Hayes produced 750,000 records in the firm expectation of high demand.

Copies of the album were distributed across the UK. Delivery trucks rolled back down the M4 and took a consignment to Derricks Music on Church Street, Port Talbot. The first dedicated pop and rock shop in Wales[38] had been keen to pre-order the LP. On 3 December the album went on sale and from radios and Dansettes in Sandfields rang out the words of the first track – 'Drive My Car':

Yes I'm gonna be a star
Baby, you can drive my car
And maybe I'll love you

Carried on the liquid rocks from beneath Ogoni, the sounds of the Beatles spilled out into the south Wales air.[39,40]

* * *

The mid-1960s were boom years in Sandfields. Although the Steel Company of Wales was facing competition and laying off staff, employment in Port Talbot remained buoyant as men found work in the chemicals plant where the number of jobs doubled over the decade.[41] Sandfields became an iconic estate, busy with its own pubs, shops, school and lido. This was the thriving community in which the Baglan plant grew.[42]

In June 1965 the Queen and Prince Philip visited. Passing crowds of school children waving Union Jacks in the driving rain, they opened the Afan Lido with its sparkling swimming pool and sports hall. There to meet them was the manager of the lido, Graham Jenkins, who had grown up in a mining family. His brother Richard Jenkins, now better known as the actor Richard Burton, had married Elizabeth Taylor the year before. A son of the coal world, Burton was an icon of the glamour that petroleum enabled, of films and cars, jet travel and distant cities.[43]

Sandfields Comprehensive School, opened in 1965, was the first purpose-built comprehensive in Wales and had a Great Hall capable of accommodating an audience of 1,200. The school was seen as an outstanding pioneer in education, especially in theatre. Godfrey Evans, a drama specialist appointed to the English department, embedded theatre at the heart of the institution. As Kevin Matherick, the head of drama, recalled, 'People used to come in droves from across the UK to see the state of the art facilities.'[44] All of the 2,000 pupils had weekly drama lessons in their first three years and the option to take acting exams after that. The actress and former pupil Francine Morgan remembered seeing plays such as Brecht's *Mother Courage* and a *Hamlet* in the round, and that the department exuded 'an air of something extraordinary'.[45]

The homes on the Sandfields estate built in the 1950s and 1960s, many of which housed workers from the BP factory, were all equipped with indoor toilets, running water and electricity. A world of modern comforts quite distinct from the mining villages from which many of the new inhabitants had come. Down on the Aberavon seafront was the Miami Beach funfair with its Florida Restaurant – the dream of America in south Wales. Along the foot of the mountains was built the Port Talbot Bypass, raised up on stilts, soon to become part of the M4 running to London. Like the Thames Estuary, with its network of refineries and car factories, new towns and airports, the Severn Estuary was being built out of the coal age and into the oil age. These were the fruits of British social democracy, built on an empire of oil.

* * *

In 1959 Shell-Mex and BP Ltd employed the services of advertising wizard John May of the agency S. H. Benson to work on a new brief. He was to promote the benefits of oil-fired central heating in the home.[46] At the time only a tiny proportion of households had such a luxury and it was widely dismissed as being unnecessary in Britain's temperate climate. May recognised that to encourage the purchase of oil, and the installation of entire plumbing systems, the company had to sell not a product but a lifestyle, the promise of a home that was warm, clean and modern.

May suggested that a film be shot portraying that future, proposing the title *Mrs 1969* and then improving on that as *Mrs 1970*.[47] The resulting advert showed an actress as a paragon of domestic bliss, leaping exuberantly around her kitchen, answering the telephone, proudly receiving a saluting deliveryman and playing with a naked baby; sometimes she looked alert in angora, at others sleepy in her nightdress. To sell heating oil, one of the bi-products of refining, Shell and BP sold Britain – and especially the middle class – a future to aspire to.

We imagine a family moving from elsewhere in south Wales when the man started work at Baglan in the early 1960s. They rented a home from Port Talbot Borough in the Sandfields estate. A year or so after starting work, he would have crossed the threshold into a house extruded from the output of the petrochemicals factory. Although Baglan itself did not manufacture any household products, the intermediates that the factory created were transformed in other plants into the building blocks of a home. Just as the ethylene dichloride from Baglan was transformed by factories in Barry and Hayes into the *Rubber Soul* album.

On stepping into the house, the carpet underlay and the floor tiles were made from styrene monomer, the emulsion paint on the walls from vinyl acetate, the ceiling tiles from polystyrene, the electric light switches and plugs from phenolics, the cable from which the bulb hung was coated in PVC. In the front room the windows were in PVC frames, the foam in the cushions on the settee was from propylene, the cabinets of the TV, radio set and Dansette were cast from styrene monomer. The shelf of albums and rack of singles were all PVC. In the kitchen the refrigerator, washing machine, vacuum cleaner and electric iron were all partly made of polystyrene, the bowls, buckets and bottles of detergent around the sink were from polyethylene, the rubber gloves and sponges from styrene monomer, and the pipes that brought water to the sink were PVC. All the Formica-covered surfaces contained phenolics. Much of the food in the fridge and cupboards was wrapped in PVC packaging. Out in the garden the hose was from styrene monomer and drainpipes, water butt, flower pots and plant trays were all from PVC. When he turned to passionately kiss his true love, she wore a dress made of vinyl acetate and her lips were covered in propylene. Potentially almost all that he touched had poured out of Baglan.[48]

This plastic world of comfort was designed in part to draw profit from the crude pumped out of the Middle East and Africa onto the British market.[49] However, these blunt economics were hidden by the dazzling colours of wipe-clean modernity. In March 1965 Charles Evans, managing director of BP's chemicals arm BHC, visited Baglan. In his speech he declared, 'All within the company should be especially proud of the contribution our products have made to the better standard of living in Britain.'[50]

Out on the street in front of the house in Sandfields was a new car, perhaps a Ford Cortina made in the Thames Estuary at Dagenham. It may have been brought on hire purchase using his wage from Baglan, but it belonged to both him and his wife. For the 1960s saw a boom not only in car ownership but also in women driving. In 1956 Petula Clark sang 'With All My Heart' and posed beside a brand new pink Turner 950 sports car, which symbolised her independence and sexiness. Three years later, in Vince Taylor's 'Brand New Cadillac', it was the woman who was driving. In the Beatles' 'Drive My Car', she's telling her man that he can drive the car that belongs to her. This was more than about getting around, this was about getting it on.

Inexorably cars flooded the streets of Sandfields and filled the lanes of the M4 as it extended west. The lust for vehicles, and the petrol and diesel that drove them, not only meant profit for refineries such as Llandarcy and Grain, it also fuelled the booming motorcar industry and all its ancillary factories which required more naphtha. Petula Clark's pink sports car had been fitted out with leatherette[51] that was manufactured in the Distillers plastics plant in Barry.[52] By the mid-1960s Baglan was part of that supply chain funnelling materials towards the production lines of Birmingham and Coventry. The Port Talbot Steel Works was pulled out of a temporary slump by the growth in demand for metal to build Fords, Vauxhalls, Minis and much more.

BP was determined to outstrip its rivals Shell and ICI and so plunged into a spree of acquisitions, most importantly buying the chemical and plastics division of the drinks group, Distillers, in 1967. By the following year the company had plants in Grangemouth next to Edinburgh, Wilton on Teeside, Saltend near Hull, Sandbach and Oldbury near Birmingham, Carshalton in London,[53] Hythe near Southampton and the Severn Estuary Oil Complex that combined Stroud, Barry and Baglan. By the end of the Sixties BP was the UK's largest petrochemicals manufacturer and oversaw this realm from new offices at Piccadilly and Victoria in London.

The modern homes in Sandfields were kept clean and bright in part by the oil from the Niger Delta, but comforts of the petroleum world extended to only the tiniest fraction of Ogoni society. The future portrayed in *Mrs 1970* had arrived but, as was acutely obvious, it was unevenly distributed.

* * *

However, the promise of wealth from oil in the Niger Delta exacerbated the tensions within Nigeria, a state created by the British by combining over 30 distinct peoples and ethnic groups. The Igbo, in south-eastern Nigeria, declared independence and established the state of Biafra, which encompassed much of the new oil-producing region. Nigeria under the recently installed military dictator, General Yakubu 'Jack' Gowon, invaded Biafra on 6 July 1967 and a bitter war ensued in which an estimated 500,000 to two million civilians died, many of them through starvation. Images of children with bloated stomachs and emaciated bodies were shown around the world in the press. In Britain they even featured on *Blue Peter*, the children's TV programme.[54]

Tamana described the events through the eyes of his ten-year-old self:

We started hearing shelling from the nearby town of Bonny. The Nigerian army were now in Bonny shelling the Biafrans, and then warplanes started coming. They drop bombs and shoot rockets on the village. Before it is 5.00 am in the morning, all of us would go and hide in the thick forest where we dug trenches and we just stayed in those trenches.

An indication of how concerned the British were about the war is given in a document sent from Prime Minister Harold Wilson to the foreign and commonwealth secretary in July 1969, three weeks into the war. It states:

> It is of national importance that the Shell-BP investment in Nigeria, which is a major British interest, of crucial importance for our balance of payments position and for our economic recovery, should be protected. I therefore wish that everything should be done, as a matter of greatest possible urgency ... to help Shell-BP and the Federal Nigerian authorities to establish effective protection of our oil investments.[55]

Eventually Biafra surrendered to the federal government forces on 7 January 1970. General Gowon blamed part of the tension behind the war on the former colonial power, Britain, and 'the expatriates, the ex-imperialists and the oil companies'.[56] Even before the war, in 1968 and 1969, the Nigerian government had passed legislation to provide for the 'Nigerianisation' of the oil industry, and now this process was accelerated.[57] David Llewellyn, a BP executive, after a visit to Nigeria in March 1972 reported that companies with colonial or imperialist connections were deeply distrusted, and that 'under the relentless pressure to Nigerianise the few capable and experienced expatriates who have been underpinning the existing system are being squeezed out'.[58]

These demands for change took place within a period of profound shift in the international oil industry. In the shadow of the Iranian nationalisation and the subsequent coup against Mossadegh, activists throughout the oil-producing nations, and particularly in the Arab World, were acutely aware of the injustice of the global balance of power. On 14 September 1960 the Organization of the Petroleum Exporting Countries (OPEC) was established in Baghdad. Spurned by the Western oil companies, the governments of Europe and the US, the cartel lacked power of its own. However, slowly it acquired it by using the dynamics of the oil market in the 1960s and 1970s. A demonstration of its strength followed the overthrown of King Idris of Libya in 1969. The leader of the coup, Colonel Qaddafi, immediately declared his hostility to the West and to the controllers of Libyan oil, the Western oil corporations including BP, Shell and Occidental.

The battle between Iran and AIOC had demonstrated just how difficult it was for a government to challenge one corporation that had a dominant position over the sole industry that determined so much of the economy of the whole state. Libya was different. By chance more than design, the country's oil industry was in the hands of a number of foreign corporations, some of which – such as Occidental – were relatively small and could be more easily pressured than BP or Shell. Furthermore Libya, with a small population and wealth accumulated from a decade of oil production, could weather any embargo far more easily than Iran had done. As Qaddafi pointed out, 'The Libyan people who have lived without oil for five thousand years, can live without it again in order to attain their legitimate rights.'[59]

Furthermore, the late 1960s and early 1970s saw the world uneasily balanced between two superpowers, especially as the US was mired in the Vietnam War. Libya could count on selling oil to the Soviet Union and Comecon countries of Eastern Europe and on receiving in return military and diplomatic assistance. Finally, Libya did not stand alone, but was closely allied with other radical Arab states, in particular Iraq, Syria and Algeria, the last of these still filled with the anti-colonial spirit encapsulated by the writings of Frantz Fanon. Qaddafi declared war on the Western oil companies in order to assert Libyan sovereignty and to be a beacon for the liberation movements of the Third World.

The Libyan deputy prime minister, Major Abdessalam Jalloud, announced that the oil companies in Libya would have to increase the sum they gave to the Libyans or else the state would nationalise the oil rigs, pipelines and export terminals that the corporations had built. Shell, BP and the other companies were terrified, not only by the prospect of expropriation but also the likelihood that if Libya were successful in getting higher sums for each barrel, then other OPEC states – in particular in the Middle East – would 'play leapfrog' and demand yet higher revenues. Eric Drake, chair of BP – who had been at the centre of the storm in Iran – and David Barran, chair of Shell, together lobbied the UK foreign secretary, Alec Douglas-Home, and the US State Department under Henry Kissinger. Both declared that they were unable to support the companies for fear of enflaming Arab opinion and driving these states towards the Soviets.[60] Without the military and diplomatic support that they had had in Iran, the Western corporations folded, accepting Qaddafi's terms.

As Drake and Barran feared, several OPEC states demanded parity with the Libyan deal, particularly Iran. By April 1971 the Tehran Agreement and the Tripoli Agreement had been signed and the percentage of the sale price of a barrel of crude that went to the producer states had increased substantially. The effect of OPEC utilising its powers drew others to the organisation, and in July 1971 Nigeria joined OPEC.[61] As each new producer country became part of the organisation OPEC grew in strength and the power of the corporations waned.

The two agreements were to hold for five years but in October 1973 OPEC demanded a meeting in Vienna to renegotiate. When the Western oil companies refused to comply OPEC unilaterally declared a 70 per cent increase in the price of oil per barrel from $2.90 to $5.11. The confidence of the OPEC ministers perfectly illustrated how the balance of power between the multinationals and the producer states had shifted, fundamentally in the latter's favour. Further oil price rises were on the way.

Part of the impetus for Qaddafi's struggle against America and its corporations had come from the desire to alter the Nixon administration's policy of supporting Israel. In 1967 the Israelis had defeated Syria and Egypt in the Six Day War and occupied all the territory of Palestine as well as the Sinai, creating over 300,000 Palestinian refugees in Jordan and Lebanon. Palestinian and Arab civil society was rapidly radicalised and began to take action against organisations that they perceived were supporting the 'Zionist state'. In August 1972 the Palestinian group 'Black September' blew up the BP-Shell Trans-Alpine Oil Pipeline in Italy, disrupting the flow of crude from Iran to refineries in West Germany.[62]

The governments of a number of Arab states, including those who had traditionally been staunch US allies, were extremely concerned that if they failed to try to roll back the Israeli occupation they themselves would fall to revolutions just as King Idris had. Three days prior to the planned October 1973 OPEC meeting in Vienna, Egypt and

Syria attacked Israel and the Yom Kippur War began. The Arab states, most significantly Saudi Arabia, immediately imposed a 5 per cent cut in production to be increased by 5 per cent for every month that the US refused to alter its policy of arming the Israelis. The Arab states also placed an embargo on all oil exports to the US and two states who were particularly supportive of its pro-Israel strategy, the Netherlands and Portugal.

The consequence of the threat to oil supplies to the West was dramatic.[63] Immediately prices rose in the UK and there was near panic in government. At petrol stations around Britain there were tailbacks as drivers struggled to compete in obtaining fuel. At petrol stations in Port Talbot, Gravesend, Liverpool and across the country queues of cars filled the roads. It perfectly illustrated just how dependent British society had become on oil and the oil companies.[64] There was widespread discussion of how corporations such as Shell and BP had come to own and control the modern world of cars and homes and music – the petroleum world. It seemed that unlike the Libyan people, the British people could not live without oil for a few weeks, and that in the face of having to do so their 'legitimate rights' might collapse.

On 21 October, four days after the Arab states announced their production cut, Prime Minister Ted Heath, intensely concerned about the impact of falling oil supplies on the UK's fragile economy, summoned Eric Drake of BP and Frank McFadzean – who had replaced David Barran as chair of Shell – to Chequers (the prime minister's country residence) for a formal dinner.[65] Heath pressed upon Drake and McFadzean that they should loyally assist Britain in its hour of need and prioritise crude deliveries to UK refineries, such as Llandarcy or Shell Haven, above those destined to other states. Drake resisted, emphasising that this would mean BP having to break its collective responsibilities to other European states now that the UK was a member of the European Economic Community (EEC). The companies had asked Britain for support in their struggle with Qaddafi, but now the roles were reversed they declined to assist the government. Heath was livid and 25 years later wrote in his memoirs that he 'was deeply shamed by the obstinate and unyielding reluctance of these magnates to take any action whatever to help our own country in its time of danger'.[66]

Heath went on to face an even greater threat when days later the National Union of Mineworkers (NUM) declared an overtime ban, using the 'Oil Crisis' to emphasise just how vital they still were to the British economy. On 13 December Joe Gormley, president of the NUM, said in a BBC interview:

> The supplies of cheap oil are finished for ever. They [the government] are going to need more and more of our own indigenous energy sources to fill the gap. And they won't have anybody to do that because men will not work at the coalfaces of Britain for less than forty quid a week. If the country doesn't see the sense of our argument, then it's woe betide them for the future.[67,68]

By 1 January 1974 Heath had declared a 'three-day week', with electricity for businesses rationed to three working days, in order to deal with the 'Oil Crisis'. Four weeks later the NUM escalated their overtime ban to a full-scale strike, and by March Heath had lost the snap general election he called. After the Chequers meeting there were protracted discussions between BP and ministers – with managing director, Peter Walters, at the heart of the company's team – and BP did try to alleviate the UK's predicament. But it was far too little too late for Heath, and the Conservative leader always blamed his fall partly on the intransigence of BP and Shell.

The 'Oil Crisis' entered the popular imagination as the consequence of OPEC using their 'oil weapon', just as the lyrics of Bunny Wailer's 1974 track ran:

> It's the Arabs oil weapon
> It's the Arabs oil weapon
> It's the Arabs oil weapon
> It's the Arabs oil weapon
> It's the Arabs oil weapon
> Ain't got no ology
> To keep civilization going
> And by the signs of the times
> With all your scientific research
> You never reach completion

No more trains, no boats, no planes
You gotta try creation.

Indeed, the four years since the overthrow of King Idris had seen
the formal posted global oil price rise from $1.80 a barrel to $11.65
a barrel by January 1974. The cost of producing crude, in Saudi for
example, remained constant at 12 cents a barrel, and a far greater percentage of the profit now went to the producer states rather than the
oil companies. Henry Kissinger, the US secretary of state, described the
shift as 'one of the most pivotal events in the history of the century'.[69]

* * *

These years did show the very significant strengthening of the power
of the oil-producing states in relation to BP, Shell and all the oil multinationals and Western states. The oil price rise arguably enabled and
encouraged the producer states to nationalise the assets of foreign corporations. However, the oil companies were not – over the long term
– adversely affected, for they had sufficient political muscle to navigate
the change that was thrust upon them and ensure that the UK, as well
as other consumer states, carried the burden of that change. Not only
did this mean placing their own interests before Britain's – much to
Heath's fury – but they also benefited from the overall rise in the price
of oil. Once the Arab states lifted the embargo on the US, Netherlands
and Portugal in March 1974 and the 'Oil Crisis' was deemed to have
passed, the oil price did not fall.

The rise in fuel prices did have an effect on the petrol market in
the UK and this impacted the refining sector. BP stepped back from
a planned expansion of the Kent Refinery, and elsewhere plans by
other companies for new plants were shelved.[70] However, the rise in
the price of oil had the effect of dramatically altering the profit that
could be accrued from projects already underway. Shell, BP and other
companies had planned to exploit the highly capital-intensive UK
North Sea when crude oil was trading at around $21.50[71] a barrel.
When the first oil came ashore in the autumn of 1975, crude was
trading at $52.00[72] and what had seemed risky projects were now
highly lucrative.

Furthermore, the 'Oil Crisis' burst upon a world that was already deeply concerned about the prospect of oil – and other resources – running out. In 1972 the Club of Rome had published what became a much discussed study, *The Limits to Growth*.[73] In the UK as elsewhere this catalysed debate around the need to conserve energy and prepare for a post-oil world. The oil companies were themselves not unaffected by these ideas. From Shell came the highly influential M. King Hubbert who declared that US production had peaked, while at BP the chief geologist, Harry Warman, was making gloomy forecasts about the amount of oil that remained to be discovered.[74] In June 1973, before the 'Oil Crisis', Mobil announced that in the US it was ceasing to advertise gasoline and rather was trying to persuade people to conserve energy. 'Smart drivers make gasoline last', ran one ad; 'She plans to save a gallon a week', ran another.[75]

* * *

From the mid-1950s, chastened by the experience of nearly losing their key Iranian assets, the board of BP had striven to gain oil prospects across the globe. On 3 April 1959 John Pattison, Eric Drake and Harold Snow, under the chairmanship of Maurice Bridgeman, purchased drilling rights over 312 square miles of Alaskan tundra. The lease was brought from the US federal authorities in Washington, DC, for Alaska had not yet been formed as a state.

This swathe of the North Slope, facing the Arctic Ocean, was Inupiat land. A vast plain that bloomed with flowers in summer, home to caribou, fox and 200 species of birds. A land hunted and fished for thousands of years with husky dogs and whaling boats. But to the men who made the purchase, this land – which they had never visited – was effectively *terra nullius*, a land without people.

After nearly a decade's search, led by BP's geologists Harry Warman, Peter Kent and Alwyne Thomas, oil was discovered in commercial quantities in 1969. Then came a long struggle in which BP strove to construct an 800-mile pipeline to ship crude from its North Slope fields to the port of Valdez from where it could be shipped to the refineries of the US west coast. Fierce resistance to their plans was put up by the Inupiat indigenous peoples working in alliance with a number of environmental groups. Costs escalated and the entire BP Alaskan strategy, running since 1959, was beginning to look unwise. Suddenly the 'Oil Crisis' shifted power in BP's direction. Kissinger urged Nixon to send a message to Congress that events had 'brought home our vulnerability' and they should drop opposition to the Alaska pipeline. The Trans-Alaska Pipeline Authorization Act was finally signed into law on 16 November 1973. After nearly four years of legal battles the project was a go,[76] and the First Nations and environmentalists were pushed aside.[77]

Four months later, in April 1974, John Browne was moved from BP's New York office to San Francisco in order to focus on the construction of the Alaska Pipeline which started eleven months later.[78]

4

Dirt Behind the Daydream

Dirt behind the daydream
Dirt behind the daydream
The happy ever after
Is at the end of the rainbow
There may be oil
(Now looking out for pleasure)
Under Rockall
(It's at the end of the rainbow)
There may be oil
(The happy ever after).

Gang of Four, 'Ether', 1979

Just before noon on Boxing Day 1965, Ronnie Farrow, the head chef
on BP's *Sea Gem* drilling rig, tossed and turned in his bunk. He was
working night shifts in the southern North Sea and was desperately
tired. The high winds and the screech of metal upon metal made it
impossible to sleep. Suddenly the rig lurched violently to one side and
he was thrown across the room onto the bunk of his cabin-mate. Both
men, clad only in their underpants, ran to see what was happening.
The rig had heeled over alarmingly and pans, food and crockery were
strewn across the galley floor. Farrow dashed back to the cabin, grabbed
a coat and went up on deck where he knew there was a lifeboat by
the radio-operator's shack. It had gone. To his amazement, he saw the
radio shack floating on the sea, and the lifeboat alongside with the
radio-operator sitting in it. Farrow yelled at him, but he seemed dazed,
staring into space.[1]

By now the rig was listing at a steep angle and, desperate to reach
the boat, Farrow jumped into the sea, only to be lifted up again by
a wave which swept him beneath a metal companionway, trapping
his legs under a heavy object. As the rig tilted further, the water rose

around him. He had almost given up hope when Ken Forsythe, one of the drilling crew, grabbed him by the arm and pulled him to safety. By a stroke of fortune, the object that had trapped his legs was a life raft. They quickly inflated it and jumped in. More men clambered aboard until it was packed with wet, shivering bodies.

After two hours the rig collapsed entirely. By sheer luck a passing freighter, the *Baltrover*, saw the *Sea Gem* foundering and went to assist the survivors amid freezing waves 20 feet high. A helicopter from RAF Leconfield, near Beverley, flew through a snowstorm to aid the rescue. Thirteen of the 32-man crew drowned 40 miles off the East Yorkshire coast. Another man died of hypothermia the following day. Among the dead was the radio-operator.

Forty years later the *Sea Gem* derrickman, Kevin Topham, recalled the night: 'The rig seemed safe enough, but evidently it wasn't. I'd been in a number of hot spots in the armed forces and this was another one of those sequences of events you have to get through. I remember it very clearly of course. Some people still have nightmares about it, I'm told.'[2]

In the afternoon of that cold, bright Boxing Day, David Steel, the head of BP's Exploration and Production division in London, was informed of events. Head office was closed for the holiday, so he was at home in Onslow Square, Chelsea. By the evening BP had given its response to the press, and the news was on the BBC.

Steel was known as a man who was modest, patient, intensely conscientious and seemingly at ease with all social classes. In World War II he served as a tank commander fighting in the fall of France, the North African desert and the invasion of Italy, followed in the rear by Mac Macalister's mobile field hospital. Steel was awarded a Distinguished Service Order and Military Cross for bravery, and summed up his experience succinctly: 'Tank shot up six times – not wounded.' Perhaps, like Kevin Topham, he too saw the *Sea Gem* disaster that day through the lens of his military experience. For offshore drilling was in its infancy and the North Sea had all the dangers and camaraderie of warfare.

Even before the rise of OPEC the board of BP was desperate to obtain reliable new sources of crude. The company was struggling in Iran, only a decade after the coup against Mossadegh. Steel was

THE DEE & FORTH ESTUARIES

St Fergus
Gas Terminal

Fraserburgh

Peterhead

Cruden Bay
Pump Station

Aberdeen Bay
Wind Farm

ABERDEEN

Dyce
Airport

Dee

Netherley
Pump Station

Stonehaven

Forties
Pipeline System

Don

Aberdeen Gas
Compressor

Dee

North Esk

Balnabriech
Pump Station

Lossiemouth

Eden

Kirkcaldy

Mossmoran
Chemical Plant

Hound Point
Terminal

EDINBURGH

Edinburgh
Airport

Dalmeny
Tank Farm

South Esk

M90

Dunfermline

Forth

Union Canal

M9

Kinneil
Stabilization
Terminal

Forties
Pipeline System

National
Gas Pipeline
System

Grangemouth
Refinery

Falkirk

M8

Almond

Avonbridge
Gas Compressor

M9

Stirling

M9

M876

M80

Carron

*Forth &
Clyde Canal*

M80

M73

Forth

Tummel

0   5 miles
0   5   10 km

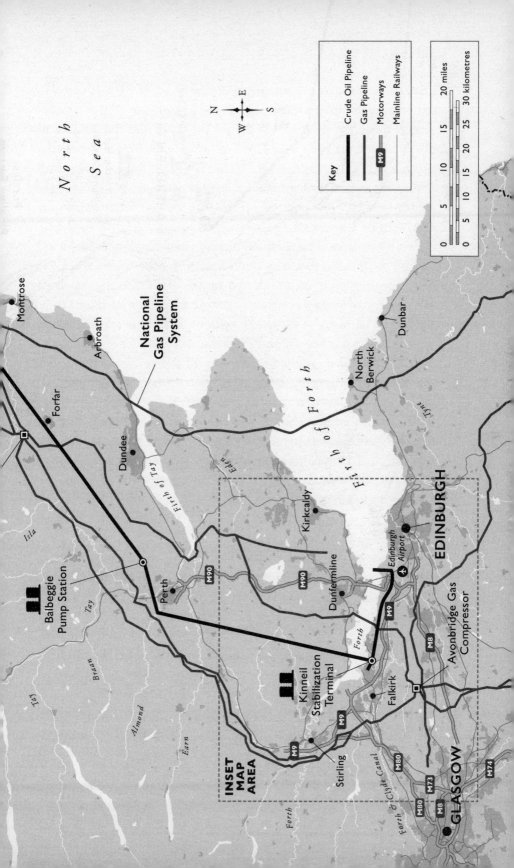

involved in tortuous negotiations with the Shah, as BP attempted to maintain terms highly favourable to the corporation. These circumstances drove the purchase of licences in Alaska and the pressure to develop resources in Nigeria. However, in 1959 Shell had discovered a massive gas field in the northern Netherlands and this had opened up unforeseen opportunities. If the geology beneath the North Sea held similar resources, and especially if this was the case in the British sea area, then BP's future might look brighter.

Five years later a competitive bidding round was opened for oil and gas exploration blocks in the UK sector of the North Sea. Each 'block' constituted of an area of 250 square kilometres over which an oil company, or usually a consortium of oil companies, purchased from the British Ministry of Power the right to drill for oil and gas for six years. Many executives within BP felt that as a national champion, the company should be granted rights over all the 32 blocks that it bid for. Maurice Bridgeman, chairman of the corporation, hoped that BP could expect its position as a wholly British company, majority owned by the British government, to count in its favour in licence awards.[3] One BP manager, writing to the company's chief geologist, thought that they had a 'moral right' to favourable treatment. Angus Beckett, the permanent under-secretary at the Ministry of Power,[4] later remembered ensuring that BP received the most sought-after blocks in the face of stiff competition from its rivals.[5] The corporation was awarded 22 licences in September 1964.[6]

Immediately BP set to work. The 'most sought-after block', a section of seabed drawn onto the abstraction of an Admiralty chart, was Block 48/6, some 40 miles east of the Yorkshire coast. No technology existed to drill for oil or gas this far offshore and in this depth of water, so a new drilling rig, the Sea Gem, was constructed by leasing a barge from a French company and converting it in Le Havre. It was then kitted out with a drill tower and accommodation units in Smith's Dock shipbuilding yard, Middlesbrough[7] and towed to Block 48/6. On 17 September 1965 the Sea Gem hit gas and the energy prospects of the UK shifted overnight. Three and a half months later came the Boxing Day disaster when BP's technological innovation sank beneath the waves taking the lives of 14 men.

The rig was constructed in France and England, but the culture of offshore work in the UK North Sea, right from these first moments, was essentially American. The techniques, and many of the staff on the rigs, were imported from the US Gulf Coast, bringing with them the experience of Louisiana. Life on the oil platforms was a version of the American South – well fed, intensely male, primarily white and entirely non-unionised. It echoed the world of the military, with its constant threat of danger. A world set apart from the civilian realm, away from communities, away from women and children, and far away from the comforts of life at head office. These were the conditions of the men, ultimately under the command of David Steel, who advanced into a new territory for BP, and created a new colony for the UK.

*   *   *

At Britannic House, BP's head offices in the City of London, senior management gathered on the morning after Boxing Day 1965 to consider the consequences of the disaster off the Yorkshire coast the previous afternoon. Eighteen men were recovering in hospital in Hull, but work continued in Northern Ireland on the construction of the *Sea Quest*, a much larger sister to *Sea Gem*. The *Sea Quest* was being built from 150,000 tons of welded steel at the Harland and Wolff Yard in Belfast.[8] It was due to be launched in twelve days' time and towed to explore the North Sea blocks over which BP had purchased rights. Despite the disaster, the campaign to conquer this new territory would not cease.[9]

The *Sea Quest* spent the next four and half years in the waters north-east of Aberdeen, but progress was slow as BP was convinced that there was little chance of finding commercially viable oil in the northern North Sea.[10] The seismic exploration carried on and was supported by supply boats working out of the harbours along Scotland's north-east coast – Dundee, Peterhead, but most of all Aberdeen. This, the 'Granite City', had been slowly slipping into decline, in part linked to the falling North Sea fish stocks. It was a town sliding into poverty, until suddenly the lure of oil and gas offshore quickened its pace.

The cold of Aberdeen was a universe away from the new Britannic House in London – or Britannic Tower is it was more commonly known. This 35-storey block was commissioned by BP under Maurice

Bridgeman and opened in 1967. It was the tallest skyscraper in the City, a part of New York airlifted to London and striking in its extravagance. It had a high lobby with marble pillars, acres of panelled conference halls and 'state rooms', a self-contained village of canteens, shops and a pub, and a garage beneath with a fleet of company Jaguars. In front was a wide piazza with a flagpole flying the green shield of British Petroleum. In the middle of the piazza a metal object like a contemporary sculpture commemorated the source of much of BP's wealth: an old 'Christmas Tree' well head from the Iranian fields, which had first flowed with oil in 1911, and since that year had carried a total of 50 million barrels.

Despite the modern appearance of the new headquarters, the company that moved in from its old imperial home retained its conservative culture. On the top two floors were the offices of the chairman, the deputy chairman and the managing directors, along with the boardroom, the chairman's dining room and the directors' dining room – known jokingly among the staff as the 'Golden Trough'. Beneath them, both physically and hierarchically, was the senior luncheon club's dining room into which female staff or female guests were not permitted. The staff manager commented at the time that it was unnecessary to promulgate this rule because it was a well-known 'tribal custom'.[11] In the lobby of the tower there were two lifts, one which stopped at all the floors, and one which went directly to the chairman and board's two floors at the top. The latter was an entirely male preserve.

BP under Bridgeman was unaffected by the changes in corporate culture that had taken place in the US, and by the shifts in the wider culture that were happening across the UK. The year after the new offices were opened, 850 machinists at Ford's Dagenham plant in the Thames Estuary[12] went on strike for equal pay and Barbara Castle MP was appointed first secretary of state. BP was not alone in using its weight in the economy to preserve the dominance of men in positions of power. The story was the same at Shell and in the finance houses of the City that the oil companies depended upon. It wasn't until March 1973 that the first women were allowed to trade on the London Stock Exchange.

* * *

We imagine the lift ride of 35 floors up Britannic Tower to emerge into what are now the offices of Winston & Strawn. The fabric of the room has been utterly changed since Bob Horton, the last BP chairman to use this building, vacated in 1991. However, the view out of the floor to ceiling windows, from a room that is higher than the top of St Pauls' Cathedral, would remain much as it was then. New blocks such as the Gherkin have arisen, and the distant towers of Canary Wharf would be seen away to the south, but the scene still gives a sense of being able to survey London, from Westminster to the City. The capital of the UK spreads out beyond.

On Monday 1 December 1975, David Steel settled behind the desk. It must have felt like a fitting birthday present – he'd been 59 two days before. After 25 years with BP he'd finally made it to the top. It had been clear since January that he would succeed Sir Eric Drake to the post of chairman. He was no stranger to the room, he had regularly come there for meetings since Maurice Bridgeman, Drake's predecessor, took up residence eight years previously. As deputy chairman since 1972, Steel's office had been on the floor below, during which time he too rode the lift to the top and tucked into the Golden Trough.

Soon after taking up his new position, Steel held his first board meeting as chair. Seated around him at the table were five other executive directors: Monty Pennell, the deputy chairman; Robin Adam, the finance director; William Fraser;[13] Christopher Laidlaw; and the man who would succeed him, Peter Walters. All of these men were in their mid-fifties, except for Walters who was only 44. The rising star.[14]

Also at the table were the six non-executive directors, who were not involved in the company's daily operations but helping to guide its future strategy. There was Samuel Elworthy, a retired senior general; Michael Verey from finance; Lindsay Alexander from shipping; and the Earl of Inchcape representing the Burmah Oil Company. Alongside them were the two non-executives appointed by the British government: Sir Denis Greenhill, the former head of the Foreign Office and part of MI6; and Tom Jackson, the general secretary of the Union of Postal Workers.

As the British state owned 68 per cent of BP's shares,[15] it had the right to appoint two directors. These two men were supposed to act at arms length from the government but they reported from the meeting to the prime minister's Cabinet Office, and in particular to the chancellor of the exchequer, Denis Healey, and the minister for energy, Tony Benn.

Benn is of course legendary as an icon of the British Left, and a standard bearer of English socialism for half a century. He was also minister of energy, who on behalf of the UK state had to lead the fight over North Sea oil against arguably the most powerful capitalist organisations, the international oil companies. By the time this battle had begun, Benn was in his early fifties and had behind him 25 years as an MP and five years as a minister. He was already known for his intense hard work, his independence of mind and the courage to be outspoken.[16] Critics on his own side saw him as an unruly maverick.[17]

His mother was a feminist and theologian. His father had been a Labour MP and minister for air in the Attlee government, where he oversaw the establishment of London Airport, later called Heathrow. As a teenager Tony Benn spent much of his time at Stansgate on the Essex side of the Thames Estuary, just seven miles north of the Shell Haven and Coryton oil refineries. Later, as an MP, Benn was profoundly committed to British parliamentary democracy and a harsh opponent of anything he perceived would undermine it, such as the EEC or private corporations.

Benn was an active champion of 'open government' and an assiduous keeper of diaries. These were published in tranches throughout his career and provide a mesmerising insight into the workings of government, revealed through his frank, and often irreverent, assessment of the events he witnessed. Their existence, alongside the diaries of his colleagues such as Barbara Castle and Roy Jenkins, highlights the fact that none of his counterparts in corporate life, such as Steel or Walters, have ever published their diaries.

David Steel's[18] first board meeting as chairman would have been as orderly as one might expect. But on coming into office Steel, who had spent much of his career in battle with the likes of the Shah of Iran, was effectively at war with the UK government and the minister chosen to act on its behalf on energy matters, Tony Benn.

* * *

On 23 December 1969 the first substantial oil find in the North Sea was made by Phillips Petroleum in the Norwegian sector, at the Ekofisk field. It led to an immediate change of pace in the British sector. By August 1970 the *Sea Quest* was drilling in BP's block 21/10, 110 miles east of Aberdeen. Here oil was finally struck. On 7 October BP announced to the press that it had discovered what was named the Forties Field, the first giant find in UK waters.[19]

The discovery of substantial reserves in the British North Sea changed the fortunes of BP, Shell, north-east Scotland and the British state. This event was far more significant than finding gas at West Sole four and a half years previously. Quite suddenly the dire situation of Britain's balance of payments, which had long plagued governments and overshadowed events such as those in 1950s Iran and then the Biafran War, looked as though it would be remedied. As more fields were discovered, the recognition took hold that the country would become an oil exporter. Laurance Reed, a Conservative MP,[20] announced in July 1973 that: 'North Sea Oil arrives in time to save us from relegation to the third division ... The 1980s will be Britain's decade. We shall become one of the most influential of nations.'[21] And Prime Minister Harold Wilson declared in the general election campaign the following February that 'by 1985 the Labour Secretary of State for Energy will be chairman of OPEC'.[22] As part of this miraculous change, the power of the post of minister for energy was dramatically increased.[23]

In June 1975 a referendum was called as to whether the UK should withdraw from the EEC, the precursor of the EU. Harold Wilson was for remaining within the EEC, while Tony Benn, then secretary of state for industry, campaigned for withdrawal. The prime minister's side won. There followed a fearsome battle between the left and the right of the Labour Party over Benn's position. Wilson wanted Benn moved from his ministerial post. The left and sections of the trades unions saw this as symbolising a victory for the business leaders who had opposed Benn's strategies at the Department of Industry. Some trades unions considered strike action in defence of Benn's position. Although furious at Wilson's manoeuvring, Benn eventually accepted demotion to the post of secretary of state for energy. The *Guardian* reported the result as 'Wilson gives Benn's head to the City'. However, it was widely recognised that Benn, the outspoken critic of international capital, now had a large say over what was potentially the UK's most important economic sector.

A week after becoming energy minister, Benn travelled with his wife Caroline down the Thames on a launch organised by BP and attended a ceremony next to the Kent Refinery on Grain in the estuary. On that day, 18 June 1975, the first oil from the UK North Sea came ashore, extracted by the US companies Hamilton Oil and Texaco and the UK mining group RTZ. Benn turned a valve on board the tanker newly arrived from the Argyll oilfield and proclaimed that the date from then on should be a day of national celebration.[24]

In his diaries he noted he'd had to participate in the ceremony in the company of 'a complete cross section of the international capitalist and Tory establishment and their wives'. These included Fred Hamilton, head of Hamilton Oil; Elliot Richardson, the US ambassador; and Vere Harmsworth, 'from the filthy *Daily Mail*'. Benn recalled: 'It was a bright, hot day and even the Isle of Grain, the most ravaged desolate, industrial landscape in the Medway, looked quite beautiful.'[25] Still working at the canteen in the bowels of the Kent Refinery, far from the dignitaries and press, was Rose Hammond, shortly to retire at 76.

Five months later, on 3 November, Benn and Steel were at a more momentous event near Aberdeen, where before civil servants, politicians, oil executives and journalists the Queen and Prince Philip opened the pipeline that brought oil directly from the Forties Field.

Benn recorded the day in his diaries:

After breakfast at the Skean Dhu Hotel, a film was shown of the Forties and then Harold Wilson arrived with Jim Callaghan (Foreign Secretary), Sir Eric Drake of BP and a lot of others, and we all drove to BP's headquarters at Dyce. The first thing I noticed was that the workers who actually bring the oil ashore were kept behind a barbed wire fence and just allowed to wave at us as we drove by. We arrived at a huge tent, constructed at a cost of £40,000, and laid with an extravagant red carpet. The tent was about the size of two football pitches and held 1000 people, most of whom had been brought up from London. We were given a cup of coffee as we waited for the Queen to arrive.[26]

After the Queen had shaken hands with the rows of dignitaries, Harold Wilson proclaimed that the event was "'the turning point" beloved of historians and journalists ... the beginning of the end of our dependency on overseas supplies of energy'. His words echoed the experience of his predecessor Ted Heath only two years before, when he felt the future of Britain depended upon foreign oil, and crucially the assistance of Shell and BP in providing it.

Benn had a more caustic view of the proceedings:

> To be frank, the day was a complete waste of time and money, and when you see the Queen in action, everything else is just absorbed into this frozen feudal hierarchy. All the old big-wigs are brought out into the open as if they were somehow responsible for a great industrial achievement, while the workers are presented as natives and barbarians who can be greeted but have to be kept at a distance. It is a disgrace that a Labour Government should allow this to continue. I also felt that this great Scottish occasion was just an opportunity for the London Establishment to come up and lord it over the Scots.[27]

Benn also noted that 'at lunch I sat next to Mrs Steel the wife of David Steel, who is to succeed Drake as Chairman of BP'. This day was the grand finale of the career of Sir Eric Drake, who had gone through the battle of Abadan, the launch of first oil at Port Harcourt and the struggle with Heath. Three weeks afterwards he handed the office at the top of Britannic Tower to David Steel and the battle between the new chairman of BP and the UK energy minister began.

* * *

Fifteen miles north of Aberdeen we turn off the A90 main road and make for the village of Cruden Bay. A cluster of houses leads down to the harbour sheltered by the headland of Slains Castle[28] and Goats Hillock. The sea is calm, benign, rollers on the sandy beach, oyster-catchers at the mouth of the burn. There's absolutely no sign of what we are looking for. Cruden Bay has been a name interwoven with our working lives. For here is where a steel tube, three feet in diameter, comes ashore after running across the seabed for 105 miles. The Forties Pipeline System carries 40 per cent of the UK's crude oil. And there's no sign of it.

We drive a little further onto the higher land, fields of grazing for cattle and sheep and big farms that must feel so remote in the freezing north-easterlies of winter. We find a pipeline marker at the side of the lane, 100 yards from the cliff edge. Before us the expanse of the grey sea, behind us, a small fenced-off compound, the first of four pump stations that push thick black crude 130 miles south. There's almost no security. The two big storage tanks are protected by two lines of fences, and their banality. There's a tiny hut at the entrance, grey pebbledash, a man in his late sixties, no armed police.

We drive south. Skeins of greylag geese in the grey sky, rain on the windscreen. We scan the fields for pipeline markers and cross the wide estuary of the River Ythan. Soon we are passing Menie Links, Donald Trump's golf course on the outskirts of Aberdeen. The Forties Pipeline System plunges under the River Don and swings to the west of the city. We grind through the traffic jams of the suburbs. On the southern side of town we cross the River Dee, its waters flowing from the Cairngorms to the harbour where we will later encounter Vattenfall. On the western horizon are those mountains celebrated by the writer Nan Shepherd. The Forties Pipeline System snakes around the village in which she lived on the banks of the Dee.

The next pump station is near Muirskie Wood just south of Aberdeen. Here at Netherley is where the pipeline had to be excavated over Christmas 2018. All flow was stopped and the diggers worked through the day and night. South again, the tube passes under Cowie Water and Bervie Water and along the Howe of the Mearns, land of the revered writer Lewis Grassic Gibbon. Beneath the River North Esk and on the banks of the South Esk is the next compound near Brechin. On, past Forfar into the Sidelaw Hills to Balbeggie and the last pump station. The Forties Pipeline System runs west of Perth and dives under the Tay and the River Earn.

On and on, west of Kinross, west of the Cleish Hills. Over to our right beyond Cowdenbeath and Dunfermline we see the flare stacks of Mossmoran Chemical Works, owned by Shell and ExxonMobil, fuelled by North Sea gas. Finally, near Culross, we meet the Firth of Forth and the pipeline runs beneath the wide estuary to the southern bank and the Kinneil Terminal on the eastern perimeter of Grange-mouth Refinery.

We have travelled the river of oil that links the estuaries of the Dee and the Forth. It has been flowing for 45 years. Looking at the map of all the oil fields it drains this is arguably the largest single machine in the UK.

* * *

On 19 December 1975, a month after the official opening of the Forties Pipeline System, Benn wrote:

> To BP ... for an hour and a quarter's briefing from their top people. They talked about BP as a British international company – there were 650 companies in seventy countries with 170,000 employees, 15,000 of them in the UK. They dealt with 165 million tons of crude every year and the value of capital employed was £4,300 million. They went through all the figures. We got on to BP–Government relations and British National Oil Company (BNOC)[29] and in effect they repeated what they had said before that they couldn't possibly cooperate with the Government until we sold off the shares that the Bank of England held. They argued that Government ownership was a great handicap to them. I've got to take that seriously but I don't think we can drop below 51 per cent and I did make that clear.[30]

At the point at which this meeting in Britannic Tower took place, between the 'top people' of a multinational corporation and the energy minister of its host nation, both sides were in a febrile mood.

Despite the promise of North Sea oil, the British economy was still in rapid decline, with rising national debt and unemployment. The Labour cabinet was split between those, such as Chancellor Denis Healey, who wanted to pursue a monetarist policy, and those such as Tony Benn who advocated national companies that were largely publically owned, a greater role for the unions in economic policy, a big project of reinvestment and import controls.

The post-war social democratic society of the UK was underpinned by Britain's political and military power that helped create a world of cheap oil imports and high profitability for the private oil companies. By the late 1960s this order was being turned on its head. British imperial strength – illustrated by the shrinking scale of its armed forces and crises such as Abadan and Suez – was rapidly waning, oil-producing states were demanding a higher price for the crude produced by foreign companies working in their fields, and this was hitting the profitability of corporations such as BP and Shell. The 'Oil Crisis' of autumn 1973 had two years later fuelled a crisis in the UK's social democratic government and a crisis in Steel's multinational corporation. The fates of both seemed intimately intertwined.

Into this landscape came a miraculous gift, the October 1970 discovery of oil in the Forties Field, a concession 100 per cent owned by BP and operated by the company. This find preceded the 'Oil Crisis' by 36 months, so the commercial and political significance of the UK's North Sea reserves only became truly apparent in late 1973.[31]

For the UK government rapid exploitation of North Sea oil and gas meant not only taxation revenues for the Exchequer, but more significantly the ability to reduce the crude imports from the Middle East which had leapt alarmingly in cost.

This unforeseen shift in circumstances suddenly offered a solution to Britain's balance of payments problem. The gift of the North Sea offered a magical escape route from the country's plight. Little wonder that the *Daily Express* cartoonist Michael Cummings portrayed the new Energy Minister Benn as an Arab sheikh sitting astride a North Sea rig with the prime minister answering to his beck and call.[32]

However, the UK's oil, and the power of the energy minister, despite being built on the exploitation of sovereign geology, depended upon the role of private oil companies, in particular Esso, Amoco, Shell and BP. For these companies the discovery of oil in the UK sector also offered a miracle solution to their own woes. Here was a resource which, although expensive and technologically challenging to extract, lay in a state whose political stability could be relied upon over the long term, with governments that would be supportive and indeed compliant – in contrast to Iran, Libya or Nigeria.

The Forties discovery in October 1970 took place four months into Edward Heath's Conservative government[33] and was followed by a frenzied rush to drill in other northern North Sea licensing blocks.[34] The creation of the legal and fiscal conditions under which private oil companies could extract from the North Sea had been underway since 1965, but the Forties find was catalytic. Striking oil in a field that many had assumed would only deliver gas radically increased the scale of the capital that the corporations were prepared to commit to the UK sector and the scale of the profit that they wished to generate in return. The companies scrambled to persuade the government to agree to terms that would maximise their profit.

In the two Conservative MPs who were secretaries of state for industry between 1970 and 1974 they found helpful allies. The first, John Davies, had been a senior accountant at BP for 20 years and latterly managing director of Shell-Mex and BP Ltd, then director general of the Confederation of British Industry in 1965, before becoming a minister.[35] The second, Peter Walker, was avowedly on the right of the Conservative Party.

However, for BP and Shell there was a bigger prize to be gained through creating a regime in the North Sea that was to their benefit. In the UK they could build a model on their own terms that they hoped could be exported to other oil provinces around the world. The oil companies could shape Britain's newest colony, the North Sea, into a form that most suited their own ends – just as the Royal Niger Company had shaped what became Nigeria and the Anglo-Persian Oil Company shaped modern Iran.

The North Sea, so cold, grey and far from the imagination of all but a handful of UK citizens, became a crucible of a New Britain just as powerful an the petroleum city of London after World War II. As happened so starkly in the 1940s, the oil corporations were centrally engaged in creating a new order in which they would be able to prosper, and in doing so they were shaping the future economy, politics and indeed culture of Britain.

John Browne, future CEO of BP, told us of working on BP's Forties Field in 1976 and 1977[36] while based in an office next door to the

Britannic Tower. He explained, 'I think there was enormous pride in the North Sea ... it was so unexpected, it was a jewel in the New Britain as it were ... the Queen kept inaugurating things.'[37]

* * *

In the decade since the first UK sector licensing round the development of the North Sea had unfolded in a way that was entirely favourable to BP, Shell and multinationals from the US and France. However, the tables of fortune began to turn, against BP in particular. In December 1974 the Burmah Oil Company suddenly revealed it was on the brink of collapse. Its demise would have impacted on jobs across the UK[38] and the Labour government was determined to prevent this. Burmah had, over 60 years previously, become the owner of 20 per cent of BP's shares. So in order to rescue the company the government, in the form of the Bank of England, stepped in to purchase these shares and thereby inject capital into Burmah.

Overnight BP became 68 per cent owned by the British state. The company was being taken in exactly the opposite direction to that which Drake, Steel and the rest of the board desired, and placing it potentially under increasing control of the UK energy minister. This was not such a concern to BP in the winter of 1974 when the energy minister was Eric Varley, on the right of the Labour Party,[39] but six months later it was a different matter, for the minister was now Tony Benn, left-wing bogeyman of the City.

Five weeks after taking office Benn recorded in his diary:

At 3.30 Sir Eric Drake of British Petroleum came to see me and I went out of my way to be charming. He said the government holdings of British Petroleum shares must be kept below 50 per cent because it would destroy the credibility of the company in the United States, in New Zealand and elsewhere – BP operates in 80 countries.[40] Therefore, he wanted the BP Burmah shares sold off in the open market but not to foreign governments. Well, I'm not accepting that.[41]

His diaries continue: 'I had contemplated giving Drake the chairmanship of BNOC but he was so negative and hostile that I changed my

mind. I'm glad I saw him and it was probably a good thing to be on rea-
sonable terms with him, though he is the most Tory of Tories.'[42] Drake
was famous for rubbing people up the wrong way due to his forceful
views. As James Bamberg, the BP historian, wrote in an obituary: 'If
he had self doubts, he did not show them.'[43]

BNOC was delineated by Eric Varley, Benn's predecessor as energy
minister, and established by an Act of Parliament in 1975. The aim was
to create a state-owned company that would be empowered to take
the licenses for blocks in the UK sector, build offshore platforms and
extract oil just as the private companies were doing. Indeed, BNOC
was created to compete against BP, Shell and the other corporations
following the model established in oil-producing states such as Iran,
Nigeria or Libya, and as a mirror to companies being created by other
North Sea countries, such as Statoil (now Equinor) in Norway and
DONG (now Ørsted) in Denmark.[44]

Building a corporation to rival BP or Shell was a complex undertaking, expensive in terms of time, capital and labour. It needed to be
able to hire the best staff from the private oil companies. Norway recognised that the only way Statoil would be strong enough to compete
in licensing bidding rounds, and to have the expertise to extract oil,
was to ensure that the opening up of the Norwegian sector was taken
at a measured pace. Oslo instituted a 'depletion policy' to make sure
the rate at which the oil was pumped out of the seabed was carefully
managed. For it is in the interests of private capital to extract oil as
quickly as possible and thereby generate the highest profits, whereas it
may be in the interests of the citizens of a state to prolong the process
as far as possible ensuring a steady and enduring stream of revenues.
The Norwegians' policy meant that their oil fields lasted longer and
contributed longer to the national coffers.

So successful was this strategy that the Norwegian method became
recognised in many countries as a model of good practice whereby the
state rather than multinational corporations could gain the maximum
benefit from oil production. These were not the tactics pursued in the
UK where successive governments following the first licensing round
in 1964 agreed to a pace of rapid development being driven by the
corporations. This meant that the UK's oil fields were drained at a
breakneck pace and that effectively BNOC, born without teeth, was

not given the time to develop sufficient strength to take on its rivals. Part of Benn's plan was to slow down the pace of development.

BNOC was also to be a key tool in 'participation', a strategy that Benn pursued with vigour throughout his four years as energy minister. 'Participation' was a legal principle that had evolved in the struggle between Middle Eastern states and the multinationals, and utilised in many countries such as Iran and Nigeria. It meant that the state would make it a requirement of any private consortium extracting oil that a percentage of that consortium would be owned by the state, so that the nation would 'participate' in the business of producing oil. Consequently the government not only drew revenue from the oil fields, but also had a greater degree of control over how the fields were operated. It allowed the state to acquire intelligence, to train civil servants with relevant skills and to shape future plans for the sector. This gave government a far more powerful role than if it merely collected a tax on every barrel extracted, which left the state blind, unable to truly see the inner workings of the oil province and its possible development.

Benn was determined to enforce participation on all the companies working in the UK North Sea. The companies, including BP, Shell and Esso, were determined to resist him. It is clear that the industry's collective strategy was to try and sit Benn out and wait for the appointment of a more pliable minister. Benn's diaries are a long record of meetings at which company executives foot drag and complain, and at which the minister becomes increasingly frustrated.

In one vivid recollection Benn described a meeting about participation with Norman Rubash, president of Amoco (Europe), an American oil company later bought by BP. The executive and his lawyer were essentially refusing to agree to the terms that Shell and Esso had committed to. Benn wrote that he said:

'I cannot go beyond the Shell-Esso arrangements.' Rubash looked absolutely white. 'That is it. You are dealing with Her Majesty's Government and these participation talks are intended to make a real difference. You are not dealing with a sheikh in the 1940s, you are dealing with the British Government in the 1970s' ... I was boiling with rage. I felt like the president of a banana republic dealing with a multinational company. I'll never forget that experience with Amoco.

David Steel, with Monty Pennell at his side, had a particular weapon that BP could wield. Repeatedly they said to Benn that there was no way that they could move on the issue of participation until the government sold the BP shares it had acquired from Burmah. At meetings in Britannic House, at Chequers and at the Department of Energy it was clear that the chairman and the minister were pulling in entirely opposite directions, one wanting more public control, the other less.

For the minister the principle of participation was not only about drawing more control to the state, but also about enforcing a more democratic culture. Benn was deeply committed to strengthening the role of trades unions in industry, and to expanding industrial democracy in all parts of the economy. Here was a relatively new sector, the offshore industry, that for a decade had followed American culture in which the companies resisted the unionisation of the workforce on the rigs and in the construction yards. Benn wanted this changed, Steel was determined that it would not.

Together the management of BP, Shell, Esso and a host of smaller companies ensured that this new industrial realm of the North Sea, seen as the future hope of New Britain, was organised in a manner which broke with a century-long tradition of industrial relations, and pointed to the world that would become the norm after the Thatcher government achieved power. This was a laboratory of Thatcherism.[45]

*　*　*

Much as Steel wanted to resist the demands of Benn, he knew that BP also needed North Sea oil for its own survival. Between 27 and 29 June 1976 a mammoth round of negotiations took place at the Civil Service College in Sunningdale, lasting 27 hours with only six hours' break for sleep. Finally Steel did agree to a level of participation and to assist in the evolution of BNOC.

But then, quite suddenly, the tables turned in favour of the BP board. Just as 18 months earlier the collapse of Burmah had tilted things in favour of the government, now only two days after the Sunningdale negotiations Benn was at a pivotal cabinet meeting. Prime Minister

Callaghan revealed the depth of the crisis in the British economy and the news that Chancellor Denis Healey was in negotiations with the International Monetary Fund (IMF) for a several billion dollar loan.

This was the first time since the IMF was established in the wake of World War II that a first-world country had come to the point of bankruptcy and asked to be bailed out. It was clear that when the news became public it would be a cause of national humiliation and a massive blow to the Labour government. The IMF demanded changes to the UK economy as a condition of the loan, and for the next six months the business of cabinet was dominated by the debate over how to meet them. Healey and others around him proposed deep cuts in public spending, and a significant scaling back of the welfare state. Around Benn gathered another faction, with its 'alternative strategy' proposing import controls and only a modest series of cuts.

Between the two sides came the question of the government's holding in BP. The chancellor's team was firmly behind the sale of 17 per cent of the state's shares, as a way of raising income and thereby not having to implement some of the public spending cuts. The minister of energy was determinedly set against the sale. Meanwhile Steel, who though not in the cabinet meeting room was kept closely informed of events, quickly saw that here was an opportunity to obtain the reduction in government control that he, and Eric Drake before him, had so long campaigned for.

The debate in cabinet raged until in December Healey's side finally won and the IMF agreed to the $3.9 billion loan partly on the condition that the government would sell 17 per cent of its stake in BP. It was the largest loan the IMF had made in its 30-year history. Six months later, on 24 June 1977, the share sale went through, generating £535 million. It was the largest equity disposal any company had ever made. The government still held 51 per cent of BP, but Benn saw it as a massive defeat, recording in his diary:

> We have handed some of the most valuable assets of this country to the Shah [the National Iranian Oil Corporation was reported to be trying to buy 1 per cent of BP shares] to the Americans and to private shareholders, and I am ashamed to be a member of the Cabinet that has done this ... We have provided a blueprint for

selling off public assets in the future and we will have no argument against it. It is an outrage.[46]

The government holding had not gone below 51 per cent as Benn had so long insisted, but his observation about the wider significance of the sale was prescient. Nick Butler, who was at the time working in the economics team in Britannic Tower, observed to us:

This was the first example anywhere in the world of public sector shares being sold on the private market. It gave [Conservative shadow ministers] Nicholas Ridley and Keith Joseph the idea of privatisation. The idea that you could sell off shares in public sector companies, simultaneously raise money and rid the state of nationalised industries. It showed Mrs Thatcher what could be done.[47]

In that same year of 1977, David Steel received a knighthood at the behest of Callaghan. The IMF crisis had given Steel the opportunity to win what he'd fought two and a half years to achieve. In setting a precedent, the sale of 1977 affirmed an alignment between BP and the shadow Conservative government. Two years later Thatcher came to power and immediately the new Tory energy minister, David Howell, and Steel set to work on a further sale of BP shares, raising £290 million for the government and reducing the states' holding to 46 per cent less than six months later.

Working alongside Nigel Lawson, who was first minister for energy and then chancellor, CEO of BP Peter Walters later completed the task that Steel had begun. In September 1983 he oversaw the sale of a further 14.5 per cent of the government shareholding, and then four years subsequently a final massive 31.5 per cent. It was once again a record-breaking share offer, generating £5.5 billion for the government and reducing its stake to zero. In just over a decade BP had been entirely privatised, and it had set the trend for all the other privatisations, such as British Telecom and British Gas, that symbolised Thatcher's neoliberal revolution.[48]

The June 1977 sale of 16 per cent of the government's shareholding was not the end of Benn's career as minister of energy, but it marked a turning point. Up until then the government stake gave him a strong hand in negotiations over the North Sea, but now the threat that Benn would oversee the complete nationalisation of the company receded.

As the Labour government's position weakened in the late 1970s the BP board flexed its muscles evermore. On Wednesday 21 February 1979 the minister of energy's diaries gave their last entry on the company: 'I approved a paper saying that we'd move to monthly or even weekly approvals of flaring to get some control of BP. I must say it made me totally sick. They have no loyalty to this country at all.' Two months later Benn was out of office, replaced by David Howell the Tory energy minister. The oil companies had blown hard and waited, until finally Tony Benn had fallen.

Fifteen years previously, during the first offshore oil licensing round, the chairman of BP, Maurice Bridgeman, had felt that the corporation should get special treatment because it was a British company. On Benn's first meeting at Britannic Tower, Steel had asserted that it was a 'British international' company. Over the following five years BP had fought the minister of the elected British government week in and week out. Step by step BP asserted its independence as an international oil company and moved to leave Britain behind.

* * *

Dirt behind the daydream
Dirt behind the daydream
The happy ever after
Is at the end of the rainbow
There may be oil
(Now looking out for pleasure)
Under Rockall
(It's at the end of the rainbow)
There may be oil
(The happy ever after)

You could say Jon King is a beneficiary of oil and Margaret Thatcher. The singer songwriter from the post-punk band Gang of Four, and now a Vice Media executive, made his mark critiquing neoliberal politics and economics for a cultural audience.

Nowhere more so than in the song 'Ether' which criticised North Sea crude as the 'dirt behind the daydream' – providing the cash to bankroll an assault on trades unions and others who opposed privatisation. 'I think everyone on the Left believed the only reason that we could have such a crumbling and useless society was North Sea oil, because North Sea oil paid for unemployment', says King sitting in a Soho café looking every bit as trim and elegantly dressed as you would want a post-punk rock star to be.

The 62-year-old son of an electrician, King was born in south London and educated at Sevenoaks School. There he met Andy Gill with whom he formed Gang of Four in 1976 when they were both at Leeds University. The group was once described by *Rolling Stone* magazine as 'probably the best politically motivated band in rock 'n' roll' and name-checked as an influence by Kurt Cobain, Michael Stipe and others.

We ask him about oil and his songs:

I wrote the lyrics of 'Ether' because there was this idea at the time that you (with oil) could turn everything around: Britain could become an island Saudi Arabia ... There was this promise of utopia there. And of course what Thatcher did in the eighties was to use the money to fund unemployment. The grotesque difference between our country and Norway, where the Norwegians had the sovereign wealth fund. They actually did the most intelligent thing probably any government's ever done, which is to invest in stocks outside of Norway. We just spent it on crushing the trades unions and destroying the working class, the organised working class.

King continues: 'It was a deliberate thing. Thatcher must have thought "marvellous" because she could never have crushed the miners without that money. The deindustrialisation of Britain is the result of oil. Then from a song perspective, it just seemed one of those things where you saw the naked wrongness of how our society was, it just crystallised in that.'

King is also convinced that the oil industry was behind the 'organised destruction' of mass public transport, most notably seen in US cities such as Los Angeles:

Obviously you can't sell cars without people wanting to use them ...
Then the musical experience. It became about individual brands on
their journeys with other people. The road movie became a function
and music is like road music. A car is where you get laid for the first
time. In one place you have these intense experiences and escape.
Having a full tank of petrol means freedom.

* * *

We consider the battleground. The former office of the Chairman
of BP, in the former Britannic Tower. Here was the place of a set of
struggles that entered company mythology. Thirty years later John
Browne, future CEO of BP, wrote in his memoirs: 'The company was
now seen as so important that Tony Benn, then Secretary of State for
Energy, was intent on nationalising BP, particularly as the government
already owned a controlling interest. I was too low down in the organisation, and too busy travelling back and forth to the North Sea, to
understand the boardroom efforts to keep Mr Benn at bay.'[49]

The course of events in the UK in the mid-1970s was not inevitable.
Britain could have followed the path that Norway pioneered. This saw
the Norwegians take the pace of licensing slowly, allowing sufficient
time to enable their national oil company, the equivalent of BNOC,
to build up strength and expertise. They developed a set of offshore
working conditions quite distinct from the US model that Britain
imported into its sector, conditions which took strong union presence
as a norm. And they established a sovereign wealth fund. Norway's
fund, built up steadily from investing income from oil revenues, was
worth \$1.2 trillion in November 2020 and was the world's largest.[50]

If the British governments had pursued the idea that Benn was
promoting, the UK's sovereign wealth fund would today be of a bigger
scale and would underpin a far higher level of social provision than
the current government sees fit to support.[51] A section of the Labour
cabinet under Callaghan wanted to resist IMF pressure to cut public
expenditure and to maintain and improve the provision of the welfare
state. This is largely what Norway did in the decades that followed.
Had Britain utilised the miraculous gift of oil in the same way, the
future of this country, the years in which we are now living, would have
looked very different.

It was Britain's fortune to find oil in the North Sea and to have at least some parts of its government feel that this resource should be licensed and operated by a publically owned company. After all this was a time when the resources of water, electricity and gas, and the infrastructure of railways, buses, airports, aeroplanes and docks, were all owned and managed by the state. After 25 years of struggle most other major producer states, such as Iran, Nigeria or Libya, had built national oil companies. Only Canada and the US disdained such a strategy. The UK chose to follow the model created by American capital, and continued down the path, which it had set out upon in the 1940s, of embracing American culture.

It was Britain's misfortune to find oil at a time when the UK's oil companies, losing control of sovereign states in the Arab World and imbibing a corporate philosophy from America, were determined to fight hard to ensure that North Sea oil should be extracted by private capital and that the companies should be evermore independent from the state. There's a tendency to blame successive UK governments for the failure to spend the treasure of oil wisely, such as by setting up a sovereign wealth fund, but this story ignores the fact that the corporations fought a determined war against such measures by the state.

When David Steel died on 9 August 2004 at his home in Brittany, one of his obituaries was headlined 'Sir David Steel – BP chairman who triumphed over Tony Benn'. It read: 'For nearly two years Steel and his colleagues slogged it out against Tony Benn, then the minister responsible for Energy, but in the event BP won, largely thanks to Steel's quiet obstinacy, a legal training and experience that had accustomed him to long drawn-out negotiations.'[52]

When Tony Benn died ten years later his departure was accompanied by an acreage of newsprint and yet there was little mention of the particular struggle that had taken place in Britannic Tower. [53]

5
Only One Road to Paradise

Went out walking I recall
Me and my best girl along the wall
In the long grass side by side
Where the big ships go gliding by, go gliding by
Skylark singing in the sun
Something told me she's the one
When I looked down into her eyes
I saw pictures of paradise, of paradise, of paradise
People talking but they don't know
Where I've been to or where I'll go
World keep turning all things change
I love two girls I ain't ashamed, I ain't ashamed
All your loving it thrills me so.
I can't go with you I just can't go
A thousand highways you don't live twice
Only one road to paradise, to paradise, to paradise.

—Wilko Johnson, 'Paradise', 1977

For two millennia everything that moved across farmed land was pulled by horses and oxen or pushed by women and men. At Binney Farm by the Isle of Grain, where Rose Hammond worked in the 1950s, all that was grown required horsepower or sail power to move it to market. The use of horses continued for a surprisingly long time after the arrival of the petrol-fuelled combustion engine in the 1880s. In the village of Higham, on the Hoo Peninsular to the west of Grain, the last carthorse, Boxer, continued to work until 1974. But in parallel with the world of horses evolved the world of tractors and cars, fuelled by diesel and petrol refined from the rocks of Iran, Nigeria and other nations. As well as becoming dependent upon oil for traction, farms also became dependent upon oil for fertilisers, pesticides and the wonders they promised.[1]

The most legendary pesticides were the 'drins' – aldrin, dieldrin, eldrin and heptachlor. These chemicals were manufactured from naphtha[2] and marketed in a host of forms under names including Soil Pest Killer, Ant Doom and Shelltox. The drins were designed in laboratories with the express intent to kill insects on a mass scale, in particular locusts, grasshoppers, nematodes and boll weevils. They were targeted at species that were deemed to be 'pests', creatures that dented the productivity of farming.

The drins had been invented by a private company Julius Hyman & Co. at the Rocky Mountain Military Arsenal at Denver in Colorado as part of the US army machine of World War II. Entirely separately in Axis Europe, the Shell laboratory in Amsterdam under German occupation had developed similar pesticides.[3] In 1950 Shell acquired the rights to be sole marketing agent for the drins globally, and the right to manufacture the chemicals. At the point of doing so concerns about the toxicity of these products were already being voiced. The previous year C. J. Briejer, head of the Dutch Government Department of Plant Pathology, had criticised indiscriminate use of these insecticides.[4]

But such concerns did not halt Shell in its embrace of the drins that offered to create a new future for farming. Pests on crops were to become a thing of the past. Pesticides combined with fertilisers were to bring farming into the modern world. As Ben van Beurden, CEO of Shell who had once been the head of Shell Chemicals, reminisced to us in 2018: 'We used to talk of how "we are an energy company and we provide things that are essential for modern life".'[5]

For Shell, marketing the drins not only enabled the corporation to sell the future, but it also generated substantial profits. Stephen Howarth and Joost Jonker wrote in the official history of the company: 'the 'drins represented the Group's ideal chemical product. They were well protected by patents, hydrocarbon-based, and research-intensive; they were modern and glamorous in offering an instant, technical solution to ancient scourges of mankind; they fitted in well with other manufacturing processes; and they combined low volume with stupendously high margins.'[6]

European plants to manufacture the chemicals were established at Pernis next to Rotterdam in the Netherlands, in the UK and in Belgium, Germany, France and Italy. The products themselves were aggressively marketed to British farmers and widely taken up. Not only were the drins sprayed on crops but workers were encouraged to dress seed with the insecticide before it was sown. Around the Thames Estuary, as elsewhere across Britain, in the 1960s and early 1970s it was a common site to see a light aircraft or helicopter swooping low over the fields crop spraying with pesticides. This technological fix was being employed even on farms where horses were still in use for traction.

The effectiveness of the drins was constantly researched and laboratories were established to conduct crop and animal trials. In the UK Shell's R&D facility was at Woodstock Farm on the edge of Sittingbourne in Kent,[7] a mere 15 miles across the estuary from Coryton and Shell Haven as the gull flies. New buildings were constructed, designed in 1954 by Sir Frederick Gibberd, one of Britain's foremost modernists, architect of London Airport and the Thornton R&D Laboratory on Merseyside. This was bold research, with the company proudly building the modern world.

The new R&D lab was watched by those who lived locally, one of whom later described 'in the fields to the right of Broadoak Road between Ruins Barn Road and the Woodstock Farm there were lots of square and oblong plots all growing under different tests of chemicals. It was quite a bit later when … the animal research and other strange things started happening there.'[8]

Concern about this 'farming of the future' began to grow. In 1955 an article in the British journal *Chemical Age* was critical of Shell at Sittingbourne testing the drins on a yearling Hereford cow that died within five hours.[9] Two years later a pair of British vets clearly linked dieldrin with the death of song birds. The following year a toxicology unit was set up on the Sittingbourne site.[10]

In June 1962 the pioneering essays of American biologist Rachel Carson were serialised in the *New Yorker* magazine. When published together these pieces became the bestselling book *Silent Spring*. Carson had studied the impact of the drins on the animal life of the north-east US states and built on the work of C. J. Briejer. Her findings were a sensation. All of a sudden the chemicals that had promised to deliver a cornucopia for farmers were revealed to be agents of death for entire ecosystems, products that would poison the web of life.

Carson described the impact of the drins not only in the US but also in the UK. From 1960 onwards there were alarming reports. A landowner in Norfolk wrote: 'My place is like a battlefield ... innumerable corpses, including masses of small birds – chaffinches, greenfinches, linnets, hedge sparrows, house sparrows ... the destruction of wildlife is quite pitiful.'[11] The issue was heard by a House of Commons committee in the spring of 1961 and evidence was called from farmers, government agencies and wildlife NGOs (non-governmental organisations) such as the Royal Society for the Protection of Birds and the British Trust for Ornithology. One witness said, 'pigeons are suddenly dropping out of the sky dead'. Others described 'huge bonfires on which the bodies of the birds were burned'. Animals were also affected. At least 1,300 foxes died between November 1959 and April 1960. Some were seen 'wandering in circles, dazed and half blind, before dying in convulsions.'[12] The committee recommended that 'the Minister of Agriculture should secure the immediate prohibition for the use as seed dressing of ... dieldrin, aldrin or heptachlor'.

Alarm over the impact on birds and mammals was followed by the recognition that streams and rivers were being poisoned, fish dying and entire ecosystems heavily impacted. The drins were moving into the food chain and thinning the eggshells of birds of prey. The breeding populations of sparrowhawks, peregrine falcons and other raptors plummeted.[13]

The naturalist J. A. Baker wrote *The Peregrine*, built from his intense study of the bird and its world in the northern quarter of the Thames Estuary. He followed these creatures with obsessive attention in the late 1950s and early 1960s, in the valleys of the Chelmer and the Blackwater as they run to the North Sea. This is the land around Stansgate where Tony Benn had grown up in the decade before, just a few miles from the refineries of Coryton and Shell Haven, 20 miles north of Sittingbourne.

Towards the end of Baker's study he noted the devastating impact that pesticides had on his beloved raptor and wrote:

For ten years I followed the peregrine. I was possessed by it. It was grail to me. Now it has gone. The long pursuit is over. Few peregrines are left, there will be fewer, they may not survive. Many die on their backs, clutching insanely at the sky in their last convulsions, withered and burned away by the filthy, insidious pollen of farm chemicals. Before it is too late, I have tried to recapture the extraordinary beauty of this bird and convey the wonder of the land he lived in, a land to me as profuse and glorious as Africa. It is a dying world, like Mars, but glowing still.[14]

In 1963 the Nature Conservancy, a governmental body, concluded that the insecticides were by their nature uncontrollable and recommended a total ban on aldrin, dieldrin, heptachlor and eldrin.[15] The company fought a rearguard action. It is noticeable that although Shell was the sole producer of the chemicals that were coming in for such potent criticism, the company itself avoided ever being named. After a prolonged fight a restriction on just one of the drins was implemented in 1974. Thirteen years after the Commons committee first recommended a partial ban, the use of dieldrin was outlawed in the UK by Minister for the Environment Geoffrey Rippon. Aldrin and heptachlor for use as pesticides were banned by 1989.[16]

This was not the end of the production of the drins, for Shell continued to manufacture and sell them, although no longer in Western Europe and North America. As with oil production a two-tier colonial operation was instituted, one set of practices in the Global North and another set in the states and former colonies of the Global South. Only by the mid-1990s was the use of the drins banned in most countries. Finally, in 2001, the Stockholm Convention on Persistent Organic Pollutants instituted a global ban on the production of aldrin, eldrin and dieldrin. It had been 40 years since the House of Commons heard of 'pigeons falling from the sky dead' but up until the 1990s Shell continued to produce these chemicals.

Throughout this battle of the 1960s and 1970s, Shell produced – as it had since the 1930s – its *Shell County Guides*, and plentiful volumes in the series *The Shell Book of* … . Aimed at the 'average motorist' these publications reflected a way of looking at the world. They echoed the Festival of Britain gazetteers that encouraged the everyman to visit the nation by car. One of the most telling is *The Shell Book of Rural Britain*, released in 1978. It is copiously illustrated with 152 photographs. Only six images include a tractor or a motorcar, whereas the book is saturated with pictures of horses ploughing or pulling carts, of men and women building fences, making cartwheels, or churning butter. There is no mention of the crop-spraying helicopters, nor the seed dressed with drins. A fantasy of the British countryside was sold to the Shell readers and drivers. As the company sold a future to farmers, so it sold a past to car drivers.

* * *

For nine miles through the south-west corner of London runs the River Wandle, a tributary to the Thames within the tidal limit of the estuary. What was once a fast-flowing chalk stream whose clear waters drained the North Downs is now a canalised waterway that passes through the boroughs of Croydon, Sutton, Merton and Wandsworth. The largest contributor to the flow of the Wandle today is the Beddington sewage works, especially in the summer months when the springs in the Surrey Hills run low but the toilets and the baths of the suburbs continue to be emptied.

The Wandle has languished in this sad state since the 1960s,[17] but in spite of the chains with which it is bound, the life of the river thrives – its banks are lined with crack willows, water crowfoot blossoms white in the stream, kingfishers fly the river's length and brown trout are caught here. That so many species should have survived is in large part down to the extraordinary level of environmental activism that has flourished along the Wandle valley over the last five decades. Hundreds of residents have been involved in a whole panoply of groups from Wandsworth Friends of the Earth to the Wandle Festival, from the Wandle Delta Network to the Wandsworth Society.

The first of these organisations, the Wandle Group, was born in response to the impact of the oil industry. In 1967, under chairman Maurice Bridgeman and head of chemicals John Hunter, BP brought the Distillers works at Carshalton on the banks of the Wandle. The plant used naphtha feedstock from refineries such as Coryton and processed it into solvents, intermediates and plasticisers which in turn were sold to paint, pharmaceutical and food industries who used them to produce washing machine finishes, perfumes, cigarette papers, soft drinks, toys, shoes and so on. It was a sister to other BP petrochemicals works at Stroud and Hythe, Baglan and Barry.

Less than two years later, in February 1969, the BP plant had a major pollution spill that killed wildlife all down the length of the river and poured into the Thames. The local newspapers reported extensively on the poisoning of the watercourse and the Wandle Group was formed in response.[18] The incident was merely the final catalyst, for concern had been growing among residents throughout the valley. Over the previous decade the Wandle had been frequently covered in a thick layer of foam. Photographs show a white froth on the river's surface at times a foot deep. This scum was the result of household detergents pouring into the stream in large part via the mains sewage system that flowed through the Beddington works.

In the late 1940s Shell, BP and other oil companies, as part of the process of experimenting with what could be created from the waste products of refining, had invented detergents. Initially these were used only in industrial plants, but from the early 1950s they were marketed directly at the domestic household. In 1959 BP's film *Mrs 1970*[19] portrayed the kitchen of the future. Not only was this heavily automated and decked out in plastic, but the 'housewife' was equipped with a full armoury of detergents and household cleaners. Here was the home of the modern world but it seems that little concern was paid to the fact that this utopia of cleanliness produced dirty rivers topped with foam and starved of fish.[20]

The detergent that poured down the Wandle was joined in the Thames by similar pollution from hundreds of other tributaries flowing out of Oxfordshire and Essex, Berkshire and Kent. The effect on the main river was disastrous. Jeffrey Harrison and Peter Grant described how 'boat crews have told us that the problem was especially marked in the dry summer of 1959 when craft moored near the Woolwich Ferry were completely submerged in mountains of foam stirred up by the continual movement of the ferry'. In the worst years 'a film of detergent covered the water of the Inner Thames'.[21] These two biologists explained that 'the effect of detergents is to greatly reduce the rate at which oxygen can be transferred by aeration systems', so that detergents 'provided the last straw that brought about the completely anaerobic state of the river in the 1950s'.[22] Effectively the Thames from Tower Bridge to the North Sea was biologically dead, starved of oxygen, meaning that only eels, which breathe at the surface, could survive in it. The mighty River Thames had been suffocated, and the same fate had befallen the Severn, the Mersey and the Forth.

* * *

At 9.11 on the morning of 18 March 1967, on the Seven Stones rocks between the Isles of Scilly and Land's End, the SS *Torrey Canyon* ran aground. She was one of a new class of VLCCs, similar to the MV *British Admiral*, with a cargo of 119,328 tons of oil from Kuwait bound for the BP Angle Bay Terminal in Milford Haven. She had been chartered by BP from the Barracuda Tanker Company of Bermuda and her load was destined for the refinery of Llandarcy and the petro-chemicals plant of Baglan. But the oil was never delivered. On striking the rocks the hull was ruptured and a black slick poured out into the clear blue seas.

The following week, as staff at the new Britannic Tower quickly addressed the task of resupplying Llandarcy and deflecting attention from the owners of the ship's cargo, the story of the stricken tanker was captured by the cameras of the BBC. The nation watched as the ship was broken up by storms, as the RAF bombed the wreck and tried to set light to the oil with napalm, and as the beaches of Cornwall, Devon and even Brittany were covered in black crude. The corpses of tens of thousands of seabirds were washed ashore and local volunteers tried to clean the sands and the bodies of guillemots, shags and razorbills. The army was deployed both in Britain and in France.[23] It was soon rec-ognised as the UK's largest oil spill in peacetime. However, the scale of the disaster was never seen, as it could have been, as a consequence of the corporate pursuit of profit through economies of scale that led to the construction of the world's largest vessels.

* * *

Through TV, radio and the press the *Torrey Canyon* provoked a national debate as oil entered the long conflict between two distinct visions of Britain. Between Britain as rural, natural and traditional and Britain as industrial, technological and modern. Less than four years before,

Prime Minister Harold Wilson had declared the country needed 'the white heat of technological revolution'. Yet he himself famously spent his holidays on the idyllic the Isles of Scilly. This idylll was a thousand miles from the world of Vince Taylor's 'Brand New Cadillac'.

Since the 1930s both BP and Shell, and often together as the marketing company Shell-Mex and BP Ltd, had sold themselves as able to answer both visions. They were the sports car carrying the driver, with a *Shell County Guide* on the back seat, fast and efficiently into a land of oasthouses and villages, woodlands and farms. The land of *The Shell Book of Rural Britain*.

But by the late 1960s these two visions clashed: catalysed by the corpses of the drins, the rivers foaming with detergent and the oil on the beaches, the British environmental movement grew rapidly. In America the same movement, enraged by an oil disaster off Santa Barbara, was becoming powerful and politically effective, driving forward the establishment of the Environmental Protection Agency as a federal body with teeth. Soon two key North American organisations set up chapters in Britain: Friends of the Earth and Greenpeace.[24] Although the environmental movement was not yet explicitly anti-oil, BP and Shell well understood that they needed to engage with the critique of industrialism that it posed. Two strategies evolved: show concern and co-opt critics through collaboration.

By the mid-1960s there was so much evidence pointing to the impacts of detergents that concern was expressed by the companies who produced the raw chemicals. In 1966 Shell released *The River Must Live*,[25] one of the first clearly environmental films to be produced by the oil industry. In due course, biodegradable 'soft' detergents were introduced by manufacturers and the problem subsided.[26]

Both companies began to publish materials that showed how they too were anxious about the impacts of modern industry on the environment. Four years after Shell's *The River Must Live* came BP's contribution to European Conservation Year 1970, the film *The Shadow of Progress*. Like *The Island* made by Peter Pickering of DATA films only 18 years before, *The Shadow of Progress* was edited in the cutting rooms of Soho by Derek Williams of Greenpark Productions. Whereas *The Island* was filled with an optimism that portrayed the refinery as the heart of New Britain, this 1970s film was steeped in an anxiety about the future of the Earth.

Shell produced a similar film entitled *Environment in the Balance* and financed the publication of *Energy to Use or Abuse?*, a book by John Davies who was an associate of the Intermediate Technology Development Group that had grown from E. F. Schumacher's ideas expressed in *Small is Beautiful*.[27] Not everyone within the companies was happy with these strategies. BP managers in the UK, France, Germany and Switzerland were opposed to the showing of *The Shadow of Progress*, but Eric Drake, the chairman, supported Derek Williams against these internal critics and advocated the film and BP's commitment to 'good conservation practice'.[28]

Co-option of, or collaboration with, environmental organisations was inevitably more politically fraught. However, Shell achieved this with the World Wide Fund for Nature (WWF). The company sponsored a booklet, branded with the WWF panda and the Shell logo, that described the world's most endangered species. Inside the glossy pages there were blank gaps for any child to glue in the plastic 3-D cards that showed the animals 'in motion' and were given away free on the forecourt each time their parents filled up at a Shell garage. So close was the link between Shell and the WWF that John Loudon, soon after retiring as chairman of Shell, became president of the WWF in 1977.

* * *

Seventy-year-old Wilko, with those trademark staring eyes, still exudes a wild punk energy but he is also affable and warm. It's remarkable that he's here at all, meeting us in the dressing room of the Corn Exchange venue in Cambridge. Several years previously he announced he had lethal pancreatic cancer with only months to live. He took to the road on an inevitably sold-out tour, only be reprieved by an innovative operation at Cambridge's Addenbrooke's Hospital. As if to prove his new strength, he whips off his *Oil City* T-shirt and, laughing loudly, reveals a series of deep angry scars snaking down his chest and stomach. The surgeon will be in the audience tonight with his free ticket.

John Wilkinson – later known as Wilko Johnson – was born on Canvey Island, in the Thames Estuary. In 1959, when Wilko was twelve, his father started working at the British Gas Canvey Methane Terminal, importing gas from Louisiana. And Wilko bought his first guitar, a Telecaster, in Southend.[29] Construction of the Coryton oil refinery, half a mile away, had been going on for nine years. 'When they were building the things. You could see these towers rising up on the horizon. I remember my mum telling me the biggest tower was a cat cracker. What a thing to tell a little kid! And later on I found out there was indeed a pub right next door called the Cat Cracker. This rather pleased me for it meant I had not made up this story.'[30]

The world of Wilko's childhood was enveloped in the gas terminal of Canvey and the refineries of Coryton, Shell Haven and Kent on the horizon. The plants operated round the clock, the rhythm of southern Essex determined by the men and women coming on and off work in these vast machines. There was an intense camaraderie among those on each shift, many of whom lived as neighbours in the surrounding villages and towns – Stanford-le-Hope, Corringham, Basildon and Canvey. After work, social life gravitated around clubs like the Pegasus Club with their myriad activity 'sections'. They filled the pages of the 1960s editions of the *Coryton Broadsheet*. Here were notices of a trip to *Robinson Crusoe* at the Palace Theatre in Westcliffe-on-Sea, and photos of the Pegasus Club Drama Section performing at Woodlands Boys Secondary School. And every year there was the Christmas children's party, held at the refinery, when buses brought kids from Southend, Grays, Canvey and beyond.

After leaving Westcliffe High School for Boys, and before going to university in Newcastle to read Anglo-Saxon, Wilko took an office job in London in 1963.

I was so miserable commuting. I remember being on the platform on Benfleet Station in the early morning. It was horrible and I saw the refinery a mile away. I imagined some kind of wonderful life. You are 16, you feel awful and over there it seems heroic. I used to call it Babylon. In the night time it was amazing, these huge flares and all the lights on the towers.[31]

By the age of 24 Wilko had been in several bands, including The Roamers and The Heap, had travelled overland to Goa in India and returned to Canvey complete with regulation hippy hair. That summer of 1971 he met Lee Brilleaux, and the legendary pub rock band Dr Feelgood was born. 'We used to sit around getting stoned and talking and getting into this kind of fantasy trip: the idea of driving round in a Chevrolet and wearing these cheap suits. I think it was Lee who started calling Canvey, *Oil City*. Outside of New York it was probably the best skyline I have ever seen. With these flames coming out of the stacks.'

Perhaps this slice of America on the shores of the Thames infused Wilko's cutting R&B guitar, which emulated Mick Green of Johnny Kidd & the Pirates[32] and channelled the sound of the Southern States. Undoubtedly the presence of the refineries pervaded every part of Dr Feelgood's work. The title of their first album, *Down by the Jetty*, referred to the loading jetty at the terminal where the oil came ashore. A crude tanker featured on the cover.

With Wilko's aggressive guitar playing, Lee Brilleaux's manic vocal delivery, the band's shabby suits and their adrenalin performances, Dr Feelgood developed a cult following in the clubs of Southend and Basildon, playing to audiences that included many refinery workers. After three years of gigging they got a record deal with United Artists and began recording *Down by the Jetty* on 8 June 1974.[33]

* * *

By the early 1970s, before the 'Oil Crisis', the rapid increase in fuel use in Britain, combined with the relative cheapness of crude from the Middle East and Africa, meant that refining in the UK could generate intense profits for the oil corporations. Building on the three existing Thames Estuary refineries, plans were drawn up for more plants, one in Cliffe on the Kent shore and two on Canvey Island on the Essex shore.

The corporations and British government expected that the local population, so enmeshed with the existing industry, would welcome the prospect of further refinery jobs. But Canvey rose in revolt. As Wilko describes: 'We found out there was a proposal to build two refineries and cover the west of the island. There were two companies, the Americans, Occidental, and the Italian company, ENI. The western half of the island, towards Shell Haven was empty, still farmland, marshes and unmade roads.' Opposition stirred. 'There were people who had grown up there and a lot of people who had moved down from east London, and suddenly they were going to have these fucking things on their doorstep, and the value of their houses destroyed. It was like a paradise under threat.'

Canvey had a long history of independence – the community that had grown up on the low-lying marshes after World War I were plot-landers, families who had left London and come to squat land in the estuary, building their own bungalow homes. By the 1960s the place was more regulated, with tarmac roads and mains water, but the spirit of independent defiance still ran strong.

Canvey was nearly cut off from the rest of Essex every high tide as the waters of the North Sea ran up Benfleet Creek and East Haven Creek filling the muddy channels. The only connection with the mainland was via a narrow bridge. In order to build and operate the planned refineries, a far grander roadway onto the island had to be constructed. This project, supported by Edward Heath's Conservative government and Essex County Council, was in part financed by the oil companies. Soon a smooth concrete arc sailed over the creek and the marshes and funnelled traffic off the A13 directly onto the island. Dignitaries gathered to cut the ribbon at the bridge opening ceremony in April 1973. But they were faced with an angry crowd led by a Conservative councillor for Canvey armed with a megaphone.

There had never been a demonstration like it on the island. The Essex police overreacted, swamping the event. The number of officers almost equalled the number of protesters. Wilko and his brother Malcolm encouraged people to storm the stage where the mayor stood. 'You can imagine Mr Plod watching us lead the crowd to the podium. "Long-haired anarchist agitators!" I was arrested mid-slogan and taken to the police cell at Benfleet to be followed 15 minutes later by a very flustered Malcolm.'[34] The heavy-handed policing and arrests catalysed the formation of the Canvey Island Oil Refineries Resistance Group or CIORRG.

The following six months were filled with protest. There was a demonstration by 5,000 people in Canvey, with a mock funeral procession through the heart of the community. The refinery construction sites were picketed and lorry drivers persuaded not to deliver materials.

There was a march by 200 'angry mums' in London, down Whitehall through Parliament Square to the Department of the Environment and the Minister for the Environment Geoffrey Rippon. An armada of small boats, with 700 islanders on board, took to the Thames punching the tide six hours up river from Canvey to Westminster. They delivered a petition to Prime Minister Heath at Downing Street. These events were widely reported in the press and on TV. A documentary was filmed as part of the *Man Alive* TV series.[35]

CIORRG also took the struggle directly to the corporations, as Wilko describes: 'We were getting pretty effective by then in our protests. We had invaded the oil company's offices in London with two coach loads of women and kids and stuff like that.' The occupation of ENI and then Occidental provoked the desired response from the latter. 'The oil company sent people down to Canvey saying they would make the refinery really nice and paint everything. And that prat, his name was Hammer, Armand Hammer. He came.'

Armand Hammer, the chairman of Occidental,[36] was 75, an inveterate dealmaker and a giant in the oil world. In 1965 he'd purchased an extremely lucrative concession in Libya, and by 1967 was supplying 25 per cent of Europe's crude needs.[37,38] Libyan oil was shipped to refineries such as Coryton, Shell Haven and Kent. Two years later King Idris of Libya was overthrown by the army coup led by Muammar Qaddafi. In January 1970, Qaddafi's Revolutionary Command Council launched an offensive to rewrite the concessions with Western oil companies on terms more favourable to the Libyan people. Occidental was the focus of their attack and it threatened to put Hammer out of business. After ten months of negotiations he achieved a deal and his company survived.

By the time the battle with CIORRG broke out, Occidental was not only building the Canvey Refinery, it was also investing heavily in the exploration of the UK North Sea and had struck oil at the Piper Field. Soon after, it began planning to build a crude terminal at Flotta in Orkney. Little wonder that Heath's government, and especially the Minister for Energy Peter Walker MP, was so supportive of Hammer at Canvey.

Wilko continues: 'Armand Hammer was going to come down and see us. We were all sitting in this house on Canvey and it was great.

This bunch of people who'd never been together before. We had to wait until we got a phone call that Hammer was in this motel in Basildon.' Once the call came the CIORRG group went to meet the chairman of Occidental. 'They were sitting there and going on about all they were going to do to Canvey Island, to make it acceptable to people. I remember saying: "We don't care what colour you paint it, we don't fucking want it, all right".'

The oil magnate was renowned for letting people hear what they wanted to hear. 'They were batting this thing backwards and forwards. They were trying to make some arrangement to make us limit our protests. But they were talking bollocks. They were going to build it and I knew there was nothing we could do. This Hammer he come up to me, and he said: "Keep on punching".' Wilko laughs out loud and continues: 'We made some agreement not to invade their offices in London again. But of course we did exactly that. And this little flunky who looked after Hammer came up to me and said "You agreed not to do this" and I said: "We were fucking liars weren't we".'

It looked inevitable that Occidental and ENI would steamroller ahead with their plans. But then quite suddenly the balance of power shifted in CIORRG's favour. On 6 October 1973 the Yom Kippur War broke out between Egypt, Syria and Israel. There was already a shortage in world oil supply so the war, and the Arab states' oil embargo against the US and several European countries, made the cost of crude rocket. This in turn meant the price at the petrol pump spiralled, drivers could not afford to fill up and demand for fuel in Britain plummeted. Production at Shell Haven, Coryton and Kent had to be quickly scaled back, and work on the Occidental and ENI plants came to a rapid halt.

By the time Dr Feelgood had finished recording *Down by the Jetty* the oil companies had written off their investments and CIORRG was victorious.

Occidental built this huge chimney, this jetty right out into the river. They built huge tanks and then this economic upset in the oil industry and all of a sudden they just left. One minute we had government ministers coming down saying it was absolutely essential to have new refineries, that we must build this thing. And then there's this upset in the desert and they were gone. And now it's all there rusting.

The struggle between the oil companies and several Arab states brought the radical transformation of this land and these communities in the Thames Estuary to an abrupt halt.

Wilko's son, a guitarist with his own Southend band, hurries into the dressing room where we are talking and reminds his dad he will be on stage soon. Of course tonight he'll play 'Paradise' with its knowing line:

World keep turning all things change

PART II

1979–2008

6

And If You Tolerate This
Your Children Will Be Next

The future teaches you to be alone
The present to be afraid and cold
And if you tolerate this your children will be next
And if you tolerate this your children will be next
Will be next, will be next, will be next
— Manic Street Preachers, 'If You Tolerate This
Your Children Will Be Next', 1998

Feet slipping on the soft sand, hands grabbing the marram grass, we make it to the top of the dunes. It's early morning and the bay stretches before us. Silver light on the calm waters. The silhouettes of the Sunday sleeping city of Swansea and Mumbles Head. The pale cliffs of north Devon several miles south across the Severn Sea and a coaster labouring against the tide up the channel towards Bristol.

The long sweep of sandy beach at Aberavon Prom is dotted with a few dog walkers and joggers. Beyond it on the horizon an outline of the steel works, steam drifting up from its towers and chimneys. At this distance, a couple of miles upwind, the plant is silent and gentle. Almost toy-like. Between us and it are the roofs of Sandfields, the western quarter of Port Talbot. Little stirs in the streets between the two-storey houses. Only the occasional passing car or a pair of cyclists, their voices carrying in the Sunday quiet.

Behind the houses, behind the clouds of factory steam, are the mountains Mynydd Emrich and Mynydd Dinas. Their soft forms, clothed in bright green grazing and dense woods, look down on the town and the industry camped out on the narrow strip of sand running between the foot of the hills and the sea.

OIL IMPORTS TO BRITAIN

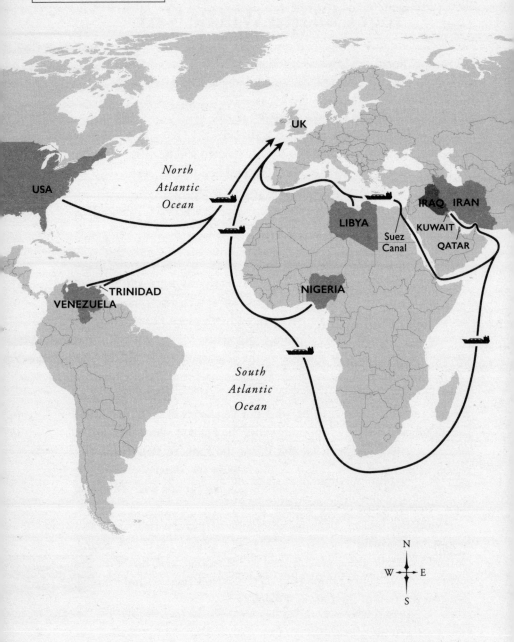

UK

USA

North Atlantic Ocean

TRINIDAD

VENEZUELA

LIBYA

Suez Canal

NIGERIA

IRAQ IRAN

KUWAIT

QATAR

South Atlantic Ocean

N
W · E
S

0	1,000	2,000	3,000 miles

0	2,000	4,000 kilometres

We turn to look behind us, across Baglan Moor. In the distance is the solid cylinder of a gas-fired power station chimney and beyond, the bridge of the M4 motorway sailing high over the River Neath. We check the OS map, the chimney is a mile away. Between it and us is a dark plain. Here and there are rusting metal structures. Two yellow JCB diggers sit immobile in the distance. Most of the land is covered in low scrub, rough grass and hunks of masonry. Immediately in front of us the perimeter barrier runs around the entire site. There are seven miles of grey concrete posts, chain link fencing and rusty barbed wire. Nearby on a brick wall is the faded name of a band painted in white letters – New Order.

This is what we have come to see, the site of the former BP Baglan petrochemicals factory, but the scale of it, the immensity of this plain stretching into the distance, is overwhelming. We have read that some parts of the soil of this area are contaminated with toxins to a depth of ten feet. The plant ran for 41 years, and at one point it was the largest petrochemical works in Europe. It operated night and day. For perhaps 380,000 hours it spewed the effluent of plastics into the land, into the seas and into the skies.

We look over the roofs of Sandfields, the second largest housing estate in Wales. The construction of 3,500 new homes for workers was begun in 1948. It was mostly complete by the time the smoke began to pour from the chimneys of the petrochemical plant. Two generations were no doubt thankful for the employment created but breathed in the refuse of the oil company. On mornings such as this the west wind took that junk from the works and dumped it in the lungs and on the skins, the houses and cars of those who lived on the land between the sea and the foot of the mountains.[1]

We walk through the streets of Sandfields and stop off at Franco's café on the seafront near the Afan Lido. Over coffee we look at a map and graphs and unpick how Britain was positioned within the world of oil during the late 1960s.

Here are the resources in Iran, Nigeria, Libya, Kuwait and Iraq that the UK was dependent upon. And here are the lines showing the routes of the supertankers bringing that crude to Britain. We trace the voyage of MV *British Admiral* from Nigeria and the SS *Torrey Canyon* coming from Kuwait bound for Angle Bay in Pembrokeshire, shipwrecked off

Land's End. The black tide of oil was blown along the Channel and up the Severn Sea. It covered the beaches of north Devon only 25 miles south of where we are sitting.

Oil production in these six states had leapt by the mid-1960s and oil consumption in Britain increased correspondingly. The financial pressure on the corporations to grow new markets drove Occidental, ENI and Total to attempt to build refineries on Canvey Island. That same pressure transformed the world of Sandfields and Baglan.

By 1968 the Baglan petrochemicals factory was already said to be the largest in Europe, but BP was concerned about its profitability. John Hunter, the head of the company's petrochemicals division in an office on Piccadilly, believed that the heart of the plant, the catalytic cracker opened only a few years before, was already too small to be viable, declaring it to be an 'economic weakness that must be eliminated'.[2] A grander scheme was conceived and agreed upon in the boardroom of the new Britannic Tower.

On 28 October 1968, in a press conference[3] simultaneously held in BP's London Head Office and in the Dragon Hotel Swansea that was linked by closed-circuit TV, a new goliath of a cracker was announced. The project had an estimated capital cost of £60 million; £22.5 million of which was provided by government grants with £37.5 million coming from BP[4] – although the latter was itself 68 per cent owned by the state. It was claimed as the largest investment that BP had made in the UK, and certainly it was a massive provision of public funds to underpin the petrochemicals industry. Hunter said the project would 'tax, at all stages, (the) manpower and experience' of BP.[5]

The giant new machine for the breaking and burning of crude would be the centrepiece for the future of south Wales. Ten new plants within the factory would manufacture different types of raw materials, or 'intermediates' as they are called, which other factories elsewhere would turn into everything from plastic flower pots to foam for chair cushions, and from medicines to cosmetics and paints. The construction phase would create over 3,000 jobs[6] and once the factory was running as planned it would employ 1,500 more staff. This in turn would lead to house building in Sandfields and a new housing estate on Baglan Moors.[7]

The expansion of the factory was at the heart of the expansion of what we call the Severn Estuary oil complex. In this web of pipelines, roads and railways, Baglan was fed by Angle Bay and Llandarcy, and Baglan in turn fed the chemicals it produced to BP's factories in Barry and Stroud. The complex itself supplied BP works further afield including ones in Carshalton in London, Oldbury in Birmingham and Sandbach in Cheshire. BP had slowly acquired all of these plants in the 1960s as it endeavoured to become the UK's largest petrochemical company, outstripping the mighty ICI.

The Severn Estuary oil complex was intended as a mirror to the network of BP factories that stretched down the north-east coast, from Grangemouth near Edinburgh to Wilton on Teeside and Saltend near Hull, all linked by a chemical products pipeline.[8] This British realm of plastic was administered by new BP offices at Piccadilly and Victoria in London.

A year after the announcement of Baglan's fivefold growth, as construction work on the plant expansion was underway, the Afan Lido hosted the Afan Festival of Progressive Music in December 1969. Top of the bill were Pentangle and Pink Floyd.

* * *

During the building of the original plant seven years previously there had been protests against the factory. On 25 April 1963[9] – five months before the official opening – the *Daily Mail* reported 'Soot turns whites black in Golden Grove'. For the pollution from the new flare stack had ruined clothes on the washing lines of residents in Sandfields, especially in the streets around Golden Grove, and deposited grit across gardens, streets and roofs. Forty women marched on the factory, carrying sacks of grit swept up from around their homes, to demand action. As the history of Baglan published by BP noted: 'Management met the group at the factory and invited the ladies to the canteen for a cup of tea. Assurances were given by the company that the problem ... would not reoccur; this was accepted by the residents.'[10] This demonstration by these women of Sandfields was perhaps the first protest against the oil industry in the UK, a decade before the actions that took place in Canvey. In the same month of April 1963 a residents' association was set up in the village of Baglan on the northern edge of the plant, which pledged 'To wage war against the local authority' and demanded compensation for the impact of the works.[11]

As the newly expanded factory came into production in the early 1970s there was again strong opposition among local residents. News of some of it reached national press and TV. The *Sunday Times* printed the words of the Rev. A. R. Wintle, curate of Baglan's St Catherine's Church that overlooked the factory:

It has become virtually impossible to open bedroom windows during the hours of darkness ... Nightly ... enormous clouds of vile vapours ascend from the stacks and towers of BP and drift towards the defenceless dwellings of a sleeping population nearby. The air is heavy with indescribable smells; the whole district reeks with the odours produced by the uncaring ingenuity of modern technological man.[12]

On arriving in his new post as the Baglan works general manager, Ray Knowland had to go head-to-head with the Rev. Wintle on live radio.[13]

Despite reassurances from the management, the chemical plant had a powerful impact on the immediate environment. Lynne Rees, who attended Sandfields Comprehensive from 1969 to 1976, writes in her memoir that the school was 'a cough and choke away from BP, and I remember the warning sirens to stay inside until an all-clear' was given. She recalls that because of the pollution 'washing turning black or even disintegrating on the garden line'. And 'I remember a monstrous open-mouthed effluent pipe that used to cut through the sand dunes from the works onto the high-tide line of the beach'.[14,15]

As with the resistance to the refineries being built in Canvey, the protests at Baglan in the early 1970s tapped into the growth of the wider British environmental movement. However, unlike at Canvey the resistance did not halt the construction of the works. Perhaps this was because of the strength of BP; unlike Occidental it was a 'national institution' and had powerful links with officials in Port Talbot, the county of Glamorgan and Westminster. Or perhaps it was because the company already employed over 6,000 personnel[16] directly and indirectly across south Wales, from Angle Bay to Barry, and consequently the wages it paid underpinned thousands of households. Or perhaps it was because at night the lights of the plant reflected in the waters of Swansea Bay, shimmering white like the skyscrapers of Manhattan, proclaiming themselves as the inevitable future of Wales, the future of Britain.

We look up from our coffees and through the window to survey the two miles of one of Wales' finest bathing spots and wonder at what remains here. Perhaps one of the motives for the building of the Afan Lido, opened by the Queen in 1965, was that it meant that residents of Port Talbot would not have to bathe in a toxic sea. Today this sunny beach does not look like the blackened land at the site of the old factory, but it is implausible that 40 years of toxins have not left their mark here too.[17]

* * *

In November 1971[18] one of the plants of the expanded factory came into production making vinyl chloride monomer – known as VCM – a key constituent of the plastic PVC. Just over a year later news came that labour unions at a petrochemical works in the US had raised the alarm after three men died from angioscarcoma, a rare form of liver cancer possibly linked to the manufacture of VCM.

The joint shop stewards' committee from the three key unions at Baglan met with Peter Sharrock, the works general manager.[19] Through collective action they rejected his initial offer of 'danger money', opposed the higher levels of protective clothing they were offered and refused to work in certain high-hazard areas. They managed to force changes that reduced the levels of toxins and have monitors installed to ensure those levels were maintained.[20] In August 1973, after months of union representatives negotiating with Sharrock, the head of BP Chemicals, John Hunter, came down from London to reassure the workforce, saying 'I think all one can say is that since January when this particular situation came to light, a great deal of progress has been made and more will be done.'[21,22]

After protests by Baglan residents, BP financed a team that toured the neighbourhoods monitoring air quality, responding to complaints and demonstrating that the company was concerned about the community. The pollution of Swansea Bay, the fouling of the air around the homes of Sandfields and Baglan and the inconvenience of 'stay indoors' alarms could be accommodated by the company with relative ease.

But the issue of cancer hung like a spectre over Baglan. In June 1989 the headmaster of Sandfields Comprehensive School, David Winfield, drew attention to the fact that out of 70 staff, seven had suffered from cancer and three had died, while six pupils had suffered from various forms of cancer since 1984. Here was the school which had been so famed in the 1960s for its progressive drama section, now becoming known as a place of lethal pollution. Winfield pushed the West Glamorgan Health Authority to investigate the causes of the cancer and the links to the Baglan factory a few hundred yards from the playground.[23] The health authority's subsequent report reassuringly claimed that the cancer rate was not higher in Baglan than for anywhere else in West Glamorgan, while BP responded by saying, 'detailed monitoring of the air around the factory, gave no indication of any environmental problems caused by the plant'.[24]

But the financial implications of cancer for BP were alarming. The fate of companies such as H&N Ltd who were asbestos manufacturers was widely understood. They had resisted protests about asbestosis, but eventually the health impacts were incontrovertible, the use and manufacture of the substance was banned, and the company was drowned in a flood of lawsuits. Shell had narrowly escaped censure over the drins. If BP, or any of the oil majors, was found to be knowingly manufacturing chemicals that caused cancer in its workforce and in the neighbouring residential areas, then it too could face monumental legal struggles.

* * *

That there should have been a robust response from the trades unions to the initial threat of cancer from VCM should come as no surprise. The Baglan workforce had a long history of fighting for decent working conditions.

From the moment when construction began in 1961, the number of jobs at the factory grew and contracted in relation to the physical scale of the factory itself. The complexity of Baglan, with its twelve different plants, was mirrored by the complexity of the trades unions, with different elements of the workforce represented by different unions including the Transport and General Workers' Union (TGWU), the Electrical, Electronic, Telecommunications and Plumbing Union (EETPU) and the Amalgamated Engineering Union (AEU).[25]

The expansion of the factory announced in October 1968 appeared to fit within the era of collaboration between unions and management as signalled by the signing of a special manpower productivity agreement between BP and the TGWU, the EETPU and the AEU.[26] However, the early 1970s ushered in an entirely new spirit as the company tried to bring novel operating methods into the workforce. Furthermore, disputes at Baglan became intertwined with the political struggles in wider British society as Edward Heath's Conservative government attempted to break with the fraying consensus of the 1960s.

Those turbulent times were highlighted in June 1970, in the very month that Heath came to power, with a five-week strike by crane drivers[27] on the Baglan construction site. By March 1971 the first issue of *Construction Worker*, a bulletin from the Industrial Fraction of International Socialism, Swansea Branch, was being circulated at the plant.[28] It read:

On a broader front we are confronted with a massive political attack, launched on the whole Trade Union movement, in the shape of the Industrial Relations Bill. The point is on this construction we are making a new beginning – we are having to rethink the question of organisation, at the same time Baglan is helping to lead the way in this area, for the creation of a united working class opposition to the current Tory attack. A call for a General Strike is imperative. Nothing talks faster than lost work, especially if they [BP] have to pay for it.[29]

Over the following two years there were repeated labour disputes on the site of the expanding factory. It was said that the contracting companies lost control of the giant project, and certainly it ran massively over budget and behind schedule. Two of the ten new plants were nearly two years late being commissioned. There were spiralling costs for BP[30] and a sense of chaos on the factory site. To add to this, in

September 1972 came an arson attack on the factory causing £12,000 worth of damage.[31]

The politicisation of the workforce at Baglan – the labour of both BP and of the many companies working under contract – was taking place during an upsurge in radicalism across south Wales. As BP was struggling to bring into being the ambitious Baglan plans, the National Union of Mineworkers (NUM) in South Wales was involved in an industrial struggle on a scale not seen since the General Strike of 1926.

The 1971–2 miners' strike across Britain reached its zenith on 10 February 1972 at the Battle of Saltley Gate in Birmingham where miners from south Wales pits were joined by comrades from Yorkshire and together they forced the government to back down. Not only was this struggle taking place within the close community of Sandfields and Port Talbot, where BP Chemicals households lived next door to mining households, but Baglan was also involved with the fate of the strike.

The factory was a major user of electricity so production was affected by any disruption to the power supply, mainly from coal-fired power stations, during the miners' strike in 1971–2. Consequently it was in the interests of BP, from Peter Sharrock, the Baglan works general manager, up to Eric Drake, chair of the company, to see Heath's government come to a speedy settlement with the miners, and thereby ensure the reliability of the power supply.

However, in April 1973[32] a new power station plant was opened on the Baglan site, powered by fuel oil delivered from Llandarcy. This provided guaranteed electricity for the factory, meaning that Baglan was insulated from future supply disruptions and was able to export electricity to the National Grid. Consequently it could be an actor in support of the government side in the event of any future battle with the miners.

The 'Oil Crisis' of October 1973 only raised the political temperature further. It had long been held that coal was a dying industry, and that King Coal would be usurped by King Oil, but the skyrocketing price of crude and concerns over the security of the UK's oil supplies raised questions about this presumed future in which the Severn Estuary oil complex was set to play such a key role. Suddenly coal was once again powerful, and its political muscle was demonstrated in the second miners' strike of 1973–4.

In November 1973 the NUM imposed an overtime ban that impacted on the UK's electricity supply, and coincidentally disputes in Baglan reached a new crescendo. The AEU, EETPU and TGWU made a claim for higher salary and shift differentials. The BP management pushed back and all 1,200 industrial staff were suspended without pay for a week. In this bitter dispute there were sometimes a hundred pickets at the gates and police were called. Production was brought to nearly a complete standstill. On 7 December there was a return to work and the factory began to function again, but the strike had taken place during a UK national shortage of chemicals and so it was widely covered in the media and created a reputation for Baglan's workforce.[33]

A month later the miners decided that they needed to move quickly to avoid a protracted struggle with the Heath government over pay and conditions. From the perspective of the NUM the biggest threat to their success was longer daylight hours and warmer spring weather reducing the importance of the electricity supply and strengthening the ability of power stations to hold out. Factories such as Baglan, now able to export power to the National Grid,[34] were a block to the miners' strategy of strangling Britain's electricity supply. The NUM decided to go for broke and 91 per cent of the members balloted supported an all-out strike. South Wales miners were the most militant in the country, voting 93 per cent for the strike and 7 per cent against.[35] The action began on 5 February 1974. Two days later Prime Minister Ted Heath declared the general election, fighting under the slogan 'Who governs Britain?'

Heath's subsequent defeat at the polls by Wilson was the miners' victory and a success for all organised labour. Once again the trades unions had demonstrated their power to shape the politics of Britain. The communities of south Wales were at the centre of this victory. The vanguard of the struggle was the NUM but the strength of the miners depended upon the support of neighbours, local businesses, families and friends. The network of ties within Sandfields and similar housing estates was a key part of this wellspring of defiance. BP operated at the heart of this zone of labour militancy, just as it had done in Abadan.

Four years later, in 1978, came the Winter of Discontent. The events between December 1978 and February 1979 became iconic in British history. The strikes by public sector workers, ranging from refuse collectors to gravediggers, generated TV footage which created the image of a Labour government at the mercy of the unions and unable to govern. The stories of bodies unburied and mountains of rubbish in the streets helped catapult Margaret Thatcher to power in the subsequent general election and were utilised by the Tories again and again over the next four decades as a warning to voters against supporting the Labour Party.

Pivotal among the strikes was the role of oil tanker drivers who, as members of the TGWU, went on an overtime ban on 18 December 1978 demanding a 40 per cent pay rise. Many of these drivers had been delivering fuel from BP and Shell refineries such as Llandarcy in south Wales and Shell Haven in the Thames Estuary. The Labour cabinet discussed plans for 'Operation Drumstick' in which the army would take over from tanker drivers, but it became clear that this would require a state of emergency to be declared. On 3 January a full strike began with 2,000 drivers stopping work[36] and eight days later the action was made official by the TGWU. Further emergency plans were discussed by the cabinet, as 80 per cent of the nation's goods were delivered by road meaning that essential supplies were soon running short. Eventually, after nearly three weeks, the drivers accepted a 20 per cent pay rise and the strike began to end.[37]

The crucible of Britain's political direction lay in the struggle between its financial sector and its industrial areas. The oil companies played a pivotal role in this fight not only in the City's battle over the development of the North Sea but also in manufacturing heartlands such as south Wales. The pushback of capital against union power, and Labour's drive to nationalise assets, took place in Baglan too.

7

Tuireadh: Lament

Tuireadh, piece for clarinet and string quartet,
by James MacMillan, 1991. Dedicated to the victims
of the Piper Alpha disaster and their families.

In the bright sunshine we stand on the flagstone terrace at the front
of the house, before us the lawn falls away, the grass neatly trimmed
between the bright colours of the herbaceous borders, down to the
tennis court. The foreground is dominated by a great weeping ash
tree and beyond it the view stretches across the Weald to the outline
of the North Downs. The three and half acres of garden are full of
people. Couples, mostly grey-haired, gaze at the ornamental ponds
and mutter about the topiary, before retiring to the conservatory
for tea and sponge cake served by the Filipino staff. It is a National
Gardens Scheme open day and by mid-afternoon 350 people have
passed through the gates donating their five pounds entrance fee to
charity. This garden, mentioned in the *Daily Telegraph* and featured
in *Period Living*, surrounds the weekend home of Sir Peter and Lady
Meryl Walters. The design of the grounds is the work of Lady Meryl,
assisted by her team of seven gardeners.

We've come in the hope of finding some way of talking, if only
briefly, to Sir Peter, former chairman of BP from 1981 to 1990, who
was intimately involved in the struggles against Edward Heath and
Tony Benn. He is now 86. We have tried to contact him but have been
told he does not want to be interviewed. We felt compelled to try a
bit harder without being too intrusive. Standing in the garden before a
pair of gravestones to dogs that have passed on, we chat to the wife of
one of the staff. 'The dachshunds were called Ribena and Luca, short
for Lucozade. He was given them as a present when he left the board
of GlaxoSmithKline [where he was a non-executive director]. They
love their dachshunds. They've got four.' She tells us, 'He's poorly now,

NORTH SEA OIL & GAS

0   50   100 miles
0   50   100   150 kilometres

Atlantic

Ocean

Clair
Ridge

Sullom
Voe

Schiehallion

*Shetland
Islands*

Foinaven

Lerwick

**Limit of UK
Offshore
Territory**

*UK
WEST OF
SHETLAND*

*Orkney
Islands*

Kirkwall

St Fergus
Gas Terminal

Inverness

Peterhead
Power Station

ABERDEEN

Cruden Bay
Pump Station

S C O T L A N D

Dundee

Kinneil
Stabilization
Terminal

EDINBURGH

Key

Forties Oil Pipeline System

F   FUKA Gas Pipeline System

S   SAGE Gas Pipeline System

SI SIRGE Gas Pipeline System

M   Gas pipeline from St. Fergus
Gas Terminal to Miller Gas Field

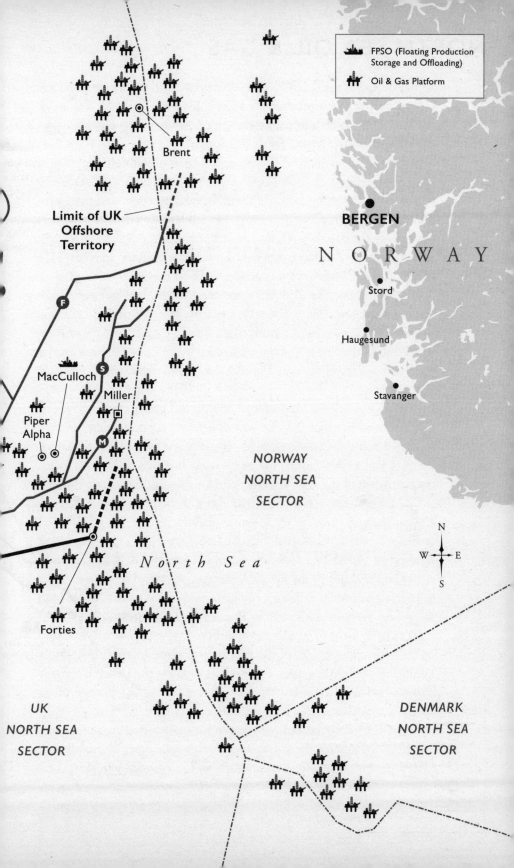

FPSO (Floating Production
Storage and Offloading)

Oil & Gas Platform

Brent

BERGEN

NORWAY

Stord

Haugesund

Stavanger

Limit of UK
Offshore
Territory

F

MacCulloch

S

Miller

Piper
Alpha

M

NORWAY
NORTH SEA
SECTOR

North Sea

N
W   E
S

Forties

UK
NORTH SEA
SECTOR

DENMARK
NORTH SEA
SECTOR

he rarely comes out. Lady Walters still goes up to the house in Kensington during the week, but he doesn't anymore.' It looks like we're too late. There is certainly no sign of him. We sit in the Japanese Zen garden and contemplate a failed mission.

How strange to think of a man, once a titan of industry, reduced to life in a few rooms. A man whose actions helped shape the country in which we've grown up, now at the moment of his decline. A man who had commanded one of the most technologically advanced organisations on the planet, spending his closing days in a former rectory surrounded by a perfect English Edwardian country garden. This corner of the Home Counties remains forever empire.

On purchasing the Old Vicarage the Walters had planted 3,000 trees to mask the traffic on the A24 roaring through the valley below on its way to and from the south coast. Now, 25 years later, the mature woodland shuts off most of the explosions in the petrol engines on the road, just as the blinds over the windows of the house shut out most of the public.

Walters was a keen supporter of Margaret Thatcher who gave him a knighthood in 1984. 'Yes, he was *definitely* a supporter of the Conservative Party, *definitely*, explicitly so', said one former BP colleague. He was also a president of the right-wing Institute of Directors and later a trustee of the now-controversial Institute of Economic Affairs.

Walters was the son a policeman, born in Birmingham in 1931, a city caught up in a boom of motorcar manufacturing. At 13 he started at King Edward's Grammar School. Soon after his father was killed fighting in World War II. Later he said: 'I had two decisions. I could either let the death of my father defeat me, or I could go on and make something of myself. I chose to make something of myself.' After each school day he helped his mother run the family off-licence.

In 1949 he went to Birmingham University, part of the cohort aided by the education policies of the new Labour government. Intriguingly he studied not mathematics, engineering or law – then the usual foundations for an oil industry career – but commerce. At the age of 23 he took a post in Anglo-Iranian, still reeling from the Iran crisis, in the year the company changed its name to British Petroleum.[1] He advanced quickly up the corporation. After nine years in Supply and Development in London and a couple in the BP office in New York, he was promoted in 1965 to be head of BP North America Inc. at the young age of 34. His predecessor had been a decade older when he was appointed to the same post.

An internal survey of BP middle managers conducted the year after Walters went to New York found that they were satisfied with the rates of pay and the generous welfare arrangements – sports facilities, social clubs, subsidised lunches – and sickness and retirement benefits. As with the world around the workers at Baglan or the expat compounds in Iran and Nigeria, the company in the 1960s provided the staff at central office with their own version of the welfare state. Furthermore, the survey revealed that 'affairs in BP were said to be conducted in a gentlemanly, pleasant fashion', but that there was a lack of drive. The typical BP recruit was, it reported, 'the kind of person if you met him at a cocktail party you'd say – awfully nice chap'.[2]

Describing himself as coming from an intelligent working-class background, Walters was conscious of the fact that so many of his peers and seniors came from a different class altogether. In the early 1960s, BP under chairman Maurice Bridgeman had an executive board made up of six men. All but one had attended Cambridge University and none had studied anything like commerce. As the official history of BP describes, Walters declared that he believed in 'an elitism based on merit' and he 'frowned on the old-boy network and systems of personal patronage that Bridgeman operated'.[3] This was just the kind of belief that would chime with the views of Margaret Thatcher, herself also the child of a shopkeeper.

Once in New York, working from a skyscraper on Madison Avenue, Walters' most important task was to push forward the development of BP's concession on the Alaskan North Slope. During his three years at the helm in New York, as BP seismic teams surveyed the Alaskan rivers and mountains, the pressure from head office in London to achieve a successful discovery became stronger as the company's position in Iran once again became weaker.

Perhaps it was the struggle to defend BP in the US from its competitors Exxon and Shell. Or perhaps it was the experience of life in New York in the mid-1960s, away from the stuffiness of London. Either way, his imagination about the company shifted and Walters came to believe that BP should change its culture to mirror that of its American rivals. The consumption of oil in Britain, in cars, planes and plastics, had held the country under the spell of American culture. Now the great producers of oil, the corporations, were themselves becoming enchanted.

* * *

In May 1967 Walters was transferred from New York back to London and promoted to run BP's global logistics, overseeing the transportation of crude and oil products around the world. He arrived in the City, which he had last worked in two years previously, stepped into the new Britannic Tower and took the lift to the upper floors.

Walters' opposition to the dominant strategic thinking in BP had hardened. Assisted by the rapid expansion of computing power, which enabled sophisticated modelling, there had been a relentless drive towards greater integration of the different departments in the company, between exploration, refining, supply, marketing and finance. This was a drive to see these departments – each of which was a massive bureaucratic entity in itself – as just bits of one gargantuan whole, rather than distinct areas whose profitability should be judged discretely.[4] To Walters it seemed that soon BP's entire operation would be run by a mainframe computer and it would be increasingly difficult to assess which parts of the company were making a profit and which a loss. He was convinced that centralised control diminished the responsibility, the motivation and the commercial acumen of the executives running the different departments.

The following month, Walters had a catalytic experience. He was mowing the lawn at his new home in Highgate, north London one Saturday morning. It was a couple of days after the outbreak of the Six Day War in the Middle East. He was called into the house to take an urgent call from BP's head of tanker chartering who told him that the Greek shipping tycoon Aristotle Onassis had abruptly cancelled all existing chartering arrangements. Onassis was offering BP his entire tanker fleet, but at rates double what they had been the day before. BP had until noon to give an answer, and it was up to Walters to give it. Tens of millions of dollars was riding on the decision. He recognised with a sudden clarity that no centralised computer programme could help him now, only his commercial judgement. He phoned back. Yes, take the offer, he said and returned to mowing his lawn. Events quickly proved the decision right; by Monday tanker rates were four times what they had been on the previous Friday.[5]

From that day on Walters became a campaigner, pushing against computer-backed centralisation and for the deintegrating of BP departments. 'That set me to thinking about the whole way we did business. I realised that the proponents of further integration were moving in the wrong direction. They were handing over to machines what should be management judgements.'[6] This moment informed what Walters pushed for within the company: 'It's nice to have some integration, obviously, but it's not something we would pay a premium for.' He explained later, 'For me there is no strategy that is divorced from profitability.' He became famous for telling managers, 'There are no sacred cows in BP' and 'you tell me which things make economic sense and which do not, and I'll tell you which we'll keep and which we won't.'[7]

* * *

The Six Day War in 1967 was described by Walters as providing him with a seismic shock. It shifted the understanding of this 36-year-old man, who went on to change the way he managed the BP Supply and Development Department to which he'd just been appointed head. These shifts not only boosted his profile in the company, but as his was one of the most powerful departments in BP they affected the whole corporation. In 1973, at the early age of 41, he was appointed to the board, a decade younger than his peers.

Twelve years after the shock of 1967 came the Iranian Revolution. Decades of resistance to the brutal regime of the Shah, put in place through the coup of 1954, came to a head with rioting in numerous towns and cities in 1978. The oil fields in which Western companies operated were hit by strikes and acts of political violence against foreign staff. BP and Shell, along with Exxon, withdrew most of their workers from Iran. The family compounds, such as that in which John Browne had spent his teenage years, were evacuated.

However, whereas the epicentre of the uprising in 1951 had been AIOC's refinery town of Abadan, now the core of revolt was in the religious city of Qom. For the new Iranian Revolution was led by the clerics, and particularly Ayatollah Khomeini. The secret services of the UK and US, which had played such a key role in overthrowing Mossadegh and installing the Shah three decades before, were unprepared for, and lacked intelligence about, an Islamic revolution. Eventually, after the revolt had swept across the entire country, the Shah and his family fled.

The position that BP, Shell and others had held in Iran and the control of Iranian oil was suddenly overturned. In an echo of 1951, all of the companies' assets were nationalised. Oil production came to an almost complete standstill overnight. But there was to be no turning back, no reinstating of the Shah and return of the Western corporations.

The direct impact on BP was far less dramatic than in the early 1950s. However, the revolution was quickly followed by the Iran–Iraq War which broke out in 1981. As warplanes attacked the refineries and oil fields of two of the world's largest crude producers, the threat to supply was immediate and the global oil price leapt. This second Oil Shock utterly transformed the industry.

For seven decades the structure of the international oil corporations had remained roughly the same – each one owned oilfields, tankers, refineries and petrol stations. Largely each company passed the crude that it extracted through its own industrial metabolism. This was the essence of the integrated oil company, built on having control of oil reserves, a share of the gasoline market and the structures in between. This house, which had begun to creak, collapsed as the Oil Shock hit.[8]

Since the 1950s the Western oil companies were under great pressure to keep oil production in Middle Eastern and African states as high as possible. Partly this was driven by their internal desires to see return on capital generated as quickly as possible, and partly on the need to ensure that the governments of those states generated high revenues from oil, while obscuring the fact that their percentage of the income from the extraction of their own resources was relatively small. It was in the oil majors' interests to build refineries and petrol stations to ensure that the high levels of crude produced were absorbed by an expanding market.

BP and Shell were keen not to reveal the scale of the profits that they were making on the crude they extracted, for fear that revelations of this would encourage oil-producing states to demand a higher share of the revenues. Consequently it was in their interests to have refineries and marketing arms that produced low incomes so that when the whole company was accounted for its integrated profits could be made to appear reasonably modest. The companies were zealous about not separating out the accounts of different divisions. However, when after the Iranian Revolution the companies had lost yet another section of oil field production, divisions such as refining were revealed as loss-making.

There was a new style of management brought in, focused on markets, and Walters led the way. The pressures outside the corporations forced a new phase of deintegration upon them. A leap that Walters had been proposing since 1967 was coming into being and he was in a position to steer the changes through BP. In 1980 he had been appointed deputy chairman and it was clear he was destined to replace David Steel as chairman when the latter reached retirement age in November 1981.

The breakdown of the integrated structure of BP was not merely a managerial exercise, but involved a revolution in the culture of the company. Since World War I refineries were retained because 'That's what BP did' and tankers were built in British shipyards because 'That's what BP had always done'. Now every division, every given practice, was brought into question. Walters stuck to his mantra, 'There are no sacred cows'. If a unit could not make commercial sense on its own account, if it could not generate sufficient return on capital employed, then it would be shut down or sold off, plants would be closed and staff made redundant.

As one BP insider told us, 'This was a big shift in practice. Big companies in Britain had not before made large numbers of staff redundant en masse. The market came to dominant philosophy in BP. Before the 1980s the principal organisational spine of the company was "do anything to avoid nationalisation of assets", now it became "the market".' The lessons that Walters had learned in the US during the 1960s were finally being employed by one of Britain's largest corporations in the early 1980s. The same process was taking place in its rival Shell.[9]

In line with his 'no sacred cows' principle Walters oversaw a number of UK plant closures, setting in train a process that rolled out over the following three decades. Smaller chemical plants such as those in Stroud and Sandbach[10] were shut down, but most dramatic was the end of BP's Kent Refinery on the Isle of Grain. This closure on 29 August 1982 took place under the chairmanship of Walters, and under the gaze of Minister for Energy Nigel Lawson, Chancellor Geoffrey Howe and Prime Minister Margaret Thatcher,

With hindsight there is a powerful symbolism in this ending. The plant had been visited by the young Queen Elizabeth and her consort only months after the coronation 27 years earlier. Shortly before the Suez Crisis, Britain's power in the Persian Gulf was still second only to the Americans, and the Kent Refinery was constructed explicitly to process crude from Iran. Now, less than three decades on, BP and Shell controlled no crude in Iran or Iraq, and were only the junior partners in Kuwait, the United Arab Emirates and Oman. BP and Shell had been forced to retreat from the Middle East and now, guided by the new principles of the 'market', it was leaving behind the Thames Estuary. When we spoke to John Vetori in the Pegasus Club about the closure of Coryton in Essex, where he had been a manager, he talked of his shock the day when the flare over the Kent Refinery, which was so visible in the southern sky, suddenly went out.

* * *

In March 1984 the NUM called on its members to stop work. As with the strikes of the early 1970s, the miners focused on the proven tactic of cutting off Britain's electricity supply. While coal-fired power stations ground to a halt, the strength of the Thatcher government to fight the NUM depended upon the reliability of oil-fired power stations. Oil was central to the Conservatives' game plan as to how to avoid a repeat of the defeats of 1972 and 1974. A network of the oil-fired power stations existed around the country, but foremost among them were Grain, Kingsnorth and Littlebrook in the Thames Estuary. In 1972 and 1974 BP's Kent Refinery had attempted to provide fuel to two of these plants, but by 1984 that refinery had closed. In the 1972 strike Kent miners had successfully worked with union officials within the Shell Haven refinery to stop the delivery of oil to Littlebrook.[11]

Would the oil industry workers support the miners again in the latest strike?[12]

Many in the workforce within Shell's and BP's UK plants, and the drivers in the company road tanker fleets,[13] were supportive of the NUM's stand against the Thatcher government, but inevitably workers were conflicted as to whether to act in solidarity with their union comrades or to avoid risking conflict with the corporations that employed them. This clash of loyalties existed not only in mining areas, such as south Wales, but also in places further removed from the coal strike. Within the branch of the TGWU at Shell's Stanlow Refinery on Merseyside on 3 May 1984 there was a vote as to whether to show solidarity with the miners by shutting down the pipeline that provided fuel oil to the Ince power station half a mile east of the refinery.[14] Had this proposal been passed, mirroring tactics utilised by workers in French refineries, it would have involved Shell staff acting directly in support of the NUM and no doubt clashing with their own employers.

But action was not taken, and throughout the year-long struggle Baglan's power station continued to export electricity into the National Grid while Llandarcy maintained supplies of fuel oil to Pembroke oil-fired power station.[15] However, at the Shell Haven and Coryton refineries in December 1984 there was some action taken by tanker drivers to block supplies to power stations such as Littlebrook, Grain or Kingsnorth. And at the Texaco depot at Purfleet, also in the Thames Estuary, two tanker drivers were suspended for refusing to cross a miners' picket line.[16]

Baglan workers may have refrained from taking action to support the miners, but it seems likely that many gave neighbourly assistance to the strikers and their families whom they lived alongside in Port Talbot. For in the town the movement to back the strike was powerful and militant. In December 1984 one of the most potent events of the struggle took place at the heart of Sandfields. Norman Willis, the head of the Trades Union Congress (TUC), came to address a miners' rally at the Afan Lido.[17] The reception was hostile. Those on strike and their families were enraged by the lukewarm support they were receiving from the TUC and the Labour Party, despite months of intense hardship and some brutal policing in the coal communities. Willis was heckled, and as he spoke a noose was slowly let down from the rafters above and dangled over his head. Onto it was fixed a card reading 'Where is Ramsay McKinnock?' an allusion to Ramsay MacDonald, leader of the Labour Party and local MP for Aberavon in the 1920s, who had not wholeheartedly supported the 1926 General Strike. Neil Kinnock was the leader of the Labour Party in the 1980s, and MP for the Welsh constituency of Islwyn, and he too seemed to be qualified in his support for the NUM.[18]

Ultimately the loyalty of the workforce in the oil refineries, and hence the workforce of BP and Shell, was vital to the Conservative government. When in October 1973 Prime Minister Ted Heath had demanded that Eric Drake of BP and Francis McFadzean of Shell ensure crude supplies for the UK as the 'Oil Crisis' bit, they had flatly refused to agree. The chairmen effectively left the prime minister at the mercy of the NUM who were already warming towards a second coal strike.

By contrast, in the 1984 New Year Honours list Thatcher awarded Peter Walters, chair of BP, a knighthood.[19] Walters stood by the Tory government in its hour of need and ensured that oil flowed from the refineries to the power stations. This support was not left to chance. In early 1984,[20] prior to the strike, oil company staff were in close discussions with John Wooley and David Bridger[21] of the Central Electricity Generating Board (CEGB) to discuss supply. BP Shipping developed a plan to deliver the carbon dioxide gas required for the nuclear generation process to the CEGB's nuclear power plants by sea in order to bypass any pickets at the plant's land entrances.[22,23]

By 3 March 1985 the miners were defeated and the workforce returned to the pits having won no concessions. The legendary strength of the NUM had been broken and the power of the Thatcher government was now apparently unassailable. The air of despondency that settled over south Wales was coupled by a blow that came five weeks after the strike ended. On 10 April 1985 Sir Peter Walters declared the closure of BP's Llandarcy Refinery after 64 years of production.[24]

Not only would this mean that around 2,000[25] workers would be made redundant during the 19 years[26] it would take to wind down the plant, but these layoffs would be added to the rising number of unemployed. Furthermore, the closure symbolised the weakening of BP's commitment to the wider Severn Estuary area. Inevitably, the closure of Llandarcy's import facility – the Angle Bay Ocean Terminal – was announced a few months later and it was widely feared that both the Barry and Baglan factories would soon shut. The most distant satellite plant of the oil province, in Stroud, had been closed under Walters three years earlier in 1982.

The 15 years of intense political action by industrial workers in south Wales – between 1970 and 1985 – was echoed in other parts of the UK. Once again BP and Shell were deeply embroiled in these struggles. Company staff included those working offshore, in refineries and chemical works, in gas terminals and oil depots, in shipping, pipelines and road tankers, and the myriad offices that administered this sprawling network. Taken as a whole the industry constituted the largest single workforce in the UK – substantially more than the coal industry. It was in the interests of the management of BP and Shell and the British state to prevent militancy within the oil and gas industry workforce and all the sectors that were linked to it, from car manufacturers to road hauliers. After all, in a wide range of states, from Russia to Iran, oil workers had been at the forefront of political action or revolutionary change.

* * *

With the benefit of 40 years hindsight Guy Standing, professor of development studies at SOAS, University of London, summed up the revolutionary aspect of Margaret Thatcher's government in blunt terms: 'Her biggest economic decision had been to privatise the North Sea oil operations and splurge the oil taxation revenues on income tax cuts and current spending. This was in contrast to Norway, which retained government stakes ... and invested the revenues from taxes and dividends in a national capital fund.'[27]

In the wake of the Conservative election victory in 1979, the government set about a radical redirection of policy over the offshore colony of the North Sea. Under Chancellors of the Exchequer Geoffrey Howe and then Nigel Lawson, and Secretaries of State for Energy Nigel Lawson and then Peter Walker, the administration undertook a strategy of privatisation that was a dramatic departure from the previous nine governments, both Conservative and Labour, since the early 1960s. It shifted the UK away from the model being pursued by other European states such as Norway and Denmark – and away from the model of OPEC states such as Iran and Nigeria – and towards the model of the US and Canada. This redirection of policy ran contrary to the aspirations embedded in Wilson's comment only a decade before, that in the 1980s the UK minister of energy would be head of OPEC. The state was to withdraw as far as possible from the national asset of the North Sea and to effectively sell it off to private corporations and their shareholders. Not only was the sale of the UK holding in BP shares pursued at pace, but other state assets were sold. BNOC, designed by Energy Minister Eric Varley and established in November 1975 under the Petroleum and Submarine Pipe-lines Act when Tony Benn was minister, was part privatised six years later in November 1981, under Nigel Lawson. British Gas, established in 1972 under the Heath government, was sold off under Chancellor Lawson in December 1986. This mass shareholder offer featured the iconic advertising campaign slogan by the New York company Young & Rubicam: 'If you see Sid ... Tell him!' Approximately 1.5 million people brought shares, at 135 pence each, in the £9 billion share offer, the largest ever at the time.[28]

After nine years of Thatcher's government[29] the UK state had sold all of its shares in BP (after having held them for 74 years), all of its shares in BNOC (latterly renamed Britoil PLC) and all of its shares in British Gas. These sales combined to raise many billions of pounds for the Treasury. At the same time the oil that, since the Forties discovery in October 1970, had held out such a promise of riches was finally coming on stream at scale and with it the revenues from petroleum tax and corporation tax. It has often been remarked that these sums combined to help finance the social security payments and unemployment benefits that underpinned the radical restructuring of the UK's economy in the 1980s. Not only had the model of Thatcherism been in part born within the oil industry, but now that same industry was helping to finance this Conservative revolution. Just as Jon King had pointed out to us, the North Sea oil boom helped finance the dole queues and the government's battle with the NUM.

The sale of state holdings in BP, BNOC and British Gas meant that the UK no longer had a player acting directly on its behalf in the game of the North Sea. This in turn had impacts on the way that the North

Sea was developed, in its regulation, its practices of labour relations and the speed at which the undersea resources were exploited.

It was in the interests of the private corporations extracting hydrocarbons to do so as quickly as possible and this was illustrated by the so-called 'Dash for Gas'. From the first discovery of gas beneath the North Sea by the *Sea Gem* in September 1965 until the sell-off of shares to 'Sid' 21 years later, the development of gas in the UK sector had been dominated by the state-owned British Gas. Now those assets were being developed by private corporations. Peter Walker, as energy minister, oversaw a new policy in which the UK encouraged the rapid exploitation of gas reserves in the southern and northern North Sea. UK gas output doubled between 1980 and 1995.[30]

Encouraging companies such as BP, Shell and Esso to develop gas supply depended on an equally rapid increase in UK gas demand. Beyond domestic consumption and sale of gas to industries such as petrochemicals, key to the gas market were the electricity generators. As with a swathe of other sectors, the power industry had been taken into state control under Attlee's Labour government through the 1947 Electricity Act. After various evolutions the generation and transmission of electricity was run by the CEGB. This body was broken up by Thatcher's Conservative government in 1990. Initially the CEGB was separated into the National Grid Company, Nuclear Electric, PowerGen and National Power. The last two were swiftly privatised.[31]

These now private corporations moved to dramatically increase their generating capacity and did so through building gas-fired power stations. For the better part of a century, coal had been the key fuel driving the turbines that created Britain's electricity, now the utilities prioritised gas as being cheaper and more flexible. The Dash for Gas was underway. Gas turbine power stations in 1990 provided 5 per cent of UK electricity. By 2002 the figure was just under 30 per cent.[32]

The shift from coal to gas was not without its deeper political motivations. The coal-fired power stations had been the central places of struggle in the miners strikes of the 1970s and 1984/5. Oil-fired power stations had been a key weapon against the miners in the 1980s strike. If the lights could be kept on by gas- and oil-fuelled electricity, King Coal was further dethroned. A steep decline in demand for coal as the key industrial power source led to faster pit closures and another blow against the NUM.

All of these actions by the Conservative government, which so dramatically reshaped Britain's energy economy, were followed and assisted by the private corporations. Without a ready buyer of BNOC's shares there would have been little incentive to sell it. By 1988 the entirety of the government's shareholding in BNOC had been bought by BP. The company gradually acquired the assets of the state. Throughout this period the change in BP's relation to the UK state was overseen by Peter Walters. When in March 1990 Walters retired Margaret Thatcher was still in power and, fittingly, the breakup of the CEGB was well underway.[33]

* * *

BP and Shell had filled the order books of the shipyards on Merseyside, Tyneside, Teesside, Clydeside and in Belfast from the 1940s to the mid-1960s. Since that decade they had increasingly had their tanker fleet built at yards on the European continent and in Japan and South Korea. This was not because British yards did not have the capacity for the VLCCs that were favoured by the industry from the mid-1960s. Swan Hunter, Lithgows and Harland & Wolff could all build them but their prices were often higher.[34]

Orders were also made for the construction of rigs and platforms for the North Sea. For example, the four platforms that were created to exploit the Forties Field were built at Laing's yard at Graythorpe on Teesside and at Brown & Root's yard at Nigg. However, the anti-union culture of the North Sea was pushed back from that offshore colony into the culture of the home country. The clearest example of this is seen in the Cammell Laird dispute of 1984.

The Edwardian grandeur of Liverpool's Adelphi Hotel lounge with its white stone columns, pink and cream marble panelling and huge chandeliers is an unlikely setting to meet and discuss the physical hardships of a strike on an oil rig. The hotel, once a stopover for rich passengers waiting for a Cunard liner to New York, is now the haunt of Chinese tourists on the trail of the Beatles or Norwegians visiting the footballing Mecca of Anfield.

Eddie Marnell,[35] a shopfloor union leader during the 1984 strike at Cammell Laird shipyard in nearby Birkenhead, sports a small white moustache and brown boots while his body is wrapped up in a black quilted jacket with a red rose Labour Party enamel lapel badge.

He explains the outline of the strike. Half the workforce at Cammell Laird – 800 men – were told in May 1984 they were being made redundant at the shipyard on Merseyside, triggering fears about the whole future of the business. The workers occupied the yard and four months later, despite warnings of dire consequences, 70 of them were still there. Marnell says it was no wonder that hundreds of workers took industrial action amid redundancies, anti-union sentiment and talk that their ultimate employer, British Shipbuilders, was to be privatised.

Writs were secured by the management in the High Court ordering 37 named strikers to leave. The men had barricaded themselves onto a Royal Navy destroyer, the HMS *Edinburgh*, and a British Gas rig, *AV-1*, the latter being built to work on the nearby Morecambe Bay gas field. The striking workers were eventually forcibly removed by the police – and members of the SAS if you believe some. The men were jailed in Walton Prison for a month and lost redundancy and pension rights.[36]

The former shipyard worker speaks almost in a whisper but his spirit – and sense of grievance from having been jailed for industrial action in defence of his job – remains strong. He believes the harsh treatment in a high-security prison was partly a warning to the coal miners:

> I think it was to deter the miners. They were on strike at the time and I think it was meant to say, 'if you carry on like this you're going to friggin' well end up in top security jail too. They were frightened the miners would barricade themselves underground. You couldn't threaten them with guns, you would have to go down the pit and get them out.

Marnell says Thatcher's government was determined to break the power of organised labour whether it was the miners or the ship builders. The North Sea oil industry had provided the template of a privatised non-union environment.

Marnell is keen to show us a flurry of correspondence he had with Michael – now Lord – Heseltine, when he was minister for Merseyside around the time of the occupation. Marnell pulls out twelve yellowing envelopes which appear to show little but Heseltine trying to avoid answering questions about whether the SAS were sent in to finally clear the rig:

I asked Heseltine if he had any recollection of the cabinet meeting that discussed sending us to jail or sending in the SAS. He wrote back to me saying 'I don't recall any of this. I don't think I was part of the meeting.' I then wrote back to him saying 'don't you keep a diary?' I've got his letter saying 'I don't keep a diary.' What MP in the world does not keep a diary?

The treatment of the Cammell Laird shipyard workers continues to anger Marnell, trades unions and Labour politicians today. In April 2017 the then Wallasey MP, Angela Eagle, urged the government to release all documents relating to the decision to prosecute and put right 'this clear miscarriage of justice'.

Richard Burgon, Leeds East MP and a candidate for the deputy leadership of the Labour Party in 2020, said in the same parliamentary debate: 'That era of Conservative government is becoming defined by suspicion of institutional interference and state wrongdoing. We know the names: Hillsborough, Orgreave and Cammell Laird. If that interference is extended to the prosecution of those trade unionists, do they not have the right to know?'

We later meet Lord Heseltine at 80 Victoria Street in London in the glass building where he rents a small office, just down the road from Parliament Square. On our way up to his floor in an escalator a cut has mysteriously opened up on one of Terry's fingers. When we are ushered into Heseltine's office by his friendly assistant, blood is pouring into a tissue. His lordship stares at the dripping digit but makes no comment.

Michael Heseltine[37] is all bushy white eyebrows and shiny black shoes. His trademark swept back hair is silver now but still luxurious, perched like an exotic bird on the top of his head. The political giant is wearing a navy jumper, speckled with dog hairs. His blue suit jacket is carefully folded over a nearby steel framed and leather chair. He peers over the top of strong dark-framed glasses.

His manner seems part chief executive, part Oxford don. What do you need, he seems to be signalling in a respectful and quizzical way. He is neither warm nor hostile. This is an information or business transaction of some sort.

The once deputy prime minister is coming up to his 86th birthday but is still discernibly flamboyant, outspoken and a believer in his version of a mixed economy, part government intervention but mainly free markets and privatisation. Tarzan, as the tabloids nicknamed him, has been given a new lease of political life by the Brexit debate where he has emerged as a passionate tribune for Remain.

Although Marnell and others had a bitter struggle in 1984, Cammell Laird was saved as an ongoing business. Part of its survival is said to be owed to Heseltine, who stepped in to ensure it was given a chance to bid for Royal Navy frigate contracts. Heseltine refutes this:

I didn't intervene to save Cammells. I intervened to ensure there was fair competition for frigates throughout my budget. It's a very important point, because the (non-interventionist) critics would use the language, 'You saved Cammells.' Now it did have the effect of making a major contribution to Cammells, but I only intervened when a proper competition was being undermined.

We raise the point that his memoirs mention the yard but not the fact there was this huge strike which ended with workers in jail, an event which now seems draconian. 'I've forgotten that', he says.

We question whether more could have been done to save jobs by giving more rig and platform orders to British yards from the North Sea boom. Heseltine responds:

You cannot make these judgements, unless you were there at the time and with the knowledge. I wasn't there and I didn't have the knowledge, but what you need is an apparatus, a state system, which seeks to find sensible policies. Now, you won't get it all right, but I think you'd get it a lot more right than we do. It's quite obvious that motor industry was extremely badly managed in the 1980s. The question is, 'Did the bad management lead to the growth of militant unions, or did the militant unions overwhelm the management?' My own guess, at least 50–50 bad management.

We move on to Brexit and whether he believes the Leave vote was partly driven by deindustrialisation:

Brexit to me is the combination of immigration and the collapse of 2008. Once you've mixed those two together, you've got Brexit. It's not a British phenomenon. It's all over Europe and it's all over America. It all has the same cause, it's a failure of living standards to keep uninterruptedly rising over a long period of time. Which people got used to and look forward to and assumed. Then suddenly living standards were frozen for a decade and still are, and people want change. They look around for answers and the easy one is all those immigrants. That if you can find a mix, foreign bureaucrats and anonymous headquarters in Brussels, it's a very easy speech to make. A very dishonest one, but it's nevertheless easy to make.

We ask: 'What about inequality? If you spend your time in west London and then you go to areas of south Wales, the Thames Estuary or Merseyside you can clearly see how unevenly distributed the wealth has been.'

Heseltine replies, 'I agree with that … Yes. That's going back to devolution and lack of industrial policy.'

We question whether it's sad but inevitable that large businesses, be they oil companies or steel manufacturers, will go where they can produce the commodities or the products cheapest? Heseltine replies:

Of course they will. That's their responsibility to do that. They have an obligation to the people who invest in them to maximise their returns. It's the most effective way of allocating resources, much better than politicians, who avoid the difficult decisions and subsidise the politically attractive ones. I assume you know what I'm saying.

We move on to climate change. The former minister is keen to see action on it both here and particularly in the US and China. We put forward the ideas that Britain has prospered in the past so well from fossil fuels that they've become part of our culture, thereby making it harder for us to change. Heseltine is unimpressed: 'No. In 1945 there were three-quarters of a million people involved in the coal industry. When I took responsibility there were 30,000 left and now there are virtually none. Trying to suggest that there's some sort of a cultural residue is simply an industrial romanticism.'

There is a shuffling of briefcases and a wrinkle of the eyebrows indicating it's time to go.

* * *

Popular wisdom places the blame for the destruction of the British shipbuilding industry at the door of labour militancy. That the demands of organised labour on Clydeside, Teeside, Tyneside and Merseyside simply made UK yards uncompetitive. However, the role of the Conservative government in ridding the UK of an industry that was a wellspring of socialist activism.[38]

In abandoning the UK merchant shipbuilding industry the Thatcher government had been preceded by the oil industry by two decades. Since the mid-1960s BP and Shell had had their vessels built in foreign yards, particularly in France, Italy and Germany, the key states at the heart of the EEC. Even BP, which was 68 per cent owned by the UK government, was placing its orders within the neighbouring economic block.

As the UK sector of the North Sea was developed there were repeated battles between the British government and the European Commission over the exploitation of the resource. Brussels was keen that oil from the UK sector should be in part landed at a continental port, in the Netherlands or Germany, but this was vigorously resisted by both Conservative and Labour governments and the crude from the offshore fields came ashore in Scotland and England. But the equipping and provisioning of this offshore colony was a massive undertaking and much of the work went to yards in other states – such as France, Germany, Italy and Spain.

The benefits accruing to multinational corporations from the consolidation and growth of the EEC became steadily more apparent.[39] The economic block had been born of an alliance of coal, iron and steel-manufacturing states, but slowly the significance of these industries within Western Europe declined and the backbone of the EEC became oil. This shift mirrored the change in the economic life of south Wales, transitioning from mines and blast furnaces to ocean terminals and petrochemical plants as the Severn Estuary oil complex evolved.

With the UK's entry into the EEC in 1973, oil plants in Britain became part of the Western European oil complex. On the perimeter of the Baglan site the export jetty in the River Neath was upgraded and a special tanker, *Sunny Girl*, was built to facilitate the transport of ethylene around Land's End, up the Channel and into the port of Antwerp. There the liquid chemical was pumped ashore and processed. For nearly 20 years[40] the newly expanded Baglan factory operated as a cog in a bigger EEC chemical machine.

However, the global production of petrochemicals was expanding at a pace. The states of the Persian Gulf, greatly empowered after the 'Oil Crisis' of 1973, began to construct plants that, with immeasurable quantities of cheap crude and a cheap labour imported from India, could easily undercut the price of chemicals produced in the EEC. In the following decade China and India began to flex their industrial muscles and chemicals from South East Asia came onto the global market creating a glut in these materials.

Multinationals had assisted in the breakup of national economies inside the EEC, with the argument that the EEC, and later the EU, could thereby form a powerful economic block whose pooled sovereignty could better compete in the global market. But in many places, such as Baglan, the opposite took place.[41]

There was by now a global chemicals glut and a reluctance of the European Commission to defend production inside an EEC in the face of foreign imports. Corporations such as BP and Shell moved rapidly to consolidate places of production within the EU. Instead of the company maintaining a plant in Belgium and a plant in Wales that worked in tandem, the board saw fit to close one. As in other spheres of activity BP showed no especial loyalty to Britain, and so in January

1994 the announcement came that Baglan's ethylene plant would close with the loss of 600 direct employees and contractors.

The company press statement was a bombshell for the workforce and neighbouring communities such as Sandfields and Baglan Moors, for it was well understood that the closure removed the heart of the plant and its end would surely swiftly follow. BP Chemicals Chief Executive Bryan Sanderson said, 'The job cuts are due to world-wide over capacity in the chemical industry and the run down in the car industry.' He announced a £3 million five-year package to help create work for the 300 direct employees who faced redundancies.[42]

Whereas the minister of energy, Tony Benn, had fought determinedly against the EEC to prevent UK closures, his successor two decades later accepted the end of Baglan with a sense of inevitability. The issue of the closure was raised in the House of Commons as Secretary of Trade and Industry Michael Heseltine said: 'I have looked at the detail of the announcement by BP that the company was closing older capacity plants because its new investments were coming on stream. As far as I know there will be no adverse effects on the British Balance of Payments. It would not be appropriate to intervene.'[43] No mention was made as to the location of these 'new investments' that BP had made.

Workers at Baglan expressed outrage at the political complacency of the Conservative government, especially as Heseltine had previously stated that he would 'Intervene before breakfast, lunch and dinner to help British Industry.'[44] But significantly there was no defiant action against the closures by the trades unions on the site. Here was a plant where in the 1970s the workforce had struck for better conditions, here was a community that supported workers' families through long and bitter struggles such as the miners' strikes of 1972, 1974 and 1984/5. Now there was next to no resistance to the job losses. The ease with which a multinational closed a plant such as Baglan, the ease with which the supranational government of the EU allowed this to happen, owed much to the trades union movement having been effectively disempowered by the British state in the previous decade. There was no strike at Baglan, no pickets, no protestations in the TUC, no solidarity from other trades unions and the wider labour movement. The silence of John Morris, the local MP and shadow attorney general who had been secretary of state for Wales under Wilson and Callaghan, was striking.[45]

On 24 August 1998 the Manic Street Preachers, from south Wales, released 'If You Tolerate This Your Children Will Be Next'. It sold 156,000 copies in its first week, going to No. 1 in the UK Singles Chart. The following month, on 20 September 1998, they played at the Afan Lido Leisure Centre in Sandfields. Eight weeks later, on 12 November, came the announcement of the effective closure of the whole of the Baglan plant.

* * *

We are blessed with a beautiful July day. The sky is clear blue, the air is still and surprisingly warm for Aberdeenshire. We stand before a plinth with three bronze figures on top of it – the Piper Alpha Memorial. It is 30 years since the night the oil platform exploded. Over the past weeks survivors, bereaved families, dignitaries from Aberdeen, journalists and others like ourselves have all come to pay their respects to the 167 men killed on 6 July 1988.

As we leave the North Sea Rose Garden in Hazlehead Park, we hear the rhythmic thud of a helicopter flying somewhere in the distance. We can be sure that we will be following the sound as we drive the three miles north to Aberdeen Airport at Dyce. Here the choppers shuttle back and forth from the heliport to the platforms far out to sea.

It would take the *Piper Alpha* disaster of 1988 to ignite effective resistance within the offshore workforce that led to the first union that was prepared the challenge the oil companies, the Offshore Industrial Liaison Committee.

The first explosion at 21.55 on the night of 6 July woke the day shift from their bunks and led to an immediate Mayday call. Gas had been leaking from an improperly secured flange and ignited. Further blasts occurred and flames shot 300 feet into the air; 165 of the 226 crew members died, while two men working on a safety standby vessel also lost their lives.[46]

Before one in the morning parts of the platform – one of the heaviest in the North Sea – began to collapse. It took three weeks for the fires to be put out. A judicial inquiry was critical of *Piper Alpha*'s operator, Occidental, which was found guilty of having inadequate maintenance and safety procedures, but no criminal charges were ever brought against the company.

Occidental, run by Armand Hammer, against whom the Canvey Islanders had battled, paid out $180 million in compensation to families but faced no fines or other sanctions from the offshore oil regulators. The incident compares with BP's *Deepwater Horizon* accident in the US Gulf when eleven people died but leaking oil caused serious damage to beaches and the environment. BP paid out over $60 billion in fines and liabilities.

8

Suude ne gbo gima de
The Eye of the Blind

Kpado Ogoni e kanaimon
Kpado Ogoni bon kiima kparake
Dene gbo gbara e dara ba boo
Dene gbo gbara nveeni kpooro
O dui O dui
E oa Bari zaa
Ken Saro-Wiwa giaakara eera
Suude ne gbo gima de
Adu kpooto no gbo buuna to
Eede pa Ogoni
E kanaimon ko o dui se?
O dui oo o dui o o O dui oo odui o o[1]
　　—MOSOP, 'Ken Saro-Wiwa the Star of Ogoni', 1996

Lazarus Tamana, in the carpeted hush of the Royal Festival Hall, is explaining the history of the uprising against the oil companies that has overwhelmed his town of Bodo and the land around it for the past half-century.

Ken Saro-Wiwa and the other leaders of the Ogoni people were enlightened. They were already graduates who were leaders of the community. They saw the impact of oil production on our people. They saw that the money was going to the central government. They saw that we didn't have any control over resources, even how it was spent. It angered Ogoni a lot. We were still being marginalised in terms of jobs, in terms of opportunities, yet the money or the resources that were sustaining the state was coming from us.[2]

In Ogoni the impact of the oil extraction, and particularly the laying of pipelines through communities felt like a wound. For Ogoni has the misfortune to lie close to Bonny, the point at which so much Nigerian oil was gathered for export, so pipelines which began their journey hundreds of miles away crossed Ogoni territory.

The travails of the Ogoni began to gain an international voice in the form of Ken Saro-Wiwa. The son of a trader in the town of Ke Dere, Saro-Wiwa attended university and spent the years of the Biafran War working as an official for the Nigerian state at the port in Bonny. Afterwards he authored *Sozaboy* which told the story of the war through the eyes of a young conscript, but was groundbreaking in that it did so in pidgin English. This bestseller was followed by writing scripts for the wildly successful TV sitcom *Basi and Company*, which had a following right across Nigeria.

By the late 1980s Saro-Wiwa had turned his attention to the plight of his fellow Ogoni and self-published the book *Genocide in Nigeria*. In this he clearly identified the source of the destruction of his people as the international oil companies, in particular Shell. As he articulated:

It is well known that a boil on one's nose is more painful to the afflicted than an earthquake which happens thousands of miles away killing thousands of people. I am inclined to think that this is why the Ogoni environment must matter more to me than to Shell International ensconced in its ornate offices on the banks of the Thames in London. But I cannot allow the company its smugness because its London comfort spells death to my Ogoni children and compatriots.[3]

The book, and the debate it created, helped provide support to MOSOP – the Movement for the Survival of the Ogoni People – formed in 1990.[4] As Tamana explains: 'The Ogoni people came together and formed MOSOP because the anger of the people over the years that was suppressed came out. I stayed here in the UK and then started the international campaign for MOSOP from 1991.'[5]

From this point on, conflict with the state and the oil companies was almost inevitable. In January 1993 an astonishing 50 per cent of the population of Ogoni came out in protest. Shell was in no position to

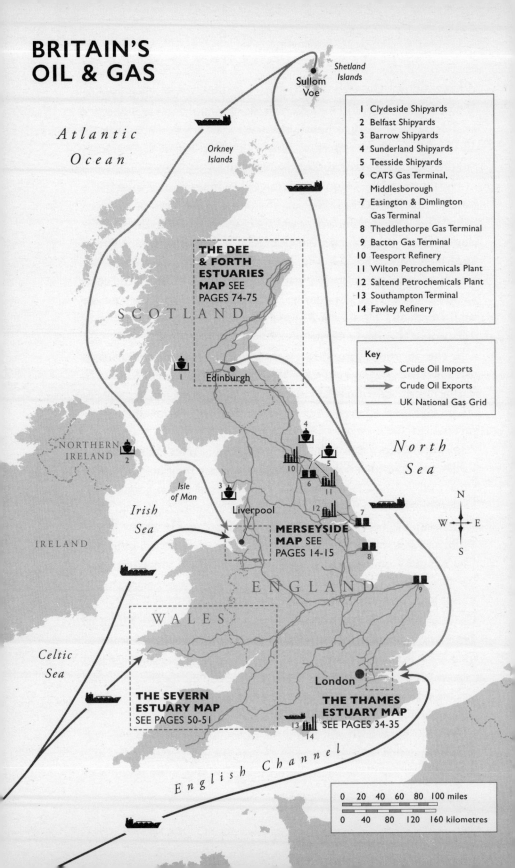

BRITAIN'S OIL & GAS

Atlantic Ocean

Orkney Islands

Shetland Islands

Sullom Voe

S C O T L A N D

THE DEE & FORTH ESTUARIES MAP SEE PAGES 74-75

Edinburgh

1 Clydeside Shipyards
2 Belfast Shipyards
3 Barrow Shipyards
4 Sunderland Shipyards
5 Teesside Shipyards
6 CATS Gas Terminal, Middlesborough
7 Easington & Dimlington Gas Terminal
8 Theddlethorpe Gas Terminal
9 Bacton Gas Terminal
10 Teesport Refinery
11 Wilton Petrochemicals Plant
12 Saltend Petrochemicals Plant
13 Southampton Terminal
14 Fawley Refinery

Key

→ Crude Oil Imports
→ Crude Oil Exports
— UK National Gas Grid

North Sea

NORTHERN IRELAND
2

Isle of Man

3
Liverpool

4
10
6   5
11
12
7

MERSEYSIDE MAP SEE PAGES 14-15

Irish Sea

IRELAND

8

E N G L A N D

9

N
W   E
S

Celtic Sea

W A L E S

London

13
14

THE SEVERN ESTUARY MAP SEE PAGES 50-51

THE THAMES ESTUARY MAP SEE PAGES 34-35

English Channel

```
0   20  40  60  80  100 miles
0   40   80  120  160 kilometres
```

do anything except withdraw its staff and mothball its operations on the territory of Ogoni. This was possibly unprecedented in the history of the oil industry. In October 1951 the Iranian state had effectively forced AIOC staff to withdraw from company assets in Abadan and beyond, but that was a sovereign government whereas Ogoni was a rebellious region.

* * *

The military ruler of Nigeria, General Sani Abacha, would not let this stand. The army and military police were dispatched to quell the rebellion and exact a brutal reprisal. Villages were razed and people slaughtered in their own homes. This operation, under the command of Lieutenant Colonel Paul Okuntimo, lasted throughout much of 1994 and 1995. In May 1994 Okuntimo wrote a memo that said: 'Shell operations still impossible unless ruthless military operations are undertaken for smooth economic activities to commence.'[6]

In some respects this brutal repression of resistance in the oilfields conformed to the pattern that had been seen in, for example, the suppression of the Ijaw uprising in 1966.[7] Under this game plan the Nigerian state imposed order while the oil companies stood back and declared that the rebellion and the state killings were not its responsibility, their concern was simply to maintain the production which created employment and generated tax revenues benefiting all of Nigeria. It was a pattern that had been perfected within the British Empire in which it was the responsibility of the colonial administration to enforce law and order and the concern of business to conduct business. The British administration in Lagos had coordinated the military suppression of any 'disturbances' in the territory of Nigeria while the Royal Niger Company carried on its orderly process of extracting resources and generating profit.

Tamana is unswerving in his critique:

That actually confirmed to us that this was a laid-out plan to intimidate the Ogoni people and to teach other Niger Delta communities, that if they dare attempt to challenge the government and Shell, that this is what would happen to them. It was basically doing that, to use the Ogoni people as an example to others. Talking about oil

extraction or campaigning against pollution in Nigeria was a no go.
The Nigerian government in collaboration with Shell were doing
every single thing they did together.

But the events in Ogoniland showed that things would be different
in the 1990s. Saro-Wiwa and MOSOP explicitly declared that the
uprising was against Shell and Chevron and not primarily against the
Nigerian state. This was not a call for Ogoni secession, as had been the
case in the Biafran War. Saro-Wiwa stated, 'I accuse the oil companies
of conducting genocide in Ogoni.' Whereas people living in the West
might have been able to say the actions of the Nigerian army are a
matter for Nigerian citizens, the revelation of the actions of a multinational corporation meant that citizens of many nations felt it was
their responsibility. Meanwhile the technology of communication was
rapidly altering. The internet was just beginning to become widely
used. Cameras that could take broadcast-quality footage suddenly
became cheaper and more portable than they had been, enabling small
production companies to make films at risk.[8]

In the film *The Drilling Fields*, broadcast on Channel 4 on 23 May
1994,[9] Saro-Wiwa claimed that Shell conducted the extraction of oil
and laying of pipelines in a way that it would never do in the UK. He
pronounced, 'I accuse Shell of practising racism in Nigeria, because
they do there what they would not do at home.'

Slowly, 40 years after oil developments impacted on Ogoni, the
global environmental movement began to wake up to the story of
Shell in Nigeria and was captivated by the charismatic figure of Ken
Saro-Wiwa. The Unrepresented Nations and Peoples Organisation[10]
in Brussels supported the Ogoni as a case study of a nation whose
existence was rendered invisible by only the Nigerian state having UN
status. In London, Anita and Gordon Roddick, directors of the Body
Shop, met Saro-Wiwa, took up his cause and accepted his invitation
to journey to Ogoni. On returning they went to meet Mark Moody-
Stuart, chairman of Shell, in his South Bank office in an attempting to
persuade him to travel to the Delta. He declined.[11]

Saro-Wiwa and the eight leading Ogoni fellow activists were eventually arrested and falsely charged and the threat that they might be
hanged galvanised the international community. In London fellow

writers such as William Boyd worked with PEN International to call for their release. Amnesty International joined the campaign, as did Greenpeace. Others such as the Body Shop, Ogoni Freedom Campaign, Delta and Platform stepped in. Prime Minister John Major and Foreign Secretary Malcolm Rifkind were lobbied intensively. There were demonstrations outside Shell's head offices, at the company's AGM in London and at petrol stations around the UK. There were constant vigils outside the Nigerian High Commission just off Whitehall.[12]

In October 1995 the judgement was passed in the military court and all of the accused were sentenced to death. There was a frantic attempt to persuade Nigeria's military head of state, General Abacha, to commute the sentences. At the Commonwealth Heads of State Summit in New Zealand, John Major – lobbied by Saro-Wiwa's son Ken Junior – was explicit in his pressuring of Abacha. Meanwhile demands were made of Shell to act to save Saro-Wiwa's life. But several times Moody-Stuart declared that Shell was powerless to act, could do nothing to save the condemned and argued that it would be wrong for a multinational corporation to intervene in the internal affairs of a host nation.[13]

On 10 November 1995, Saro-Wiwa and eight others[14] were hanged in Port Harcourt jail. Tamana recalls the day: 'All of us were shocked. We didn't know that Nigeria will go to that extent to kill these people the way they've killed them. We thought that with the intervention of all the world leaders, that they will show mercy or show clemency but that did not happen.'

The impact internationally was striking. Many newspapers led with the story on their front pages. TV coverage was extensive. John Major described the hangings as judicial murder and called for Nigeria to be suspended from the Commonwealth. This duly happened.[15] Abacha was a pariah. So was Shell.

The shock in the media was mirrored by the shock within Western oil companies. The accepted rubric of corporations being able to distance themselves from the acts of states had broken down. In global public opinion Shell was being held at least partly responsible for the death of Saro-Wiwa. Although the hangings had no discernible impact on the profits of Shell, let alone the profits of its rivals,

Moody-Stuart and John Browne[16] at BP recognised that something fundamental had shifted.[17]

* * *

The combined impact of the execution of the Ogoni activists and the defeat of Shell at the hands of Greenpeace over a controversial plan to dump the used *Brent Spar* platform at the bottom of the North Sea,[18] necessitated some deep rethinking in the corporation. How could Shell, which prided itself both on its unflappable, stately progress and its reputation to be able to foresee all oncoming storms, aided by its legendary Scenarios team, have failed so badly in its appreciation of how public perception might radically shift?

The economic strength of Shell was not in question. There was no doubt as to the viability of the company or the financial soundness of its wider investment strategy. Its position stood in stark contrast with BP, which was still struggling to overcome its financial crisis of 1991–2, when saddled with mounting debt anxious investors pressured the board who forced the resignation of the CEO, Robert Horton. At the time there was strong speculation that BP would be taken over by Shell.

Despite the furore over Saro-Wiwa there was no question of Moody-Stuart or Cor Herkströter being forced to resign by the board, no matter how much those who demonstrated on a weekly basis outside the Shell Centre called for it. Tamana describes the events:

> I'm always there. You see the Waterloo Bridge here? Anita Roddick and me are always there campaigning. The day that we'll go after Shell, we'll come here at Waterloo and we'll demonstrate. November 10 we normally have coffins, nine coffins, which sometimes we carry to the Nigerian High Commission and sometimes we just carry to the church where we normally do our activities. I'm always here.

However, 1995 had demonstrated how public opinion could turn against the corporation and how this could impact on the functioning of the company. There were widespread calls for drivers to boycott Shell petrol stations in the UK, Netherlands and Germany and in these states the operations of the corporation could be hindered by staff dis-affection with the corporation and its ability to handle public outrage.

Workers at Shell's Stanlow Refinery on Merseyside, for example, were deeply concerned about the plight of Saro-Wiwa, especially as Nigeria echoed apartheid South Africa in the 1980s when Shell staff strongly contested the company's collaboration with the regime. At head office level there was clearly public embarrassment.

The management of Shell recognised that such a climate of opinion could work against the company during its daily engagement with Whitehall and Westminster, both over UK domestic affairs and over UK support for Shell's activities around the world.

The idea that the state ensured the passive order of society while the corporation pursued profit was further thrown into question.[19] In Ogoni the Nigerian government had been unable to ensure the passivity of its citizens and Shell had had to abandon production sites. Worse still the Nigeria government could do absolutely nothing to defend the reputation of Shell in global public opinion.

Shell decided to embark upon a radical intensification of its public engagement strategies. If the state could not deliver public support in British society, then the company would set about the construction of that support in the UK; it would build its 'social licence to operate'. The term seems to have first appeared in the Shell lexicon courtesy of the public relations company Fishburn Hedges, which Shell had hired in 1998.[20]

The principle of the concept is that any company needs not only the legal licence to work in a particular locality, but also the social licence: the support of the community in that locality. Patently Shell in Ogoni had the legal licence to extract oil from blocks granted by the Nigerian government, but it did not have the licence of society in Ogoniland. Indeed, it had been forced to leave Ogoniland and the Nigerian state had not been able to enforce its will to uphold the legal licence that it had granted Shell despite massive repression. The brutality of the Nigerian army's attempted enforcement only served to undermine Shell's social licence in Germany, the Netherlands and the UK. Conse-quently Shell set out to bolster this social licence in these 'home states', and this would require more than mere advertising.

John Jennings, chairman of Shell Transport,[21] summed up the scale of the shift required: 'We no longer live in a trust me world, we now live in a show me world.' Shell determined that it would show the public in the UK, and elsewhere in Europe and America, that it cared about environmental and human rights.

* * *

Shell's response was not entirely out of the blue, for it fitted within a deeper trend that was unfolding in the way corporations had been interacting with society. The tactic of openness combined with an attempt to build a pro-company media blitz were two parts of the strategy to build the social licence to operate, to endeavour to counter negative perceptions and to mould society into a form that supports, or at least does not hinder, the workings of the company. This fuelled the move of the corporations from casual philanthropy to a muscular corporate social responsibility or CSR. John Browne was grand master of this new strategy.

In June 1995 Browne was appointed CEO of BP. He had imbibed this understanding of the social role of corporations 15 years earlier at Stanford University, California, where the participants in the executive programme were set to study iconic failures of corporate environmental and social management, such as the pollution disaster which unfolded at Love Canal in New York State during the 1970s. Browne stepped into the role of head of BP at precisely the point at which Mark Moody-Stuart at Shell was failing to respond both to the *Brent Spar* crisis and the imprisonment of Saro-Wiwa and the Ogoni 8.

Only 16 months after becoming CEO, Browne faced his own potential nightmare. In 1987 BP had begun exploration for oil in Colombia, in farmland and forest on the eastern side of the Andes, at the headwaters of the Amazon.[22] Colombia was in the midst of an intense civil war between the state and left-wing FARC (Revolutionary Armed Forces of Colombia) guerrillas.

BP's name became embroiled in the conflict after hiring guards to protect pipelines, triggering allegations the company could have become involved in human rights abuses. There was then media speculation that Colombia could turn into BP's Ogoni.[23] But a government inquiry into the allegations was eventually dropped due to lack of evidence and media focus, and public concern faded. BP's social licence in the UK had not been seriously damaged while its operations to extract and export oil from Colombia continued without hindrance.

Fourteen years later Browne wrote in his autobiography:

> The Ken Saro-Wiwa incident had started a broad international debate about business and human rights. Internally this and the specific accusations against BP were the catalyst for a major re-evaluation. Did we as a multinational have multinational accountability for human rights that would transcend national law? The United Nations Universal Declaration of Human Rights had been written for member countries agreement. Could we as a company subscribe to it?[24]

Browne was soon applying the new CSR strategy to perhaps BP's greatest challenge: climate change.[25]

*   *   *

It was the political element of climate change that had the potential to threaten the fossil fuel companies, for there was clearly a chance that could translate into policy shifts which would impact on their core businesses. In 1981 Shell commissioned a report on the 'greenhouse effect' and by 1986 it had established a Greenhouse Effect Working Group.[26] This body made a report in 1988 which estimated that 44 per cent of global carbon dioxide emissions came from oil, 38 per cent from coal and 17 per cent from natural gas. It stated: 'With fossil fuel combustion being a major source of CO_2 in the atmosphere, a forward looking approach by the energy industry is clearly desirable, seeking to play its part with governments and others in the development of appropriate measures to tackle the problem', but 'the likely time scale of possible change does not necessitate immediate remedial action'.[27]

There is an old red Nissan car out the front with mildew on the roof and bonnet and the windows next to the porch have their curtains drawn. We knock gingerly, wondering whether we were at entirely the wrong place on the outskirts of Swansea, in our search for one of Britain's most eminent scientists. But the front door suddenly opens. 'John is waiting for you', says a smiling young woman who turns out to be a cleaner called Clair.

Sir John Cadogan,[28] now 87, ushers us into his sitting room and closes the door. We are offered comfortable chairs. The walls are covered in traditional oil paintings and shelves of porcelain vases. There is a slightly musty smell and the blinds are half drawn, but Cadogan is as bright as a disco ball. 'If you want to write about the history of the oil industry you can't start at a better place than here', he says fixing us with bright blue eyes. 'D'Arcy built the first refinery down here at Llandarcy to serve the navy. I have seen the handwritten note from Churchill. He wanted it hidden behind a hill and away from the water's edge for security reasons. It's closer from Iran to come up the Bristol Channel than to go round the coast to the Thames. Milford Haven was not considered a good place then because of the lack of rail links.'

Cadogan, who was chief scientist at BP from 1979 to 1992, speaks with a soft Welsh accent. He has returned to the city of his birth, home to the Llandarcy Refinery, home also to both Dylan Thomas and Michael Heseltine. This chemist fell into the oil industry in a haphazard way:

I was a lecturer in organic chemistry in the late fifties at Kings College, London. BP headhunted me and wanted me to go and work for them but I didn't want to do that, so they hired me as a consultant. And I was a consultant for them for 20 years. But in 1979 they came back and asked whether I would consider a full-time job. By then I had been a professor in Scotland, at St Andrews and Edinburgh, and thought this was an opportunity to change my life completely. So I became chief scientist at the Research Centre in Sunbury on Thames. By the time I retired in 1992 I had a worldwide staff of 4,500 and was spending up to £400 million a year.[29]

This included a $50 million programme devoted to green issues such as increasing the efficiency on BP's huge energy usage. He was also chair of BP Solar International.

Cadogan was tasked to study climate change by the head of BP, Peter Walters.[30] A working group was established at BP's Sunbury offices[31] and in due course a report was produced that went to the board. It remains unclear what happened to that report and we thought we might find it in the BP Archive, based at the main Warwick University campus. A variety of attempts to look through the archive were blocked by BP on the grounds that everything covering the last 40 years were commercially confidential. The report and other early research work into renewables done by BP has remained beyond public reach.[32]

Cadogan explains to us that he eventually left the company because the normal BP retirement age was 60 and he wanted to do other things, but he was also aware that the axe was hanging over his department. 'The writing was on the wall. Browne was a financial genius but argued, "Why do I need to do all this R&D?" He believed in outsourcing. But when you outsource you do not have the control of the people, their loyalty when things start going wrong. All you get is the blame.'

Looking back at Cadogan's fate in the early 1990s we can see some clear ironies. Walters, who had championed the chief scientist and sanctioned the first BP study of the greenhouse effect, became an active climate sceptic after he retired from BP. He became a trustee of The Institute of Economic Affairs (IEA), a right-wing free market think tank.[33] In the early 1990s, while Walters was a trustee, the IEA embarked on a full-frontal attack on climate science, in particular through the publication of Julian Morris and Roger Bate's pamphlet *Global Warming: Apocalypse or Hot Air?* in 1994.[34]

It may well have been a shock to Walters when his protégé John Browne made an historic announcement on 19 May 1997.[35] The date was chosen with great care, just a couple of weeks after Tony Blair's equally historic New Labour general election victory and seven months before the Conference of the Parties (COP) to the UNFCCC[36] in Kyoto, Japan – the Kyoto Climate Conference. The location of his speech was also chosen with great care. Browne spoke to an invited audience at Stanford University in California; this was an alma mater, the place where he had first begun to explore the ideas of corporate social responsibility 17 years previously.[37] It was also at the heart of Silicon Valley, a symbol of the future and a battleground with BP's great rival Exxon.

Browne later described the day: 'You would expect the temperature in Northern California in mid May to be around 20°C. That day it must have been touching 30°C. I had been CEO of BP for just two years and I was now about to deliver what would turn out to be a seminal speech on climate change and alternative energy. I was aware of the blistering heat; it was as if the sun wanted to demonstrate its ultimate power'. In his speech he announced that since the link between man-made carbon emissions and global warming could not be discounted, BP had decided that it was time 'to go beyond analysis to seek solutions and take action'. He explained his view 'that business can play a key role in finding and delivering new ways of achieving both environmental stability and economic prosperity. The two must be inextricably linked.' And he committed BP 'to five specific actions to get us started on the road: first, to reduce our own carbon emissions; second, to fund research and development to create greater understanding about the issues; third, to take initiatives for joint development of activities in the developing world to reduce carbon emissions in those countries; fourth, to develop alternative fuels and energies for the future; and lastly, to contribute to the public policy debate in the search for wider global solutions.'

Some welcomed Browne's boldness, others called it 'greenwash'. But in order to drive home his point, Browne declared three days later 'If we are all to take responsibility for the future of the planet, then it falls to us to begin to take precautionary action now'.[38]

Lee Raymond, head of Exxon, was an outspoken sceptic of climate policies and his company supported a host of think tanks that attacked the science. He was the mainstay of the Global Climate Coalition, the oil industry lobby group that had fought hard to prevent Washington taking action on the issue and in any way constraining the fossil fuel industry.[39] BP had quietly withdrawn from the coalition the previous October.[40,41] Through this speech – both in its timing and its location – Browne was quite transparently attempting to break BP away from its peers in the oil industry and thereby to develop a distinct profile for the company. His aim was to ensure that caring for the climate became part of BP's evolving brand and that it should be seen in this light, particularly in the US where it held the major part of its assets.

There was extensive media coverage, and some instant words of support from politicians and NGOs. It played especially well with the Clinton administration, now in its second term, and with Vice President Al Gore who was working full tilt towards the Kyoto Climate Conference.[42]

Peter Walters was less convinced. In an interview years later he explained: 'If you're chief executive of a major company, you've got to work with the grain of the government. I'm not going to stand up and criticise them because I want something else. I don't have to bend my conscience too far to say what they'll be glad to hear.' Asked if Browne was using rhetoric to 'greenwash' BP during the Stanford speech, Walters replied: 'When I was a soldier we had an expression in the army which was, "bullshit baffles brains". Before the commanding general came to inspect, everything would be whitewashed and cleaned. I don't think the general ever knew about the standard of training or the combat readiness of the troops and all the rest', he recalled, mimicking a British general: 'What! What! What!'[43]

Browne and his team around him wanted to show that BP's new direction was not mere rhetoric as so many critics within, and outside, the oil industry believed. He wanted to improve efficiency and reduce the amount of CO_2 emitted by the sheer wastage in the company's daily production processes. On the ground this meant looking at assets such as factories like Baglan or oil platforms like the Forties. The measures were to be enacted around the world in many of the different states in which BP operated. Of course this programme did not run counter to the company's core intent to generate profit, for capital sunk into these measures not only reduced CO_2 emissions but reduced the running costs of a plant.

At the same time BP determined to make an investment in renewable energy at a level that would be meaningful and grab business press headlines. Both BP and Shell had made forays into solar and wind technology since the 'Oil Crisis' of the early 1970s. John Cadogan, BP's chief scientist, explained how he created a blueprint for significant investments in renewable energy, but like the report of the greenhouse effect this too was considered and then shelved, perhaps buried deep in the BP Archive at Warwick University. Now, barely five years after Cadogan's departure, BP began to invest heavily in renewable energy.

Factories manufacturing solar photovoltaic panels in California and Maryland were purchased by the company and the Baglan Energy Park initiative was launched in 1998.[44] Alongside these factories were high-profile solar installations. In September 1997 Browne made a further key climate speech, choosing to do so in Germany. On the same day in Berlin he launch BP's first 'solar-powered petrol station', promising that it would be the first of many if the concept proved workable.[45]

At the time the company had few assets in Germany but was keen to expand its market share. In order to do this it needed governmental approval for its intended takeover of the German petrol company Aral. These were the early days of the later famed *Energiewende* in which the state planned out its long-term transition to be powered by wind and solar. Browne was both placing BP as part of Germany's future and helping clinch a near-term business deal, enabling it to sell more oil products – and carbon – to German customers.

In part these investments followed the strategy of openness. For as with BP's response to the Colombian crisis, in tackling climate change the company sought to engage with society and particularly the NGOs. As in Germany, they found willing collaborators in the UK.

In May 1997 Greenpeace was taking direct action to prevent BP drilling for oil on the bed of the seas west of Shetland. Meanwhile Greenpeace International, based in Amsterdam, began a campaign to link climate change to exploration at the frontiers of the oil world by the likes of BP in Alaska, Australia and the Atlantic.[46] However, at the same time Greenpeace UK's Solar Electric Campaign, led by Matthew Spencer, was engaging in constructive dialogue with BP Solar, the business unit based at Sunbury.[47]

BP's commitment to this climate strategy built steadily over the next three years and reached a crescendo in July 2000 with the rebrand of the company. For in parallel with this evolution Browne had piloted a series of bold business deals. An early outlier of this process came with BP's asset swap with US company Mobil. Through this BP obtained refineries and lubricants plants in Europe, such as the Coryton Refinery in the Thames Estuary and Birkenhead engine oil factory on Merseyside. Browne was eager to go for a larger complete merger with Mobil and detailed talks were held, but ultimately they proved fruitless.

However, on 12 August 1998 BP unveiled a mega-merger with Amoco, a large US rival. Within five months the deal had been completed and the new company, BP Amoco, was launched. Despite being nominally a merger, BP was by far the dominant player and a joke did the rounds inside the company: 'How do you pronounce BP Amoco? BP – the Amoco is silent.' This £30 billion takeover was followed by the acquisition of another significant American oil corporation, Arco, announced on 2 April 1999 and completed nearly a year later in March 2000. Then there was the acquisition of Veba. Finally there was the acquisition of Burmah Castrol, a century-old British corporation based in Swindon which held the commanding stake in the UK lubricants and engine oils market,[48] completed on 24 July 2000. On that very day Browne announced BP's rebrand.

All of these deals together increased the size and carbon emissions of BP dramatically. Consequently they also impacted on the UK economy and politics, as one of Britain's largest corporations grew in capitalisation and undertook a pronounced swing towards becoming a more American company, only British in name. BP, from being in the second league of the major oil companies, jumped into the first league, alongside Shell and Exxon.[49] This rapid growth in size created a massive human relations challenge within BP as they tried to integrate the new workforce. Drawing together people who had given their working lives to Amoco or Arco, Burmah or Mobil,[50] the company needed to create a bold new brand to inspire the loyalty of this new workforce and its clients, in industry, in government, in the media and on petrol station forecourts.

On 24 July Browne unveiled the relaunch of BP with a new logo of a Helios sunflower design. Gone was the old BP shield with its allusions to the defence of Britain's realm and her empire, in came the bright yellow and green sunburst that suggested positivity and openness. Significant in relation to the company's slow departure from the UK – first observed by Tony Benn 25 years before – the official name of the corporation shifted from British Petroleum to simply BP. However, the boldest move, causing the greatest stir, was an element that built on the Stanford speech – the company's new strapline, 'Beyond Petroleum'.

The new identity of the corporation, on which $7 million was spent with a commitment to spend substantial sums on further public relations in coming years, was rolled out across the company. Brochures of many pages, created to the highest design specs and printed on high-gloss paper, extolling the objectives of the reborn corporation, were distributed to BP subsidiaries and assets from Australia to Austria. The high point of this campaign to create an internal message was the shooting of a short film explaining the rebrand and the new values of the company. Under the strict command of the BP media team it was to be watched by all company employees on the same day on VHS tapes distributed to offices around the world. In the film a range of talking heads spoke to camera about how the company was shaping the future and grasping the challenge of climate change with both hands. Among these figures was Tony Juniper, then head of the UK's Friends of the Earth, speaking to BP staff on an internal communications film, about how positive he was about Browne's message.[51]

Later on, Juniper felt let down:

> I remember standing on the top floor of BP's office looking over the City of London with the then chief executive, Lord Browne. He had seen the light, and the light was green. He told me: 'I'm convinced. we will bring in a new era at BP, we will be a sustainable business'. I was excited, but it didn't last. The rebranding looked persuasive, but it was just that – rebranding.[52]

* * *

If BP's climate strategy helped it build legitimacy and social licence within NGOs, it also assisted within Westminster and Whitehall. Despite Margaret Thatcher's pronouncement in May 1990 upon the gravity of the threat posed by global warming, the main body of Conservative MPs, both under Thatcher and her successor John Major, remained lukewarm, if not hostile, to the issue. One notable exception was John Major's secretary of state for the environment, John Gummer. New Labour was determined to strike a different note on the environment and particularly on climate matters.

With Shell reeling from its failings over *Brent Spar* and the shock of the judicial murder of Saro-Wiwa and his comrades, BP had something of an open field. Walters had clearly been a passionate ally of Thatcher, and by becoming a trustee of the IEA, her favourite think tank, he displayed his loyalty to her even after her ejection from office.[53] Successor CEOs David Simon and John Browne chose to follow a different path. As the Major government sank under its infighting over the European Union, and Blair steered Labour towards the right, so BP swung towards Labour.

The fostering of that closer relationship began while Simon was CEO and Browne was head of BP Exploration and Production. Browne wrote of his first encounter with Tony Blair in 1994:[54]

News of the success of performance contracts reached the ears of a young politician, Tony Blair. We had started talking when he was Leader of the Opposition as he was keen to learn from business. He was looking, in particular, for a way of ensuring that political commitments were delivered. Our conversations were mostly about how to get individuals and teams to meet targets. I explained what we were doing. In BP, performance contracts are just a way of agreeing what you are going to do. Because targets are written down and personalised, people feel committed. We keep the contracts short and simple, and with only a handful of meaningful targets. I have a penchant for things in fours, whereas Blair seemed to talk in fives. 'You can have five targets if you like, but ten is too many', I told him.

A pivotal figure in building the relationship between BP and Blair's shadow team was Nick Butler. In a discreet Soho club we meet the man who could be described as the foot servant of 'Blair Petroleum' as wags called the company, much to BP's irritation. Butler became involved in Labour Party politics in the mid-1970s while he was studying for an economics degree at Cambridge University. 'I remember seeing both Tony Benn and Denis Healey when they came to speak. Benn was a fascinating character.' At the time the minister of energy was in the midst of his long battle with David Steel, CEO of BP. 'He [Benn] and Frances Morrell and Francis Cripps, they thought oil in the North Sea should be government owned.' But Butler's politics were to the right of Benn's for he joined the more centrist Fabian Society and became its treasurer in 1982.[55]

In 1989 John Browne, fresh back from the US as chief executive of BP's Sohio business, was appointed head of BP Exploration and Production at the age of 41. He gathered around him a team of bright young men, including Tony Hayward, John Manzoni and Nick Butler. Two years before Butler had stood as Labour candidate for the constituency of Lincoln;[56] although he failed to get elected, it reaffirmed his political affiliations. 'I stood for Labour. I always wanted to be an MP. BP were good about it. I was told I was the first person to be a Labour candidate in the company. I was thought of as an eccentricity. John Browne was fascinated by it, and said "I wish I was your agent".'[57]

Butler became a right-hand man to Browne who described him in his autobiography as 'my trusted adviser on politics and strategy in BP for 17 years'.[58]

In July 1994 Tony Blair was elected leader of the Labour Party. From this point on Butler's long history of involvement in Labour politics came into vital service for BP. Over the next three years he cultivated ever closer links between BP and Blair, especially as the latter's grip on his party grew ever stronger. In March 1995 Blair won his campaign to change Clause IV of the Labour Party constitution and remove its stated commitment to nationalisation. For BP the spectre of the last Labour government in which Tony Benn had apparently striven to have the state take over the company was finally dispelled.[59]

Two years later New Labour won its landslide election of May 1997, and celebrated victory night with a party in the Royal Festival Hall. Immediately the close involvement of BP in the government was clear. David Simon resigned as chairman of BP and was appointed to an advisory post as minister of competitiveness in Europe at the Department of Trade and Industry and the Treasury. Bryan Sanderson, head of BP Chemicals, was made a member of the Advisory Group to the Labour Party on Industrial Competition Policy,[60] chairman of the Learning and Skills Council and joint-chair of the Asia Task Force under Margaret Beckett, secretary of state for trade and industry. The Stanford speech by Browne two weeks later served only to highlight the sense of the birth of a new era with BP and the New Labour government marching in lockstep.

That June, in the euphoric days of Blair's 'Cool Britannia', we met a close friend of Browne's. The two had been at Cambridge University together in the late 1960s and had kept in touch. He explained, 'Johnny's got green fingers'. Leaning over the dinner table he said, 'Browne gave Tony his theory of the "Contract with the British People".' It was a short aside, but in hindsight seems now to be an illustration that the BP strategy of CSR had helped form Blair's victory and had thereby decisively moved into the UK's political realm.[61]

* * *

In his Stanford speech Browne had committed BP to establishing an emissions trading system within BP to help drive energy efficiency within the company. Soon a pilot scheme was underway, guided by the chief executive. Since the age of 20 John Mogford had been a BP driller, and had worked himself steadily up the company to be head of drilling, reporting to John Browne who was head of Exploration and Production. Soon he followed the latter's footsteps and became manager of the Forties Field, based in Aberdeen. His next step up the ladder was to become head of BP Technology in London, and in this role he oversaw the creation of BP's internal emissions trading scheme.

The concept of the scheme was aimed at reducing the carbon dioxide emissions by applying a market approach to this corner of the global disaster of climate change. Each business unit within BP was given a set number of 'carbon credits' which gave them the right to pollute and they could trade these in a marketplace with other business units in order to assist the company as a whole to reduce its CO_2 emissions but to do so in a way the required the lowest expenditure of capital. Thus the BP Grangemouth Refinery recognised that it could reduce emissions by many tonnes per year by investing in efficiency measures. So, managers at Grangemouth 'sold' their carbon credits to the Forties, which enabled the latter to carry on polluting just as it had done, and then with the capital it received Grangemouth invested in the measures it had outlined. Thereby BP reduced its emissions as a whole company but did so at the lowest capital cost, and the scheme had stimulated discussion among staff about the need to reduce CO_2 and the best methods to achieve this.[62]

John Mogford lived in Barnes, west London, a mere nine miles down the Thames from his office in Sunbury, together with his wife of 25 years, Margaret. As it happened she had worked for British Gas and was then headhunted to join the Department of Industry under Secretary of State Peter Mandelson. The department was exploring ideas for developing a UK Emissions Trading Scheme (ETS) and, on an even grander scale, for promoting the concept within the EU as a means of reducing carbon emissions on a Europe-wide level. The BP design was taken up with enthusiasm by the Industry Department, the Environment Ministry and then by the secretary of state for Europe.

Whitehall under New Labour was eager to show constructive commitment to the European Union and also to do so under the abiding ethos of Blairism, also known as the 'Third Way' as expounded by Blair and President Clinton. This was a strategy to underpin a more socially just and environmentally conscious society by working with the grain of a capitalist market economy. This would create a kind of 'social neo-liberalism' in place of the more hard-line version that had underpinned government thinking since the late 1970s. The ETS fitted the bill perfectly and the UK vigorously lobbied for it in Brussels. Finally it became EU policy and the European ETS was launched in January 2005.[63]

Years later Browne told us he felt equivocal about the ETS: 'It kind of works, and it kind of doesn't work.' BP developed the prototype system in Chicago 'because we were very big in Chicago, and Chicago was the exchange that took it up', winding up by saying, 'I believed it would be great for everybody because it would further drive efficiency'.[64]

By then the ETS, which had been lauded for the first decade of its existence,[65] had become a great disappointment. Its ability to reduce CO_2 emissions had been far lower than projected, partly because a procession of industries undermined it by lobbying and winning exemptions and free credits. It stimulated a great deal of debate around climate change, CO_2 and energy efficiency, as Browne observed, but it had little real impact on the practices of EU industries.[66] Many argued that its effect has been negative as it diverted a vast amount of labour, capital and time within the EU and European national governments which should have been spent on tackling the problem at root.[67]

In effect the tale of the rise and fall of the ETS echoes the tale of the whole era of the Blair/Clinton Third Way. Great efforts were made to reduce the impacts of oil and gas extraction on the Earth's atmosphere through energy efficiency and 'working with the market'. But the structural problem beneath this activity remained – BP, Shell and other companies were still extracting carbon from the earth at an increasing rate and global demand continued to rise. These fossil fuels were then sold onto the market and inevitably global CO_2 emissions rose. In parallel the New Labour project expended, correctly, great efforts to improve the lot of the poorest in UK society through invest-ment in Sure Start schemes, refurbishing schools, housing and health care, but the fundamental problem of structural inequality in British society remained and indeed continued to grow at an alarming rate in the Blair years. Meanwhile little was done to create new jobs in the deindustrialised regions such as South Wales, Merseyside and the Thames Estuary.

* * *

It is entirely fitting that we meet with Lazarus at the Royal Festival Hall, for this was the setting of the events at the zenith of the power bloc of Browne and Blair, of BP and New Labour. We describe the scene to Lazarus. It was a sunny late April day in 2002.[68] This public hall, this 'People's Palace', had been closed off to the public. It had been hired out for the private use of BP, which had just reached the goal of being the largest company by capitalisation on the London Stock Exchange after seven years of meteoric rise under Browne.

For the second year running the Festival Hall had been utilised as the venue for BP's annual general meeting (AGM). The symbolism was perfect. It was situated just at the base of the head office of BP's great rival, Shell. The venue itself seemed to spread its arms wide to the Thames, London and nation, and BP chose to do likewise. An open corporation, displaying its achievements to the public, not only the financial heart of the capital city.

Browne was at the peak of his reputation. Four months before, *Management Today* had crowned him 'Most Admired Businessman' for the second year running. The award came as a result of a poll of his peers and indicated a deep respect for him within the finance and business sector; he had truly turned BP around. The *Financial Times* had asked to do an extensive feature on him. Nick Butler strongly advised his boss against taking up this offer but Browne agreed. The resultant piece was hagiographic and dubbed him the Sun King.[69] These halcyon days were also a high point of Tony Blair's power. In June 2001 he had achieved his long-term goal, to win two general elections. The Conservatives were in complete disarray, Labour's majority in the House of Commons had only dropped a fraction from its 1997 high point, and New Labour seemed unassailable in power.[70] Blair had supported President Bush in America's hour of need following the attack on the Twin Towers in September 2001, and had asserted the UK's position as an important global power, the US's staunchest ally. Blair's reputation on the world's stage was as yet untainted by the debacle of the Iraq War.

As Browne arrived at the Festival Hall, and the security guards parted the way through the crowds, many of whom carried placards, we looked on as the cameras captured not only the Sun King, but

176 · CRUDE BRITANNIA

also the woman at his side, his new PA Anji Hunter. Tall, blonde and slightly bashful before the press, Hunter was already an icon. She had been a friend of Tony Blair's from their school days and was at his side in the climb to become leader of the Labour Party and then prime minister. She was the gatekeeper to Blair, said to be the most powerful non-elected person in Number 10.

In 2001 Hunter had been headhunted by Browne, who offered her an annual salary of £200,000. Blair was distressed by the possible loss of his confidant and begged her to stay, at least until that vital second general election. With the offer of a doubling of her Number 10 pay packet, he managed to persuade her not to leave. But by November 2001 she was at Browne's side. Together, PA and CEO entered the Royal Festival Hall, bedecked with the iconography of the rebrand – the 'Beyond Petroleum' strapline and the Helios logo.

* * *

It is just two and a half miles north over the river from the Royal Festival Hall to Mayfair where Browne, who left BP in 2007, is now comfortably settled as chairman of a business called L1 Energy, formally registered in Luxembourg. For a man who promised to take BP 'beyond petroleum' 20 years ago it is perhaps surprising to discover that L1 Energy is largely an oil and gas company owned ultimately by Russian tycoon Mikhail Fridman. The two men met in the 1990s before Browne masterminded another of his startlingly large mergers, one involving BP's Russian arm with Moscow-based TNK, also part-owned by Fridman.[71]

L1 Energy is just one of the companies Browne chairs. Another is the UK arm of China's controversial telephony company Huawei which is under sanctions in America and at the heart of a security row in Britain.[72] Browne is also trustee of various cultural and science institutions alongside writing books, collecting pre-Colombian artefacts and watching opera in Venice where he keeps a flat.

We are asked by the young receptionist to sit on a comfortable chair in the beige-tiled entrance hall and pass the time with a cup of tea. There are pot plants, a large flat-screened TV with Sky News on but with the audio off and copies of the *Financial Times* and *Economist* on a glass table. The soft hum of air conditioning and the smell of air freshener waft over us. The photographs on the wall include a shot of Lucien Freud at work, brush in hand, stripped to the waist. It sits somewhat uneasily with the corporate formality, but of course Browne was chair of the trustees of Tate Modern.

We are keen to meet him. The youthful-looking 70 year old is for us the most interesting character that the world of oil and gas has thrown up in the past half-century. Despite the time of the meeting being muddled, we get a message that Browne has finally agreed to have a quick chat in the middle of a busy schedule: he is off to the theatre at 5.15. And here he is, a small, neat man with a beige tie, light blue shirt and shiny black leather shoes. He gives an outstretched hand and broad smile. Browne is famous for being as charming in social settings as he is ruthless in business deals.

We have been following his career for the past 30 years. He is smart, with a first in physics from Cambridge, charismatic and a great communicator. Critics claim he was a celebrity CEO who began to believe in his own omnipotence and he is certainly highly conscious of his public image. An assistant is called into his office alongside us and asked to tape proceedings even though we are doing the same thing.

We pitch into deindustrialisation and whether it's not regrettable that oil companies such as BP and Shell have hugely downsized and almost left Britain behind. Browne says there is no point in mourning that: 'I think it's inevitable. When you're in a resource business, you can only go where the resources are.' And 'left behind' communities just have to live with that on the Thames, Merseyside and South Wales? 'No, it's a consequence probably of many, many decisions to do with the use of proceeds in the hands of the government and the business environment established and the commitment to education. I think its very complex, it's not just one thing.'

We ask with reference to government how successful he was when he was brought in as the government's lead non-executive director in a bid to improve the efficiency of Whitehall departments? 'Like everything in government – good in parts. If you get five out of ten you're doing quite well, I think in government. You can ask around, it's working well in some departments and not working well in others.'

And the greatest contribution to the UK from oil would be what? 'There was this amazing brand new industry in the 1980s. It went from a little bit of onshore oil production in the Second World War to being almost a petrocurrency at some point. It was so big the North

Sea. It made a big difference: the very large amount of tax revenue generated in the 70s, 80s and early 90s.' And should that have gone into a sovereign wealth fund for future generations like they did in Norway? 'There's a question.'

Browne has to go but he agrees to a second meeting at which we go back to his earlier career and what he learned doing a master's degree at Stanford University and then working for BP in Alaska, New York and San Francisco. 'I was very much influenced by that. I really thought British business practice was lost in the Stone Age and America was way ahead in the world of management science. In understanding how to finance companies, how to get the best out of people, even how to integrate business with society.'

And the development of the North Sea was a test laboratory for American business practices? 'Absolutely. The North Sea was really built by Americans. The original building was American, the ideas were American, the execution was British because of BP and Shell. But Shell was in conjunction with Esso, later Exxon.'

And the fact that unions were discouraged if not prevented from operating? 'Well there were no unionised workforces offshore and everyone felt enormous pride in the North Sea. It was so unexpected. It was a jewel in New Britain as it were. The Queen kept inaugurating things. While we had difficulty with Anthony Wedgwood Benn and BNOC. You know Benn wanted to nationalise BP if you remember and that was resisted. And BNOC became privatised Britoil and that was eventually purchased by BP.'

So to what extent was BP caught up in the neoliberal politics of Thatcherism? 'Difficult to say. It's not as if you can divide systems into Marxist, capitalist, sub-capitalist, American, north-west European. It doesn't work like that and so I think BP was a mix of things.'

We go back to BP's selling down of its 'downstream' refineries such as Coryton on the Thames and Grangemouth in Scotland. Browne said other executives inside BP had done research on whether refineries were good business or whether you could get the same products from Rotterdam or wherever. 'We looked at our refineries and said they weren't that good, they really weren't that good in terms of the mix of products they made and their cost base. So we proceeded to sell them.'

But why sell Coryton to a company like Petroplus that quickly went bust? 'Nobody wanted them. I think Petroplus put a very good face on it at the time, you know and I don't think we sold things irresponsibly, because we had to keep certain liabilities I believe. The Grangemouth sale worked out fine. INEOS is a pretty good company', although he admits he did not know who they were when they first came along.

So in the end no regrets? 'In the end you can't buck the reality of something. A fixed asset has to be improved, so you are always looking at the cost of improving it and the benefit that comes from that and saying to yourself, "I could sell it to someone else, they could do that' so they might as well cash out at that point".'

And lastly on climate, what role does the man who tried, but effectively failed, to get BP beyond petroleum, see for the oil and gas sector that he still inhabits? He has earlier pointed out to us that L1 Energy is 70 per cent invested in gas over oil:

I am a firm believer that gas, appropriately handled, is a very important thing for climate stability. We really can't wave a magic wand and all hydrocarbons disappear I'm afraid. It would be fine if they can but they can't. Energy is the biggest platform there is and we need it to be clean, reliable, totally secure, low cost and available to everybody. There's about a billion people in the world who don't actually have any energy so we need to get it to them. We need to do everything in renewables, everything in gas, oil for heavy transportation and the rest to optimise on the margin: nuclear fission, coal, oil for non-transportation purposes.

So will the world meet the Paris Agreement goals? 'We are not going to meet them with rhetoric either', he says tartly. 'We don't have enough sufficient incentives to drive it [change] and that has to be some charge on carbon. I used to think [carbon] trading was a good idea but there are too many ways of cheating so maybe taxes are the way to go.'

A female staffer pushes open the door of Browne's office and says it is time for him to go. He ushers us out quickly and firmly, with a flash of the trademark smile.

9

Local Hero

Dire Straits – 'Going Home'
instrumental soundtrack for *Local Hero*, 1983

Mac, the young oil executive from Knox Oil in Texas, is walking the tide line with Oldsen, the naïve company rep in Aberdeen. The older man is imparting his worldly wisdom, garnered from life in Houston, the distant capital of oil:

Mac: Could you imagine a world without oil? No automobiles. – No heat. – And polish. – No ink. – And nylon. No detergents. Perspex. You wouldn't get any Perspex. No polythene. Dry cleaning fluid. Waterproofs.
Oldsen: They make dry cleaning fluid out of oil?
Mac: Did you not know that? No I didn't know that.[1]

This statement of all the benefits that the industry holds comes as the two men are privately troubling over the fate of the bay and the community of Ferness, threatened to be utterly transformed by the construction of an oil terminal.

Local Hero, in which this scene takes place, is one of only a handful of movies that touch upon the story of North Sea oil, and encapsulates the contradictions that lie at the heart of the industry – great wealth but also irreversible change. Bill Forsyth's film echoes Peter Pickering's documentary *The Island*, shot on the Isle of Grain 30 years previously. But whereas the construction of the refinery is a given in Pickering's piece, the terminal is never built in Forsyth's.

Local Hero, shot at Arisaig on the Highland west coast and at Pennan on the north-east coast, using a largely Scottish cast and written by the director of the Glaswegian hit movie *Gregory's Girl*, was not only a statement about oil but also about Scotland. A testament to the resurgence of Scots culture as distinct within the UK. Forsyth is kind on oil. The industry men are eccentric and full of contradictions. Happer, the CEO of Knox Oil, eventually visits the bay and captivated by its beauty, decides that the terminal should not be built, but rather Ferness should be a centre for research into marine ecology – an idea first mooted by the half woman half mermaid character, Marina.

Local Hero, released in February 1983, was a box office hit. Its title track, 'Going Home', by Mark Knopfler of Dire Straits, became one of the band's anthems, played at stadium concerts over the coming decade.[2] The success of the film, and its powerfully romantic image of Scotland, echoes the way in which the rise of Britain's newest offshore colony, North Sea oil, intertwined with the fate of Scotland.

* * *

The Scottish National Party (SNP) was founded in 1934, and despite its first president being the socialist Robert Cunninghame Graham, it was long perceived as a conservative force. Its Labour opponents referred to the SNP as 'Tartan Tories', but by the 1960s it was tacking back to the left and a victory in the industrial central belt of Scotland was a significant marker. In November 1967 the SNP achieved a breakthrough when Winnie Ewing won the safe Labour seat of Hamilton in north Lanarkshire.

Three years after the Hamilton by-election, BP had its breakthrough in the North Sea, striking oil in the Forties Field. Suddenly the game changed. Now the rallying cry of the SNP became: 'It's Scotland's oil!' For the economic prospects of the country had radically altered. Aberdeen and the north-east, which had languished since the 1920s, were proffered a golden future as the service base for a vast new oil colony. The Scottish Highlands, which had been relentlessly depopulated for 150 years through the forced evictions of the Clearances and economic migration, were suddenly proclaimed as a site for rig construction yards. This phenomena was the premise of *Local Hero*.

The calls for Scotland to regain its independence as a sovereign state had long foundered on the question of whether the nation could survive in a global industrial economy. Now in the oil of the 'Scottish Sector of the North Sea' was found not only a viable economic future but wealth on the scale of Kuwait. Oil became the central plank of the

SNP programme and vision for Scotland. This was also true in the wider culture through works such as 7:84's touring music-theatre piece 'The Cheviot, the Stag and the Black, Black Oil'.

In March 1973 the show began its tour of Scotland and the actors' chorus declared to the audience at the close of each show:

Have we learnt anything from the Clearances?

When the Cheviot came, only the landlords benefited

When the Stag came, only the upper-class sportsmen benefited.

Now the Black Black Oil is coming. And must come. It could benefit everybody. But if it is developed in the capitalist way, only the multinational corporations and local speculators will benefit.

We must organise, and fight ... with the help of the working class in the towns, for a government that will control the oil development for the benefit of everybody.[3]

The SNP surged ahead and in the general election of October 1974 the party polled nearly a third of all votes in Scotland and returned eleven MPs to Westminster. Even before the election, the Labour government in London was rattled by the SNP's advance. To stem the tide the Prime Minister Harold Wilson decided to revive a half-forgotten commitment to devolution, in which Scotland would remain within the UK but be granted a Scottish legislative assembly with limited powers.

In August 1974[4] a special conference was held in Glasgow in an attempt to persuade a reluctant Scottish Labour Party to back the idea of a referendum on devolution. On resigning 18 months later Wilson handed this poisoned chalice to James Callaghan who committed himself to the issue by passing the Scotland Act in 1978. It seemed that the road to greater independence opened up with every new oil discovery in the North Sea.

However, when the referendum was eventually held in March 1979, although 51.06 per cent of the voters supported a devolved assembly the turnout of 33 per cent was too low to meet the terms of the Act and therefore the Labour government left the status quo as it was.

Two months later Callaghan lost to Thatcher's Conservatives and the cause of devolution seemed dashed, pushed underground by a strongly pro-Union Tory Party.

The dream of independence did not die. In fact it was stoked by Thatcher's apparent disregard for the social if not socialist traditions of Scotland, a bias that was especially revealed by her government's use of the country as a test bed for the hated Poll Tax. Throughout the 1980s and 1990s the spirit of Scotland as a separate nation was fuelled by films such as *Local Hero* and *Braveheart*,[5] and bands such as the Proclaimers with their anthem 'I'm Gonna Be (500 Miles)'.[6] The prospective Blair government, partly as an inheritance from his predecessor John Smith, promised a new referendum if Labour returned to power. In September 1997 turnout was 60 per cent and of those 74.29 per cent voted for devolution. On 12 May 1999, the new parliament met for the first time in Edinburgh. Winnie Ewing, the oldest MSP, opened the session with the words: 'The Scottish Parliament, adjourned on the twenty-fifth day of March 1707, is hereby reconvened!'[7]

* * *

Sometime after the parliament opened we meet Tom Stewart at Kenneil Stabilization Terminal just west of Edinburgh on the Firth of Forth. Stewart is a throughput analyst on BP's Forties Pipeline System. Working in an office on the edge of the sprawling refinery and chemical works of Grangemouth, he makes predictions as to how much crude will be flowing down the Forties system the next day, the next week, the next year. He is a futurologist.

At the time Brent crude is trading at $40 a barrel and the Forties system carries about 40 per cent of the UK's production. Every 24 hours over 100 million dollars' worth of hydrocarbons passes through Kenneil, and Stewart studies this clock.

After we meet Stewart we reflect on this machine that charts the passing of time. It took 37 minutes for oil and gas to rise 11,000 feet from the Paleocene sandstone – created 57 million years ago – to the drill deck of the platform in the Forties Field. Over the next 48 hours the oil and gas passed down the pipeline to Kenneil. It took a few

184 · CRUDE BRITANNIA

hours for the gas to be separated from the oil, and the latter to be pumped to the loading terminal at Hound Point.

It took about 17 hours for the crude to fill the 300,000 tonne tanker in the Firth of Forth. The journey time of the vessel to the Coryton Refinery in the Thames Estuary varied with wind and tide.

The crude was refined into a number of products including aviation fuel. This high-grade liquid was pumped to the Buncefield depot, near Hemel Hempstead, before moving to Heathrow Airport via the West London Pipeline.

Seventy-two tonnes of the fuel filled the tanks of a Virgin Atlantic Boeing 747. Bound for New York, travelling at 555 miles per hour at an altitude of 31,000 feet above the Irish Sea, the fuel was burned.

The clock, part of which Stewart observed, most likely took under ten days to run its course, ten days for oil to move from 8,000 feet below sea level to 31,000 feet above sea level, ten days for liquid rocks to be incinerated into gas. The clock worked minute by minute, hour by hour, 365 days a year, part of a system transferring carbon from the lithosphere to the atmosphere.

Seventeen years later we try to arrange a visit to the main platform of the Forties Field.[8] On a previous occasion Terry had flown by helicopter from Dyce to the Forties,[9] then owned by BP, prior to its sale to Apache. So we begin the long and tortuous process of trying to persuade the PR man at Apache to allow us a helicopter ride to the platform. We are told we will have to undergo a safety training course and readily agree to that. At moments it looks possible, but eventually we are met with a flat 'no' and without any reason given except the company's obvious desire to control any story that might arise from the trip.

The phonecalls back and forth served to remind us that although technically the sea space 110 miles east of Aberdeen, where the Forties platforms stand, is UK territory it is in practice another country. Ever since the early days of its development in 1964 the UK North Sea, both the southern and northern halves, was treated like a new colony by Westminster and Whitehall. In this foreign land, far beyond the view of any citizen standing on an eastern shore, a different set of rules applied.[10]

* * *

By the mid-1980s, with 15 years of transformation under its belt, Aberdeen had become an oil city. Production in the UK sector was at a high of over two and a half million barrels of oil per day.[11] There was a battalion of platforms out beyond the horizon of the cold grey sea working a wealth of fields. The largest stakes in this offshore colony were held by BP and Shell, the latter doing so with its long-term North Sea partner Esso, an arm of Exxon, later ExxonMobil.[12] Wealth flowed into Aberdeen, house prices rocketed, and while other parts of the UK such as Merseyside and south Wales entered long-term decline, the city and north-east Scotland boomed. The shops on Union Street thronged with people, the bars on Belmont Street were full and *Dallas*[13] was showing on TVs in the suburban homes newly built with the famous local granite stone.

The UK offshore industry had been constructed with expertise imported from the US, but slowly the technological advances made in response to the severity of working in the North Sea began to create an industry that became a significant British exporter. The tax revenues coming in for the government were enormous: $12 billion alone in 1984–5, £11 billion the following year and still £12 billion in 2008–9. The figure, made up mainly of corporation tax, had dropped below £1 billion by 2019–20[14]

In the autumn of 1989 John Browne, only a few months into his tenure as head of BP Exploration and Production, met with Prime Minister Margaret Thatcher who, in the course of their conversations, encouraged him to invest in the USSR. Thatcher had recently met Mikhail Gorbachev, president of the Soviet Union, and concluded that he was 'a man we can do business with'. On the strength of this she strove to catalyse one of the UK's rising businessmen. Browne took the bait and dispatched a team, headed by Tom Hamilton, which initially tried to sign a deal in the Soviet Socialist Republic of Kazakhstan.

This failed, so Browne tried two other tacks. First, BP organised a visit to the UK for a select group of executives from the Soviet oil ministry, including Vagit Alekperov, the deputy energy minister.[15] In 1990, months before the fall of the Soviet Union, they were brought to Aberdeen and flown to see the technological miracles offshore. The visitors from Moscow were sceptical when BP executive Rondo Fehlberg took them to an oil field that, by Soviet standards, was impossibly clean and quiet. 'They initially refused to believe that oil was being produced there', Fehlberg said. 'They thought it was a propaganda effort by the West. We had them put their hands on pipes so they could feel the oil.'[16]

Second, the BP team went to Baku, capital of the Soviet Socialist Republic of Azerbaijan, part of the USSR. When they arrived they discovered that they had been beaten to it by Steve Remp, CEO of Ramco, a small, privately owned oil company from Aberdeen. Remp was already negotiating with the Azeri authorities attempting to get licences to explore for oil offshore in the Caspian Sea. Six months later, by December 1990, Remp had agreed to join Ramco up to a BP team which included Norway's Statoil.

In due course BP became the largest foreign investor in Azerbaijan, and throughout the rapid expansion of the oil and gas industry in the 1990s and 2000s Aberdeen played a key role. The first direct flights from Western Europe to Azerbaijan ran not from London or Paris, but from Aberdeen Airport at Dyce. The jets that took this route were packed with oil workers from the rigs shuttling back and forth, two weeks offshore in the Caspian, a night out in the newly established Lord Nelson pub in Baku, then two weeks back at home in Britain. In the searing heat of Azerbaijan it was intensely hard work, but the money was good. The working practices of the BP-operated platforms in the Caspian were directly transposed from Scotland, just as the first rigs in the UK North Sea had been influenced by the oil culture of Louisiana.

Through the portal of Aberdeen flowed men from all over the UK. Some of them came to work offshore as a result of being laid off elsewhere in the oil industry. In 1994, as the Baglan factory went through another round of closures, staff took up the offer of being retrained and found work on the rigs beyond Aberdeen. In Essex, as the Shell Haven refinery closed, and later Coryton Refinery, men found they could transfer skills to working in the North Sea. Some relocated with their families to Scotland, others commuted, two weeks on and two weeks off. Groups of riggers were a familiar sight on the Intercity trains heading south from Aberdeen. Men swapping stories over six-packs of Tennent's or Special Brew.

The extent of the UK's export of exploration expertise, channelled largely through Aberdeen, was generally ignored by the national media and politicians in London. Aberdeen, like the hidden offshore colony of the North Sea sector, became almost set apart from the British economy and culture. However, from north-east Scotland labour was dispersed across the world, not just to 'new frontiers' like the Caspian but also to more mature provinces such as the Persian Gulf, the Alaskan North Slope or the Niger Delta. The geographical reach of this export market was inevitably limited to the political range of the Western oil corporations. Parts of the globe could be forced open, as BP and others had done in the USSR, but some doors were barely ajar – such as Saudi Arabia, Iraq and Iran.

* * *

The sudden opening of the Soviet Bloc to Western capital and corporations was an unexpected gift to an array of UK and US politicians. It underpinned the Cold War triumphalism of Margaret Thatcher and George Bush Snr, and made possible the sense of inevitability that fuelled the Third Way of Tony Blair and Bill Clinton. There was an alignment of the interests of states and corporations. Shell and BP saw the former Soviet Union as a miraculous answer to the problems of dwindling oil reserves, while the UK government saw corporations as a tool to enact British foreign policy.

Civil service notes of a 1993 meeting between Foreign Secretary Douglas Hurd and BP's CEO David Simon and Chairman Lord Ashburton, sums up the relationship between company and Whitehall neatly: '[Hurd] emphasised that there were some parts of the world, such as Azerbaijan and Colombia, where the most important British interest was BP's operation. In those countries he was keen to ensure that our efforts intertwined effectively with BP's.'[17]

In Azerbaijan that intertwining was illustrated by the British Embassy in Baku. BP needed the UK to give it more support in its struggle to gain exploration licences offshore in the Caspian. In July 1992, 18 months before the meeting with Hurd, David Simon had asked Prime Minister John Major to assist the company. Subsequently it was decided that the UK would no longer communicate with the Azeri government through the embassy in Moscow but set up a new consulate in Baku. BP agreed to provide space in its offices, so the British diplomatic Union Jack was flown permanently outside the oil company local headquarters.[18] As Browne wrote in his autobiography, although BP was then an entirely private company, with no state shareholding, it 'was essential for us [BP] to be closely aligned with the UK government, as post-Soviet countries still found it easier to understand and accept government-to-government dealings'.[19]

This alignment of UK foreign policy and the needs of BP and Shell remained largely hidden from public view and barely delineated in the national media. However, towards the end of 2002 it came into the spotlight in the build-up to the Iraq War.

The State Department in Washington, and the industry lobby groups in that same city, believed the greatest prize was for US oil companies to drill in the fields of Saudi Arabia, Iran or Iraq. That possibility materialised in 2002 with the growing hostility between President George W. Bush and President Saddam Hussein.

There has been much print devoted to the question of whether or not the Iraq War was, or was not, a 'war for oil'. Tony Blair said in a television interview in February 2003, in the run-up to the conflict: 'The oil conspiracy theory is honestly one of the more absurd when you analyse it.' However, research over the 15 years that have passed since the attack on Baghdad have slowly revealed just how strongly BP and Shell, among other Western majors, had relished the prospect of the opening up of Iraqi reserves. As the minutes of a meeting on 6 November 2002 in the Foreign Office reads: 'Iraq is *the* big oil prospect. BP are desperate to get in there.'[20]

If by 2005 BP's and Shell's hopes of gaining access to Iraqi oil reserves on a 'game-changing' level had been thwarted, the Iraq War had quite definitely shifted the political game in the UK. In February 2003, on the eve of the war, the country's largest ever demonstration took place. Around a million people marched through London, many of them holding placards that read 'No War for Oil'. The march failed to thwart the plans of Tony Blair and George W. Bush but it created long-term change. First, it put a brake on the ability of prime ministers to wage extensive ground wars while ignoring parliament and public opinion. And second, it fixed the term 'No Blood for Oil' in the popular imagination and created a widespread understanding of the link between UK foreign policy and the interests of the oil companies.

*   *   *

By 2005 the violence of the occupation of Iraq and the inability of the US and allies to establish a stable Iraqi state had become clear to many. There seemed to be no end in sight to the turmoil and the endlessly rising death toll. For the oil companies this posed immense challenges while at the same time dispelling the vision of Iraq that had long been held, that it would become simply a version of Azerbaijan. This model would make it a country with a compliant government, unlikely to be democratic, which was able to make and hold to long-term commercial contracts and ensure a level of social order and compliance for foreign companies to easily construct oil wells, pipelines and export terminals.

A stable regime in Iraq had not been established, but the violent overthrow of Saddam Hussein and the brutal manner of his end was starkly clear. Just as Washington intended, this had geopolitical repercussions. Muammar Qaddafi took note and it is possible that this contributed to his sudden decision to open diplomatic channels with the West.

Qaddafi had come to power through an army coup in 1969 and in the immediate aftermath the new government nationalised the assets of all foreign companies working in Libya.[21] Although new arrangements were made and the Western companies returned, Libya remained locked in a dispute with the US, UK and France throughout the 1970s, 1980s and 1990s. This diplomatic war occasionally flared into outright conflict, such as with the downing in 1988 of a civilian Boeing 747 over Lockerbie in Scotland, blamed on Qaddafi; the subsequent bombing of Tripoli by the US at President Ronald Reagan's command; and the shooting of PC Yvonne Fletcher outside the Libyan Embassy in London in 1984. The Lockerbie disaster, in which 270 passengers were killed, caused the greatest rift between the Major and Blair governments and Qaddafi, so it was a surprise when the latter announced in 2003 that Libya would open itself to foreign weapons inspectors.

BP was quick to seize the opportunity of gaining, or regaining, access to the Libyan oil fields. On 19 June 2005 CEO John Browne accompanied by Mark Allen, former head of MI5, and one other assistant met Qaddafi in his desert tent. Two years later the CEO of BP – now Tony Hayward – met Qaddafi, this time in the presence of Tony Blair as well as Mark Allen. On 29 May 2007 a deal was signed in the shade of a tent that gave BP exploration rights over the offshore Gulf of Sirte basin and the onshore Ghadames basin.[22] As Browne later wrote: 'Rebuilding the relationship with Qaddafi was a rare modern example of where political diplomacy opened up trade. Without Tony Blair's intervention, I doubt BP would ever have been as significant a player as it turned out to be in the re-opening of the oil and gas industry of Libya.'[23]

Shell had indeed been ahead of BP in re-entering Libya. On 24 March 2004 the corporation had acquired rights to drill for gas, also offshore in the Gulf of Sirte, together with a commitment to refurbish a liquid natural gas export plant. However, Shell's advance into this new territory had also been assisted by Labour politicians, including Foreign Secretaries Jack Straw and David Miliband, Foreign Ministers Duncan Alexander and Baroness Symons and former Labour Leader Lord Kinnock.[24] Revelations of Whitehall assistance to Shell lent weight to the suspicions that the release of the Lockerbie bomber, Abdelbaset al-Megrahi, from jail in Scotland to receive a hero's welcome in Libya was driven by commercial considerations.

Qaddafi had long proved to be politically fickle. From the outset both BP and Shell were wary that their current good fortune could be rapidly turned on its head and they would summarily lose all that they had gained with little chance of recourse. Consequently they both made the strategic decision not to approach Libya in a manner similar to Azerbaijan, but in a manner similar to Iraq.

In Azerbaijan BP had essentially worked to embed itself in the evolving politics, economics and society of the new state, building a version of Aberdeen on the Caspian. The company had created extensive offices in Baku, set up housing facilities for BP staff to live on the edge of the city, made a prolonged attempt to engage in Azeri politics and indeed made efforts to foster the evolution of Azeri civil society, through sponsorship, bursaries, joint ventures and the like. In Iraq and Libya neither BP nor Shell attempted to do anything like this.

* * *

The years surrounding the fall of the Berlin Wall in 1989 saw neo-liberalism enter its triumphalist phase, crystallised in works such as Francis Fukuyama's book *The End of History and the Last Man* and the phrase always attached to Margaret Thatcher, that 'there is no alter-native' to free market capitalism. These years also entirely transformed the geography of the corporate world,[25] with the opening up of the resources and markets of the Soviet bloc, and soon after China.[26] Just as fundamental to these momentous changes, came the boom in infor-mation technology (IT), and alongside it the internet, with the birth of the world wide web in 1989.

The oil industry had long been an enthusiastic adopter of computers and IT. Both Shell and BP developed extensive computing departments from the late 1950s, drawing on the expertise of outside companies such as CEIR (UK), Ferranti and IBM.[27] By the late 1960s computer modelling had become so fundamental to both oil companies that there was a substantial backlash against it. The formidable Australian Bob Deam had been pressing for ever greater computerisation in BP, but Peter Walters, general manager in BP's Supply and Development Department, resisted.[28,29]

The steady Americanisation of BP and Shell in the 1970s went hand in hand with computerisation, especially with the rise of Silicon Valley. By mid-1970 Browne was renting capacity on Control Data Corporation's large mainframe computer CDC 6600 in Palo Alto, California for BP and using it to model oil projects in Alaska.[30] It was entirely fitting that in 1979 Browne did his business school masters at Stanford University, in the heart of Silicon Valley. This was in contrast to David Simon, his predecessor as CEO of BP, who attended the European Institute of Business Administration at Fontainebleau, just outside Paris.

Computerisation changed not only the speed and scale of data analysis but also altered the daily life of the companies. Britannic Tower under Maurice Bridgeman in the late 1960s was an edifice of hierarchy with the male directors on the top floors and an army of women in the typing pools on the lower floors. BP Chairman Robert Horton's 'Project 1990', launched in the year of its name, aimed to 'flatten out' the organisation and crucially to cut staffing levels at head office. In part this was an enactment of US business theory on the functioning of the 'modern corporation', but in part it reflected the possibilities opened up by IT. For example, staff at all levels of the company began to type on desktop computers and the days of the typing pool were numbered. The workers in the post room, who for decades had trundled file trollies from office to office, moving information from department to department, was decimated by the introduction of a company intranet and the arrival of email.[31]

The use of computers revolutionised communications within discrete parts of the company, such as inside the Shell Centre or inside Shell Haven, but also between different sections of the corporation. In March 1986 an IBM System/38 was installed at the Baglan factory to implement BP's company-wide Commercial System Project. The new machine was linked both to a network of terminals within Baglan, but also directly to the centre of the project at Scicon, a BP subsidiary in Milton Keynes.[32]

Computerisation equally transformed the relationship between the oil corporations and other companies. Up until the mid-1980s BP and Shell were largely hermetically sealed entities. Within the Shell Centre and Britannic Tower were housed most of the units essential to the functioning of each company, for example the advertising department and the accounts department. All the staff in these departments were directly on the company's payroll, benefits and pension schemes.

A decade later and the world had been transformed. At BP the accounts were undertaken by Ernst & Young, and the advertising outsourced to Ogilvy & Mather, a subsidiary of WPP. At Shell the advertising was managed by J. Walter Thompson. This new pattern gradually extended to all corners of corporate activity – for example, by the late 1990s the human resources issues for BP staff working on platforms in the UK North Sea, such as on the Forties Field, were outsourced to a company based in northern California.

This may have increased efficiency but it also radically reduced the payroll, overheads and liabilities of BP and Shell. It was a manifestation of Peter Walters' drive to introduce the 'market' into BP, echoed by Sir Peter Baxendell (chairman of Shell Transport and then Royal Dutch/Shell Group) employing the same strategy in Shell. It also spread the responsibility for the oil companies' activities more diffusely through the body of British society, creating what became described as the 'carbon web'. This web was the network of companies, banks, government departments, cultural institutions, legal firms and so on that came to make up the essence of BP and Shell, making it possible for them to carry out their core function of generating profit from the extraction and sale of oil and gas.

The carbon web was enabled by, and came to depend upon, IT and the internet. As a business model it was arguably unable to function without these technologies. But although it spread knowledge and a degree of decision making beyond the walls of BP and Shell, it also served to concentrate the power of the oil corporations geographically and alter the economic and political structure of Britain.

If in 1986 Baglan became connected to Milton Keynes by computer, within just over a decade that line from south Wales had gone dead as the Baglan factory was closed and with it Llandarcy, Barry and Angle Bay. Meanwhile Shell's and BP's battles to rebrand themselves in the late 1990s had tied them ever closer to agencies such as WPP and J. Walter Thompson. Ever more data flowed between them and by 2006 BP was spending $300 million annually on the services of WPP, who devised the 'Beyond Petroleum' campaign.[33] The link between BP in Piccadilly and WPP in Canary Wharf, five miles apart across London, grew stronger, while the link between Piccadilly and south Wales died. The same pattern was visible in Shell, as the technology and business model altered the shape of the UK body politic.

In his autobiography John Browne notes that he was fortunate when he became head of BP Exploration and Production in 1989, not just because this coincided with a geopolitical earthquake, but also because of the arrival of the internet. Indeed, computerisation and Big Data had perhaps the most profound effect on the exploration divisions of the oil companies.[34]

10
Stanlow

Eternally
This field remains
Stanlow
No heart or head or mind
No season could erase
We set you down
To care for us
Stanlow
A vision fading fast
A million hearts to warm
And so restrained
She turns away
Stanlow.

—Orchestral Manoeuvres in the Dark,
'Stanlow', 1980

Stanlow is the name of a refinery on the banks of the Mersey, its back to Chester and its face turned towards Liverpool and the Wirral. It is also the title of the only British pop song about an oil installation.

In our pursuit of the story behind the track, 'Stanlow', we come to the K-West Hotel in Hammersmith, London. It is all steel and glass on the outside, bright-coloured velour cushions and rock 'n' roll portraits on the inside. A bottle of Tanqueray Gin will knock you back £120 in the bar and the basic fish and chips comes in at £16. It's the obvious place you come to meet a pop star, in room 22 down dimly lit corridors.

A stocky figure dressed in regulation black T-shirt and jeans swings open the door and then plonks himself back down in front of hamburger and chips. 'You don't mind if I continue with this do you?' he asks, wedging his plate between a flat-screen computer and his own broad chest. Andy McCluskey[1] is the songwriter behind Orchestral Manoeuvres in the Dark, or OMD as the fans know it, and hits such as 'Enola Gay' and 'Electricity'. But why write a love song to an oil refinery with the lines, 'We set you down | To care for us | Stanlow'?

'It comes from my family connections … and I had some of that post-war dream of science solving most of our problems', he explains. 'My father worked at Stanlow, well actually Thornton Research Centre, next door. My sister worked at Thornton. She started making tea and ended up as head of photography for the Test Section before they started laying people off.'

I always had this fascination about humans and technology. And from a purely aesthetic point of view the way that Stanlow oil refinery sits at the base of the Wirral peninsula, on the marshes surrounded by sheep and fields. It is this clinking, clanking collection of pipes and flames and lights and gantries looking like some giant intergalactic mother ship that has landed in a place where there used to be a religious centre, Stanlow Abbey.

He continues: 'When we [OMD] drove back from gigs in London, Manchester and Leeds, before there were the motorways, we had to go down the link road right past Stanlow. So we knew then at 1, 2 and 3 in the morning that we were 20 minutes from home.'

McCluskey grew up in the shadow of Stanlow in the 1960s and 1970s. The plant was in its heyday and his dad would regularly bring home new cars, provided by Shell, in which improved variants of lubricant were being road tested. At one point the family motor was a Mark 1 Ford Cortina with a Lotus racing engine installed in it. 'That certainly gave it a kick. I remember my mother racing an E-Type Jag down the motorway at 110 miles an hour once.' The boy filled with the ecstasy of technology and speed.

'Stanlow', written at a time when the refinery was still in full production, and using a live recording of a diesel pump at the plant as its opening rhythm, is intriguingly melancholic. A premonition.

A vision fading fast
A million hearts to warm
And so restrained
She turns away
Stanlow.

In October 1980, when the track was released on the album *Organisation*, Stanlow and Thornton were part of a massive petrochemical complex that ran along the Wirral and up the Manchester Ship Canal. A web of interconnected plants that stretched for 40 miles along the watershed of the Mersey river. Not all of these, by any means, were owned by Shell or BP, but the two corporations provided the vital engines of this megamachine.

* * *

On 26 May 2011 the 500 Shell staff at Thornton R&D Facility were gathered together in the facilities car park and told that the unit would be closing and job losses were possible. Ed Brady, the company PR manager, explained later to the local *Chester Chronicle* that an internal review had taken place and 'found there was considerable duplication and under-utilised assets at both Thornton and Hamburg, with scope for consolidation. ... Our proposals exist to transfer work to Hamburg, this has been driven by such factors as proximity to key automotive customers.'[2]

Staff were furious. Thornton was very important to Shell, and to Britain. Among many other things it had produced the fuel and lubricants for Ferrari Formula 1 cars. Racing drivers Michael Schumacher and Fernando Alonso had both been on site. It was top-end technology and the UK was losing it.

The local Labour MP, Andrew Miller, was at the time chair of the House of Commons Science and Technology Select Committee. Prior to becoming a politician he had been a laboratory technician and an official of the Manufacturing, Science and Finance union. Given the loss of jobs, both those directly employed by Shell and 150 contractors, it was surprising that Miller did not kick up a huge fuss.[3] Nor was it raised in parliament by the Conservative chancellor of the exchequer, George Osborne, whose constituency at Tatton was close to Thornton.[4]

In January 2012 Terry wrote a piece for the *Guardian* headlined 'Shell to shut its main UK research base and transfer its work overseas'. It read: 'Hundreds of senior scientists working at the centre at Thornton in Cheshire will be scattered to other offices in a move that follows the sale of the nearby Stanlow Refinery and is seen by some as a more general retreat by Shell from the UK.' There was seemingly little interest in this hollowing out either by the media generally or Westminster politicians.

In our attempts to contact some of those laid off, we find Kenny Cunningham.[5] He invites us to his home, a neat red-brick Victorian terraced house in a fast-gentrifying quarter of Chester. From the gruff Glaswegian voice and our talks about the rough and tumble of trades unionism in Merseyside on the phone earlier, we had stereotyped Cunningham as a bruiser of a man.

But the door is opened by a small, skinny fella with thick, black-rimmed glasses, a slim pale face and silvery shoulder-length hair. We settle ourselves in his front room that is all stripped pine floorboards, Persian-style rugs and wood-burning stove. Slightly more incongruously there is a pink glass flamingo on the mantelpiece and Easter bunnies wink at us from the top of the television in the corner. Cunningham's smile is big and the offer of cups of tea warm and genuine. The conversation, once it starts, is free-flowing. He is sharp as a nail, acid-tongued and very, very funny. Sitting on the wrong side of a negotiating table from him would be hell.

Cunningham reels off scurrilous stories, such as one describing his early days at Shell when the company was under fire from anti-racist campaigners for still being very active in South Africa. 'When I got the job at Shell I was a kind of Marxist by then, my politics were kind of sorted, so I felt like I had been parachuted into the heart of the beast. Shell, the oil industry, that's the heart of capitalism and I was under the skin, kind of thing ... I was supporting Greenpeace and started arranging anti-apartheid meetings at work', he says with a loud chuckle.

Cunningham lives and worked in Cheshire but was born in Paisley, Glasgow. He enjoyed science at school and became interested in lab work, which sounded like a 'perfect job you can do with a hangover'. But after getting a job in a steel works and then a paper factory, he decided to take himself off to Stirling University to study chemistry at degree level. He also moved to Leith in Edinburgh, 'more or less at the time when Irvine Welsh was writing *Trainspotting*, which was a great piece of work', and started to apply for jobs.

In fact, Cunningham thinks literature got him his opening with Shell:

I was asked in the first interview what I was reading and I said '*War and Peace* actually'. And the Shell guy wanted me to tell him about the scene with the train and I knew it and could fill in all the details. After that he's buzzing on me because I'd just read it and he'd read it like 40 years ago. That all went well. And then I had a second interview and the next old buffer. He went on to literature, 'What have you read recently?' And I'm thinking hang on this is a set up, and exactly the same thing happened, so as soon as he said that I went out and he went, 'He's a nice guy, give him the job'.

He laughs: 'They regretted it from that moment forth.'

The laboratory worker was delighted to get a job at Shell, happy to work there for 30 years ahead of being laid off, and was even flattering about the current Dutch chief executive and a former head of chemicals, Ben van Beurden. Cunningham had come across him while working as staff rep for his division in the UK and as a member of the company's wider European Forum. 'I liked him (van Beurden). His personal manner is highly attractive, he's respectful, he's human, he's quite funny, he's obviously smart.'

But then Cunningham liked the business ethos of the Netherlands generally over that of his home country. 'The Dutch influence is exceptionally civilising on the company. It's the Dutch who decide, who run the HR [human resources] department, so sod everything else. And the engineers. They like their engineers and they really struggle with the British style of management. They were horrified by some of what they heard from us.'

Cunningham was active as a staff rep in the company when the 'reserves scandal' broke and forced the exit in disgrace of the CEO Sir Philip Watts:

> I remember that coming up at the European Forum. The issue was kind of prettied up for our consumption. It was one of the things, again where your colleagues who come from Germany and Belgium were really quite pissed off that this had happened. They felt Shell had disappointed them by behaving badly like this. That wasn't really an issue for us Brits because we had a different attitude towards the company.

So you were treated well, but did that stop the selling off and downsizing, we ask? 'No.' And do you think they were creators or victims of globalisation? 'No, they're not victims because Shell can't be victims in that sense. It's hard to find a power relationship in which you would see Shell as a victim. Exxon can mess with them, right, but that's about it. So no they're not the victims. They were following the trend. They knew what they were doing.'

So when did it first start?

> Well, if you see globalisation as part of a neoliberal experiment, which I think it is, it's just one manifestation of it. It's moving to a worldwide economy and that's exactly what Shell did. They did it carefully. They opened the first outsourcing centre in Glasgow and they tried to keep it secret. When they started closing Carrington some staff were transferred to Thornton. Then Sittingbourne [a research centre] went. Then much later Thornton itself lost out to Germany.

Despite all this, and continual problems with his UK managers as Cunningham tried to organise union activity inside Thornton, there was in his view a basic humanity about the company. 'Shell have always been squeamish about laying people off and sacking people. Their general strategy has been to pay them off or transfer them, but eventually they ran out of places to transfer people.'

* * *

On 12 January 1994 BP announced the closure of the ethylene complex, the heart of the Baglan. After a lifespan of three decades, this was the beginning of the end for the plant. Over the next ten years around 2,000 men and women were to lose their jobs. As a result of the closure of the BP Angle Bay Ocean Terminal in 1985, the shutdown of the BP Llandarcy Refinery in 1998, the sale of the BP Barry Petrochemicals Plant to INEOS and the decommissioning of all the interconnecting infrastructure of oil pipelines, around 2,000[6] BP employees were laid off and 2,000 contractors lost their livelihoods in the 1990s. The loss of well-paid jobs at Baglan, coming together with a host of other closures across south Wales, had a huge impact on communities such as Sandfields.[7]

The South Wales oil age had grown up in the shadow of the declining coal age. By the 1960s oil had begun to look like the future and for many coal looked like the past. Three decades later the future of oil processing in Port Talbot was ended by a handful of decisions made in the boardroom of Britannic House in the City of London. The profitability of the output of Baglan, and particularly ethylene, the raw material for so many industrial products, was being undercut by soaring chemical production in Asia, especially China. BP was beginning to benefit from this too. When in January 1994 CEO David Simon, CFO Byron Grote and Bryan Sanderson closed the ethylene complex, the life of Baglan was mortally threatened. In December 1995 BP formed a joint venture with Sinopec and three years later opened a 200,000 tonne a year plant at Chuanwei in Chongqing.[8] Capital spending was redeployed to China and south Wales was left behind

The shift was not approached as brutally as this might sound. BP was in the first flush of what became the 'Beyond Petroleum' initiative. Some of Browne's team felt passionately that this could be part of the company's future. And if this were the case then surely it could be part of the future of assets such as this redundant petrochemicals plant. Clearly some of the dwindling Baglan staff felt so, for they won a company Breakthrough Award for their proposal 'Baglan – Beyond Production'. The entry gave an account of the evolution of the Baglan Energy Park, turning 'a 1,000 acre brown field liability into the world's first energy park ... The deal to carry this out with General Electric creates new value to BP and to the local community from a site that faced closure.'[9] The solar building, with its convex front, was to be the new visitor centre at the gates of the complex. Here was a strategy for a high-carbon works to undergo a just transition to a low-carbon works. A transition that was largely to be undertaken within the arms of a private corporation – albeit heavily underwritten by the public purse.

Some time around the year 2000 we were driving west along the M4 passing below Mynydd Dinas and Mynydd Emroch. Away to our left, between the motorway and the mirror of Swansea Bay, sprawled the chimneys and warehouses of the Port Talbot steelworks. Beyond it were acres of wasted land where the Baglan chemical works had once stood. At the edge of the former site, close to the road, was a strange jewel that stood out from its grey surroundings. A small building with a convex front of photovoltaic panels, shining blue in the sunlight. We'd seen it before, as a display unit at the 1998 G8 Summit[10] in Birmingham, used to promote BP as an 'energy company' that was venturing 'Beyond Petroleum'. Here it was, landed in this petrochemical zone of sacrifice, like a spacecraft sent by the company with a mission to drag Sandfields into the solar future.

* * *

The UK's industrial heartlands were being steadily left behind by the likes of BP and Shell, in places such as Baglan, but the attention of Britain's 'special publics' of politicians, civil servants, journalists and NGOs may have been distracted by strategies such as the 'Beyond Petroleum' brand campaign. The zenith of this initiative came in July 2005 when, timed with the G7 Summit at Gleneagles in Scotland, John Browne launched the company's pioneering carbon capture and storage (CCS) scheme at nearby Peterhead. It came alongside a commitment to spend £4.5 billion over the next decade on wind power, solar energy, hydrogen and gas-fired power stations.

CCS was the oil companies' engineering response to the threat of climate change and one that would make fossil fuels 'clean'. Fifty years of North Sea drilling experience would be run in reverse. Rather than extracting oil and gas from the rocks beneath the platforms and pumping it onshore, carbon dioxide would be captured from the chimneys of a former coal- and oil-fired power station at Peterhead and piped offshore, down into the now empty rock cavities under the seabed and sealed in for perpetuity. The corporations, in this case, would provide the solution to the climate change problem that they had created through the profitable extraction of hydrocarbons, and they would do so in a manner that still generated a handsome return as a result of government subsidy.

Browne personally presented the Peterhead CCS scheme to the assorted heads of state at the Gleneagles G7 Summit of political leaders. Tony Blair, standing alongside him, used the event to promote the UK as a pioneer in the global battle against climate change. The brand of BP PLC was used to enhance the brand of UK PLC.

A central figure in Browne's strategy was Vivienne Cox, BP CEO of Gas, Power and Renewables, and undoubtedly one of the most senior women in the global oil industry. We are delighted to get the chance to meet her in Mayfair, west London.

Most of the period townhouses in Grosvenor Street are multi-occupied with brass plates in the doorways discreetly advertising asset management or hedge fund firms. Across the road is a showroom selling luxury yachts. This is home to the 'third finance sector'. The Square Mile had been the epicentre of the banking world throughout the early part of the British oil industry, Canary Wharf became a secondary hub from the mid-1990s and then more recently came another financial base: with myriad private equity firms that cluster in Mayfair. Nestled among them are private oil companies such as Thistle Energy, Crescent Petroleum and L1 Energy, owned by the Russian oligarch Mikhail Fridman, which has John Browne at its helm.

We are at the entrance to number 20 Grosvenor Street, a building used by KPMG,[11] one of the 'big four' accounting corporations that audit the books of other multinationals. It is a private members' club for KPMG's clients, described as 'five floors in which to enhance your conversations and deepen your business relationships'. As Cox[12] is a non-executive director of a couple of such companies, she is invited to use the premises for her meetings.

We take the lift and enter a beautifully furnished room – comfortable chairs, subdued lighting and waiter service. Cox is waiting. She is a vibrant presence in a crimson blouse, black trousers and shiny patent leather boots. She gives us a warm welcome and seems keen to talk.

Cox has played a catalytic role in the story of British oil. She was a BP lifer, joining the company at 22 and serving under a range of CEOs from Peter Walters to David Simon and John Browne. It is almost a *Zelig*-like career, being present at so many key junctures of the company's evolution, from a junior assistant during the final privatisation of BP in 1987, to establishing the Integrated Supply and Trading division, to being chief executive of Gas, Power and Renewables, later renamed BP Alternative Energy.[13] She was also pivotal in putting teeth into the corporation's initiative to go 'Beyond Petroleum'.

She left the company in 2009 at the same time as the Alternative Energy arm had its budget cut and its separate headquarters shut by the group chief executive, Tony Hayward. Cox had been the first woman to become a member of BP's Executive Committee and she was one of five people in the running to succeed John Browne as CEO in 2008. You could count on one hand the number of female executives in UK oil over the past 80 years at such a senior level. Cox was made Veuve Clicquot Business Woman of the Year in 2006[14] and was awarded a CBE in 2016 for services to the UK economy and sustainability.

A decade after leaving the company, she remains frustrated over the closure of BP Alternative Energy: 'I was disappointed that BP was not prepared to support the renewable energy business to the same extent but my legacy is all those people who left the company and are now running their own businesses.' She is clear that BP going into renewable energy was 'a good idea. But it is about content and timing. 'We were a bit too early. In my mind it was always a deeply strategic move that would take time and was about building another leg for BP businesses because it was inevitable that the pressure on fossil fuels would come – exactly in the way it is coming now.'

She believes her division was a challenge to BP for a number of reasons:

Renewables has a completely different culture to it. And that was one of the difficult issues running Alternative Energy inside an oil company. There was an allocation of capital problem but also a cultural problem. Renewables attracts different people. Many people are doing it because they want to make a difference. They think they are solving a problem that needs to be solved.

After leaving BP, Cox was made chairman of the private equity-backed company Climate Change Capital and remains deeply concerned about global warming. 'I think it is an urgent, urgent problem.' But like Helen Thompson, she is unhappy about the blaming of big brands like BP and expecting everything to come right if they abandon hydro-carbons. 'I think the answer is not to pillory the oil companies. From my perspective this is quite a dangerous route being taken by some of these groups – encouraging the withdrawal of investment from oil companies, legal claims against oil companies.' She pushes home her point: 'To make the oil companies the bad guys in all this is to not engage with the real problem, which is how to reduce the consumption of fossil fuels and replace them with other sources of energy and that is an urgent, urgent problem. To blame the oil companies is fighting the wrong battle.'

Within a year of leaving BP, Cox took up a role in the civil service as lead independent director at the UK government's Department for International Development,[15] an appointment that overlapped with a post as commissioner for the UK's Airport Commission in the Department of Transport. Cox speaks enthusiastically about what her colleagues at BP have gone on to do: 'My peer group, and it's the gen-eration before as well, you know, tentacles all the way though British business life, and beyond – Rosalind Franklin Research Institute,[16] John Manzoni in government,[17] Neil Morris at the Faraday Institute.'[18] Cox herself is non-executive director or senior adviser to a dizzying array of businesses and organisations including Pearson PLC, GSK PLC, Stena AB, Vallourec SA, the Saïd Business School at Oxford University and the African Leadership Institute. Cox is also chair of the board of governors at the medical research centre the Rosalind Franklin Institute, and vice president of the Energy Institute. Her previous positions have included posts on the boards of BG PLC, Rio Tinto PLC, Eurotunnel PLC and the INSEAD business school.

We ask, gently, if the role of so many former oil executives in key British institutions – political, economic and cultural – is appropriate? It illustrates just how important this industry is to the structures of the

UK, but could it be problematic that one company has this network of alumni?

Cox laughs a little and observes: 'It is interesting how many BP executives have gone on to do big jobs.' She begins to elaborate: 'I mean I think there are two things going on, one is positive and the other is not. The positive one is the training that we got was *extraordinary*, I mean it really was.' Her eyes light up as she begins to recount her experience of the training she underwent at BP:

At the time of the generation before me, and my generation, there was very much a philosophy of a broad education, that you'd move between countries, that you'd move through parts of the company. … You'd get on-the-job-training that was very broad, but more than that the formal developmental training was extraordinary. They used to have something called Stages 1, 2 and 3 and they were gates to whittle the high-potential resource. So you got a certain number of people who went through Stage 1 and then their careers were accelerated. Then there was Stage 2. Stage 3 was basically the top, it was whittling from 100 to 50, it was part of the final selection for the executive.

Full of delight she recalls days 20 or so years before when she went through the final hurdle to join John Browne's executive team. 'One of the elements of Stage 3 was a week in the Lake District that had a lot of activities to demonstrate how capable you were at managing teams, and influencing, and all those things.' She describes the last exercise like a thriller:

We thought we'd finished. We go to dinner. And somebody comes into the room and drops some car keys on the table. So somebody goes and picks them up. The car keys fit a car that's in the car park. In the back of the car there's a map with a map reference written on it, so we go. This exercise ran all through the night. You didn't know what the problem was that you were trying to solve, and you didn't know how to solve it. It was what they call a Box 4 Problem. And there were people watching us through one-way glass, to see, during the night, how we organised ourselves, what we did, how we responded, what the interaction between us was, I mean *extraordinary*. You know, this training I got is world class still, and this was decades ago.

She returns to our question. 'So we were incredibly well trained, so there is probably a meritocratic reason why a lot of these people are now in the places they are in.' She laughs a little uncomfortably and says, 'There is also, you know, the less positive, that we all know each other, so you tend to ... It's a really close group. It's built on mutual respect and experience of working really closely together. I've got a group of people who I stay very much in contact with, cos' – again that laugh – 'because they are my friends. And these networks overlap in various places. You know, the number of times I get a call from a head hunter, and the first name that comes into my head is a BP name. I have no reservation in recommending them, because I know they are really, really high-quality people.'

There's an interesting twist to this tale. As she observes, part of the depth of BP's involvement in the UK government is because the company 'came out of the British state', but by the 1990s the corporation's position was different. She explains 'We're not British Petroleum. We're BP. The internal dialogue was: "we are a global company. The UK is a very small part of our activities. Yes, we have relationships with UK government ministers, just as we would with US, or Russian, or anybody else. And we have to take a *bit* of extra care, because we are domiciled here, but we're definitely a global company".' We mention Prime Minister Theresa May's dismissive line in autumn 2016 about 'citizens of nowhere' and wonder if BP too has become a citizen of nowhere. Cox replies, 'Maybe, but peopled by citizens of the world, because we all moved around a lot.'

When our conversation is finished we step into the street. It is a spring day. The weather calls us to be outside in the city. We head down Davies Street and weave our way through Berkeley Square, walk past the office of L1 Energy where John Browne is chairman, until we cross Piccadilly and enter Green Park.

We are both pleased with the meeting with Cox. She was insightful, reflective and quite candid. She made us want to believe everything she said – so distinct from others at her level who are invariably warm but also guarded, having signed either non-disclosure agreements or mindful of who pays a part of their future pension. What strikes us most profoundly are Cox's remarks on the role of fellow BP executives in the civil service. We talk of John Manzoni, former head of BP Refining, a good friend of Cox's who was chief executive of the civil service and permanent secretary of the Cabinet Office till February 2020, one of the most important positions in Whitehall.[19] He followed into the civil service even more senior BP executives such as Lord Browne. Despite Cox's remarks, Shell has also had a long tradition of executives moving into government roles upon retiring from business.[20]

The crossover between the oil companies and government is in many ways an inevitable consequence of the historic compromise in the 1940s when BP and Shell became national champions. Other major industries, such as coal or the railways, were nationalised and their staff became direct employees of the state. Although the oil corporations remained, in part, private capital concerns they took on the culture of the civil service and the armed forces. As John Browne wrote in his autobiography:

My induction to the company [in September 1969] had involved learning about oil exploration on a short course at the Great Fosters Hotel in Surrey ... Then my 'posting orders' came through. BP, then part-owned by the British government and seen by many as an arm of the Foreign Office, was still highly bureaucratic and posting orders meant exactly that. There was no say in the matter.[21]

Despite BP's strenuous efforts from the 1970s onwards to separate itself from the British state and follow the model of Shell, the connections between the two corporations and Whitehall remained exceptionally strong.[22] It went both ways. Lord John Kerr of Kinlochard was UK ambassador to the US from 1995 to 1997, while the storms raged over Shell's position in Nigeria. Upon the election of Labour in 1997, Kerr was brought back from the US and made head of the Diplomatic Service, and permanent under-secretary of state at the Foreign and Commonwealth Office (FCO), the most senior civil servant in the FCO. Upon retiring in 2002, aged 60, Kerr became a non-executive director of Shell Transport and Trading. His role as a non-exec was by no means ceremonial. For example, it was Kerr who headed the team coordinating the corporate restructuring of Royal Dutch/Shell in 2004. 'We non-execs perhaps took a little time to get our act together, to decide how to go about it', he told the *Financial Times*, 'but I gave up my summer last year and it wasn't for a PR bang. I mean, come on, it was not to achieve positive PR that we worked our butts off.'[23]

Kerr's move illustrates what is known as the 'revolving door', through which many civil servants move to join the boards of large companies upon retiring from Whitehall. Prior to Kerr, another permanent under-secretary at the FCO from 1982 to 1985 was Sir Antony Acland. Following this post, he too became UK ambassador to the US for five years. Directly upon retiring, Acland was appointed a non-executive director of Shell between 1991 and 1996.

Then there are the figures that move in the other direction, step out of the oil companies and glide into political roles. A bout as a contestant on the TV show *Strictly Come Dancing* and his position as leader of the Liberal Democrats placed Vince Cable MP among the unassailably 'virtuous' public figures. But earlier in his career Cable had been chief economist at Shell for seven years, during the events surrounding the judicial execution of Ken Saro-Wiwa in the summer and autumn of 1995. A decade or so after leaving Shell, Cable was secretary of state for business, innovation and skills, and president of the Board of Trade. He admits the Shell badge caused him some problems and refers to this in his autobiography, writing that he was asked on the BBC programme *Newsnight* to condemn his former employer over the company's reserves scandal that led to the exit of the then boss Sir Philip Watts, but 'I declined and said I was proud to have worked for Shell'.[24]

There are many others who, like Cable, sit in the Houses of Parliament after active service in BP, Shell or other parts of the industry.[25] They include David Lidington MP, who worked in BP when Peter Walters was CEO and was later Conservative minister for the Cabinet Office[26] and deputy prime minister to Theresa May. Liz Truss MP was a commercial manager at Shell in the days of Mark Moody-Stuart and Sir Philip Watts, and went on to become secretary of state for international trade.[27] Alan Duncan was another Shell employee, who landed the role of minister of state for Europe,[28] while Lord Andrew Robathan, a minister of state for Northern Ireland and the armed forces, had worked for BP.[29]

We exit Green Park, cross St James's Street and stroll down King Street. The sunlight dapples through the leaves of the plane trees at the centre of St James's Square. We pass Chatham House, the famous debating chamber of British foreign policy, home to a number of think tanks who, abiding by Chatham House Rules, discuss global issues. It is just a few paces from BP's head offices, and could function for the company almost like an external meeting room. The institution is sponsored by an array of corporations including Shell and BP. Executives from the two companies are regularly invited to be on panels hosted by Chatham House, shorthand for the Royal Institute of International Affairs.

Established in 1920, Chatham House has been a powerhouse of geopolitical thinking. From here came ideas on how to maintain the British Empire and later to manage the retreat from it. Alongside other institutions in Whitehall such as the Royal United Services Institute, it was central to strategic thinking in the Cold War and the ongoing defence of the realm. As Britain's imperial reach was often conducted through industrial businesses, they were key to running the empire, and the nation's life after empire. Primary among those businesses have been corporations such as BP, Shell and Rio Tinto. The brainstorming inside the rooms of Chatham House helped bolster the rationale of Britain's presence in the Middle East and Africa, and after the fall of the Soviet Union, in Russia and the Caucasus. The foreign policy thinking, Foreign Office actions and the needs of private capital concerns has determined UK relations with states such as Nigeria, Azerbaijan, Georgia, Iran, Iraq, Kuwait, United Arab Emirates, Oman, Libya and indeed Russia. We note that Chatham House has a panel of senior advisers chaired by former Prime Minister John Major, but also including a swathe of influential people with close links to oil such as Lord Browne (ex-BP CEO), Ayman Asfari (chief executive of oil services firm Petrofac), Ian Davis (non-executive director at BP), Sir John Sawers (non-exec at BP) and Lord Robertson (special adviser to BP).

Outside this Georgian townhouse we discuss what happens when the needs of the citizens of the UK run counter to the needs of these companies. Is this a way to describe the debacle over the Iraq War, vehemently opposed by so many across the country? And how does an institution such as Chatham House affect the UK's response to climate change? Shell is planning for a target of a 70 per cent reduction of CO_2 emissions by 2050, and zero carbon by 2070.[30] However, this is far less ambitious than swathes of British opinion, including the government's own Climate Change Commission that calls for zero carbon by 2050. Shell can promote its view through institutions such as Chatham House, presenting its own target as 'sensible', informed by the company's wealth of experience. This cannot help but determine what the Foreign Office and other government departments deem to be realistic. And that in turn helps to determine the actions of the UK in global climate diplomacy. As Adam Vaughan, chief reporter at the *New Scientist* put it to us bluntly when we met him: 'Every UK minister pays lip service to the Paris Agreement climate goals but few seriously think we will meet them.'

Oil industry writer and commentator Greg Muttitt[31] has another perspective, as he told us:

I think a large part of the function of Shell's plans is to define the boundaries of what the future might look like. I think all the companies do this in order to say, 'Well, we can have this much fossil fuels or this little fossil fuels. This is your range of futures. Let's keep the conversation here.' It is about putting bounds on the imagination as to what's possible and what the future could be. That's of course in the interests of those companies. It's in the interests of those who are connected with them and who benefit from them. It's in the interests of those who hold a particular kind of worldview to set out future possibilities.

We cross St James's Square and outside the head office of BP we stop to consider what Cox has told us. Does all this really matter? Is it not the case that the staff in this company are supremely well trained, supremely skilled and have a wealth of global experience that they bring to their new roles in the civil service? Is the British state not benefiting from getting such well-trained executives to act as servants to the public realm? Women and men tutored at little expense to the public purse.

The words of the ethnographer Marilyn Strathern come to mind: 'It matters what ideas we use to think other ideas.'[32] Cox has been extremely clear in how BP helped discipline her mind in a particular way. Browne too has shown this in our conversations with him. But that training also, inevitably, limits their thinking. In essence their imaginations are captured by engineering and capital. In engineering there is no problem that cannot be overcome, given enough time and energy, but there is little space for emotion, least of all grief or despair. And capital demands that one thing overrides all others, the necessity to get a return, to generate profit.

These ways of thinking have determined the way that BP, like Shell and the rest of the oil industry, has approached climate change. It is a problem that can be dealt with by technology, by wind or solar power, by biomass or Carbon Capture and Storage. And it is a problem that can only be dealt with if it makes money, if it generates a return on capital. When Shell and BP couldn't produce enough profit on making solar panels, they sold on the factories. When the Peterhead CCS project was not going to make money through UK government subsidies, it was closed down.

It is one thing when this thinking directs the behaviour of a private corporation, but when those who have been most highly disciplined in that thinking move across into the civil service, inevitably their thinking can help determine the imagination of the state. When corporate thinking and popular thinking were aligned, say in the defence of the realm in World War II, this is not necessarily a problem. But when corporations, and the state constrained by corporate thinking, start to behave in the same way, and in a manner that runs directly counter to the desires of the people over an issue such as climate change, then that is a problem. 'It matters what ideas we use to think other ideas.'

* * *

A year after Browne had triumphantly announced the Peterhead CCS scheme at the G7 Summit in Gleneagles, BP was not growing as fast as it had been. There were stirrings among the institutional investors whom Browne had wooed so assiduously for two decades. Those that questioned the CEO's golden touch found their champion in Peter Sutherland, the chairman of BP. By the autumn of 2006[33] there was a struggle between Browne and Sutherland focused on the question of his retirement. The CEO's faction wanted Browne to stay on until at least the company's centenary year in 2009. The chairman's faction wanted Browne to retire in 2008, at the age of 60, which had been common practice since the retirement of Peter Walters in 1990. Just as with Prime Ministers Thatcher and Blair, there were many around Browne who felt their leader had overstayed his time.

The year 2005 had been a run of disasters in the US. An explosion at BP's Texas City refinery had killed 15 and left 180 injured. A rupture on a company pipeline in Alaska had leaked 212,000 gallons of crude into the tundra. And two BP traders in the Midwest propane market had been found guilty of fraud. All of these events led to accusations that BP was cutting corners and not maintaining safety standards or complying with federal regulations. The company's reputation in its most vital market was damaged, its share value declined relative to its peers and shareholders were alarmed.

The end came suddenly. At 17.00 on 1 May 2007, Lord John Browne, CEO of BP, stepped out of the main entrance of the corporation's head offices in St James's Square, Piccadilly and walked the few paces across the pavement to the waiting limousine. Someone inside the company had tipped off the press. The cameras whirred and clicked as a security guard held back the scrum. After twelve years as CEO Browne had just been sacked.[34]

The Sun King's fall from grace was as dramatic as his rise to power. On 5 January 2007 Browne was on holiday in the Caribbean at the home of his friend, the banker and benefactor John Studzinsky. He received an urgent call from Roddy Kennedy, the head of BP Press and Media. The *Mail on Sunday* was about to publish a lengthy exposé of Browne's private life. A former lover, Jeff Chevalier, had clearly been handsomely paid by a newspaper, eager to attack a key member of the Blairite elite. Chevalier's testimony described Browne's London penthouse at Chelsea Harbour, dinners with Peter Mandelson, the great architect of New Labour, and the trips to Browne's apartment in Venice.

The BP CEO took the *Mail on Sunday* to the High Court and lost; not only that, he had perjured himself. He claimed that he had met Chevalier by chance in Battersea Park, but the evidence revealed that they had hooked up via a dating website, 'Suited and Booted'. It was clear that Lord Browne, trustee of Tate, former trustee of the British Museum, non-executive director of Goldman Sachs, confidant of the prime minister, had lied to the judge. From that point on his public position seemed fundamentally weakened, with the *Mail on Sunday* and other parts of the media eager for his scalp.[35]

BP always denied these issues were connected in any way, but critics pounced on the problems in America as revealing the shortcomings of the business model Browne had experienced in the US in the 1970s and studied at Stanford Business School. The core of this model ran: constantly strive to increase the company's capitalisation and global spread; expand the corporation through mergers and acquisitions; maximise the capital value of new assets by stripping out labour and perceived inefficiencies; sell on or shut down those assets that do not generate the desired returns; outsource company functions where possible; and always focus on satisfying the demands of the equity analysts who guide the institutional investors.

When we talked in 2019 at his Mayfair offices, he made no apology for his view that jobs and success could only be created by adherence to tough business decisions. Whether you chased renewable energy projects or oil schemes, it always came down ultimately to using the money you have in the most efficient way.

'A fixed asset has to be improved, so you are always looking at the cost of improving it and the benefit that comes from that, and saying to yourself "I could sell it to someone else, and they could do that".'[36] And, 'I think in business you realise that emotion doesn't have too much space when it comes to assets. When it comes to people its different. Assets are assets, you know. They are things that you can't hold on to if there's no space for it.'[37] On the process of deindustrialisation he reflected:

I think it is a natural cycle and it changes according to the comparative advantage of each country you are in ... no company is big enough to do everything, so where does it put its next dollar, where does it put its people, how does it get competitive advantage for itself? ... You have to move the business to different places because of marginal returns.

The events in America had drawn attention to the flaws in the model as applied in the US, but some of the same strategy had also been applied in the UK. The story of Coryton provides a perfect illustration. The refinery came into BP's ownership as part of a far larger merger that Browne had masterminded in 1996 in which BP and Mobil had combined their European assets.[38] In 2007 Coryton was sold off to Petroplus. But the price at which it could produce petrol was soon undercut by the importation of refined products processed more cheaply in South East Asia by the likes of BP, such as at the refinery BP co-financed in Haldia, West Bengal. Coryton closed within five years.

The workforce of southern Essex had been sacrificed to BP and the other companies' desire to generate a higher return on capital through refineries such as that in Haldia and through the destruction of tariff barriers into the EU. This pattern was repeated in the closure of factories, such as the lubricants plant in Birkenhead acquired from Mobil in 1996 and closed in 2001. Whereas at Baglan BP attempted to assist the community with a form of transition, in the case of Birkenhead no such action was taken. And in the case of Coryton, BP effectively sold on any responsibility to the communities who had served the corporation, and those people were left at the mercy of Petroplus which was ruthless in its disregard for anything other than return on capital.[39]

* * *

BP's and Shell's decade of closures of industrial plants across the country – of which Coryton and Baglan are merely exemplars – dramatically reduced the number of people employed by the two corporations in the UK. Communities in South Wales, the Thames Estuary and elsewhere were heavily impacted, and while some continued to feel loyalty to their former employers, others felt betrayed by an industry to which they had given their working lives.

The corporations claimed they were themselves victim of the ruthless roulette of globalisation where you could not stand still and be undercut by lower-cost foreign rivals. However, over the same decade the two corporations continued to portray themselves as 'national champions' who were working for the benefit of 'British interests'. Their ability to do so depended largely on maintaining the loyalty of the 'special publics' – politicians, civil servants, journalists and even some environmentalists – part of the web of allies that Cox had illustrated for us. It was a community which was based increasingly in London.

Friends of the Earth had, from their establishment in the UK, been sceptical or critical of the oil industry. In the mid-1990s Friends of the Earth joined in the collective outrage at Shell's attempts to dump the *Brent Spar* in the North Sea and the company's involvement in the murder of Ken Saro-Wiwa and his colleagues.

Meanwhile Jonathon Porritt, a former director of Friends of the Earth who was being presented by the media as the 'leader' of the UK green movement, co-founded Forum for the Future. The organisation's explicit remit was to work in 'partnership' with business, and in particular Britain's largest corporations, to pursue joint environmental goals.[40]

For a few years these alliances were a valuable asset for the CSR team at BP headed by Howard Chase. Nominally these networks supported Shell's rival, but the air of collaboration between NGOs and specific companies assisted the political position of the industry as a whole. The interchanges in the metropolis helped the corporations maintain their position as great contributors to British society while at the same time closures were impacting on communities around the country. Subtly, unintentionally, this also meant that sections of the nation's political class focused their attention away from those parts of the UK's population who lived increasingly in ex-industrial regions.

The contradictions between the missions of Friends of the Earth and Forum for the Future and the practices of BP meant that these alliances did not hold.[41] A mere two years after promoting the BP rebrand to the company's staff, Tony Juniper lent his weight to the campaign against the company's Baku–Tbilisi–Ceyhan oil pipeline. Since 1992 BP had been planning the construction of this 1,768 km oil pipe to run from offshore fields in Azerbaijan's sector of the Caspian

Sea to the Turkish Mediterranean port of Ceyhan, via the mountains of Georgia. It posed not only a threat to the ecologies of the regions it passed through but also to the global climate, for it facilitated the delivery into the Earth's atmosphere of the carbon load of crude oil from the rocks beneath the Caspian. Friends of the Earth played a key role in an international coalition of NGOs alongside the Corner House, Platform, Amnesty International and many more opposed to the pipeline. Any sense that Friends of the Earth was giving unconditional support to the rebranded BP and its 'Beyond Petroleum' initiative was dispelled. However, the attention of these NGOs was directed away from the communities in the UK impacted by the oil companies and towards communities in distant elsewheres, such as the Caucasus region.[42]

* * *

Shell or BP crude tankers would steam up the Mersey, passing the city of Liverpool to port and the shipyards of Cammell Lairds in Birkenhead to starboard. At the base of the Wirral, where the shipping channel narrowed, was Stanlow. Here their loads were pumped ashore at Bromborough and processed in the refinery.[43] The grades of the plant's products were being constantly improved by the scientists at Thornton next door.

The core of Stanlow's output was gasoline and aviation fuel which was distributed by road tankers across the north-west and by bunker barges up the Manchester Ship Canal. Half way along that waterway the vessels passed BP's Partington Dock. Occasionally they would unload there, but usually the terminal was supplied by an underground pipeline direct from Stanlow. Partington itself fed Manchester Airport, by road tanker and pipeline, and the vast Carrington petrochemicals plant right next door.

Carrington, owned by Shell, was one of the largest chemicals works in the UK, a pioneer in the development and production of plastics. It has its own R&D centre, established in 1961, and was a rival to BP's Baglan.

But as with BP in the Severn Estuary, Shell steadily retreated from Merseyside. The initial step was the oil companies ceasing to fill the dockyards of Cammell Laird with orders for tankers and offshore rigs in the 1960s, 1970s and 1980s. But the first works to close, at the behest of John Browne and the BP board, was the factory at Birkenhead in 2001, which had supplied lubricants to markets across Europe. Production was moved to a new plant in China.[44]

Then came the decline of Carrington chemical works. The first was the 2006 closure of the polyols unit, with production moved to Rotterdam, and the ethoxylates unit moved to Wilton on Teeside – 'under plans to improve the cost effectiveness of the business'.[45] In summer 2007 Shell closed all of Carrington, except for one unit left to be run by LyondellBasell.[46] On 29 March 2011 Shell announced it was selling Stanlow Refinery to the Indian company Essar. The corporation was cashing in and capital was being allocated elsewhere, leaving the machine at the heart of Merseyside which it had run for 87 years. The closure of Thornton followed two months later.

Passing through the gateway of Carrington, we drive across a great plane of concrete reminiscent of Coryton and Baglan and head for one of the few remaining buildings. In the office of the reception there are several well-built guys having a tea break. These are the men in the fire-fighting team who are assisting the demolition of the final plant. Colin McMullen explains that when he started in catering at Carrington there were 3,500 people directly employed by Shell working in seven separate units on the site. Gradually these were shut down with production moved abroad to Canada, France and elsewhere. Now there are 200 people employed on the last unit being wound down.

Another former worker tells us:

I honestly believe, and so does everybody else that worked there, it should never have closed ... because it couldn't be put in a better place ... if you think about it, we had pipelines from Stanlow, we had the Manchester Ship Canal, we had everything going for us ... the drivers could go, when they came out of the terminal, anywhere. I mean why would you close something like that?

McMullen says he doesn't blame Shell for the closure of the plant:

It's not just this industry. It's all industries. I mean I started out in the merchant navy and they did exactly the same. They decided that it was cheaper to run the ship's foreign flag and that's exactly what they did. And again since I've been here, there's been changes, and you have got to move with the changes, if you stay stagnant always, then eventually you'll get bladed or whatever.[47]

We drive away filled with a sense of how McMullen seems to accept the inevitability of the changing tides of neoliberalism. His words seem to echo the words that John Browne had said when we had met him. Except that the post-oil world of McMullen and Browne and Cox seem polls apart. Some get bladed, others don't.

* * *

We drive 15 miles to the Thornton R&D Centre where seven years before the 500 Shell staff had been told they would be laid off. Away into the distance stretch the clean lines of the purpose-built blocks designed by Sir Frederick Gibberd, the modernist architect who also created Shell's Woodstock R&D Facility dedicated to researching pesticides. Both places exude a sense of social democratic optimism. The sign at the gateway reads 'University of Chester, Thornton Science Park', and search as we might there is no indication of the 75-year history of Shell on this site. Here, in the lee of the chimneys of Stanlow, staff in white laboratory coats and tweed jackets had laboured to develop new products from crude oil and open new opportunities for profit for the company. This is where Kenny Cunningham and the father and sister of Andy McCluskey worked.

We remember a conversation with Ben van Beurden, CEO of Shell in The Hague where he admitted Shell's industrial footprint in Britain bore little resemblance to the past due to endless plant closures. He replied: 'Yeah ... that's true ... and I think that has happened, and there is no way I can position it any differently.'[48]

* * *

By the beginning of the 2010s the long arc of BP's and Shell's withdrawal from Britain had reached a climax.

The scale of this shift can best be seen by comparing the range of assets that BP and Shell owned and operated in 1940 and 1980 to what they still held in 2020 (see Table 1 on pp. 220–1).

This record of change leaves out BP's and Shell's offshore assets – the drilling rigs, oil and gas production platforms, and subsea pipelines in the North Sea. However, the UK offshore has, since the mid-1960s, been another country, abiding by different norms than those that apply onshore.

The record includes the UK's shipyards, for from the 1940s to the 1960s these yards had two key customers: the oil companies and the Royal Navy. The yards of Clydeside, Merseyside and elsewhere effectively formed arms of BP and Shell. When these two companies chose to order tankers from yards in continental Europe and the Far East, the viability of British shipbuilding was fundamentally undermined.

Table 1 is indicative, not comprehensive. It does not include a wide range of other BP and Shell assets in Britain, including petrol stations, aviation supply depots, road tanker fleets, liquid petroleum gas depots and administrative offices. Furthermore, BP and Shell are by no means the totality of the UK oil and gas industry. However, they have been the dominant players since 1940 and Table 1 shows their direction of travel, and the direction of travel of the wider sector – leaving Britain.

This vanishing act was almost entirely hidden from the general public. Of course, laid off employees at Baglan or Thornton knew that BP and Shell had withdrawn from their communities. And for sure most of these former staff would have known of how others had been made redundant at related plants, say at Carrington or Barry. For time and again the companies would explain that such and such a plant would close in order to concentrate production and safeguard jobs at another plant, only for the latter plant to be subsequently closed. Of course, closures were well covered in the local press and warranted column inches in the national papers, but few can have had a sense of the great wave of redundancies sweeping across the country. The number of people directly involved as workers in some part of the UK oil and gas industry was being reduced radically.

For the majority of the UK's consumer citizens, Shell and BP still maintained a strong position in British daily life. Their petrol stations dotted the motorways and occupied prominent locations at the heart of towns and cities, the companies' logos could be seen on cultural events, the exploits of the corporations were constantly reported in the

BRITAIN'S OIL & GAS INDUSTRY OPENING & CLOSING

	1940	1980	2020
Onshore Oil Production			
Shell			
BP			
Wytch Farm, Poole Harbour		▲	O
Refineries			
Shell			
Shell Haven		▲	⊗
Stanlow	▲	▲	O
Teesport, Teesside		▲	⊗
BP			
Coryton		▲	⊗
Grangemouth	▲	▲	O
Llandarcy	▲	▲	⊗
Crude Terminals / Depots			
BP			
Cruden Bay		▲	O
Sullom Voe		▲	O
Hound Point		▲	O
Kinneil Stabilization		▲	O
Dalmeny		▲	O
Natural Gas Terminals (utilised but not necessarily owned by BP or Shell)			
Theddlethorpe, Lincolnshire		O	O
St Fergus		O	O
Bacton, Norfolk		O	O
Easington & Dimlington, East Yorkshire		▲	O
CATS Terminal, Middlesborough		▲	O

	1940	1980	2020
Refined Product Terminals / Depots			
Shell			
Kingsbury, Birmingham		▲	O
BP			
Buncefield, Hemel Hempstead		▲	▲
Hamble, Southampton		▲	▲
Cattedown, Plymouth		▲	O
Isle of Grain			▲
LNG Terminals			
Shell			
Mossmorran		▲	▲
BP			
Isle of Grain			▲
Crude Pipelines			
Shell			
Amlwch		▲	⊗
BP			
Forties Pipeline System		▲	O
Product Pipelines			
Shell			
Stanlow – Shell Haven		▲	O
Stanlow – Carrington		▲	O
Stanlow – Manchester Airport		▲	O
BP			
Coryton – Buncefield		▲	O
Buncefield – Heathrow		▲	O
Coryton – Stansted		▲	O
Coryton – Gatwick		▲	O
Grangemouth – Wilton – Saltend, Hull		▲	O

Key: ▲ operating under Shell, BP or in joint venture with third company
O operating but sold by BP or Shell to a third company
⊗ closed

Table 1

Product Pipelines – continued

	1940	1980	2020
Joint BP & Shell			
Avonmouth – Shell Haven (A/T)	.	▲	O
Avonmouth – Bromborough (N/S)		▲	O
Old Kirkpatrick, Glasgow – Grangemouth		▲	⊗
Isle of Grain – Walton		▲	O
Thames Haven – East Anglia		▲	O
Chemical & Plastics plants			
Shell			
Carrington		▲	⊗
Wilton, Teesside		▲	⊗
Mossmorran		▲	▲
BP			
Wilton, Teeside		▲	⊗
Barry		▲	⊗
Baglan		▲	⊗
Grangemouth		▲	O
Saltend, Hull		▲	O
Stroud		▲	⊗
Carshalton		▲	⊗
Hythe		▲	⊗
Lubricants plants			
Shell			
BP			
Birkenhead		O	⊗
R & D Facilities			
Shell			
Thornton		▲	⊗
Egham, Surrey		▲	⊗
Sittingbourne		▲	⊗
Carrington		▲	⊗
BP			
Sunbury	▲	▲	▲

Shipyards & Rig Construction Yards (suppliers in which Shell & BP formed a dominant share of the order books)

	1940	1980	2020
Cammell Laird, Birkenhead	▲		
Smiths Dock, Middlesborough	▲		⊗
J.L. Thompson, Sunderland	▲		⊗
Sir J. Laing, Sunderland	▲		⊗
John Crown, Sunderland	▲	⊗	
Bartrum, Sunderland	▲	⊗	
Wm. Doxford, Sunderland	▲		⊗
Swan Hunter, Wallsend	▲		⊗
R & W Hawthorn, Hebburn	▲		⊗
Furness Shipbuilding, Teesside	▲	⊗	
Vickers-Armstrong, Barrow	▲		⊗
Grangemouth, Grangemouth	▲	⊗	
John Brown, Clydebank	▲	⊗	
Scott's Shipbuilding, Greenock	▲		⊗
A. Stephen, Glasgow	▲	⊗	
Fairfield, Glasgow	▲	⊗	
Lithgows, Port Glasgow	▲	▲	⊗
Blythswood, Glasgow	▲	⊗	
A & J Inglis, Glasgow	▲	⊗	
Harland & Wolff, Glasgow	▲	⊗	
Harland & Wolff, Belfast	▲	▲	

business media and occasionally made it onto the front pages of the mainstream papers.

However, BP and Shell had undertaken a massive, if stealthy, retreat from British society as Ben van Beurden had agreed when we met him. At the same time corporations had settled into the redoubt of the City. In contrast to the rapid shrinking of the companies' assets and their role as UK employers, the position of BP and Shell on the British capital markets continued to be dominant and indeed to grow over this 30-year period.

The primacy of finance within BP had been steadily growing for the past three decades. Key moments that had articulated this shift included David Steel's battle to begin the 'privatisation' of the company; Peter Walter's declaration that there would be 'no sacred cows'; and John Browne's strategy to woo asset managers through presenting BP's quarterly results in person. BP had been evolving into a purely financial machine rather than a producer and provider of oil. The same progress could be seen in Shell.

As John Browne pointed out: 'I once worked out that BP accounted for about £1 out of every £6 received in dividends by UK pension funds. Those dividends trebled between 1995 and 2007.'[49] The scale of BP and Shell on the London Stock Exchange and the size of holdings in them in the portfolios of British pension funds, insurance companies and unit trusts grew and grew.

When BP and Shell laid off staff, or sold on plants, they often provided generous redundancy benefits and of course did not relinquish their responsibilities to their pensioners. Steadily the numbers of people who were on Shell and BP pensions outstripped the number of staff on each of the company's payrolls. With the ranks of directly employed persons falling, and UK citizens living longer, the demographic of the wider community around the companies was shifting. This was sharply illustrated by the people who would attend each company's AGMs. They were a sea of grey heads. Many were filled with fond memories of the company they'd worked so loyally for. The industry was slowly becoming a place of nostalgia, a repository of the past rather than a manufacturer of the future.

PART III

2008–2020

11

This Bitter Earth

The London finance sector is hit by the news that Lehman Brothers of New York has collapsed. It is the fourth biggest US investment bank by capitalisation. The TV screens show staff leaving the head offices on Water Street in Lower Manhattan carrying cardboard boxes of possessions. The largest branch of Lehman's outside of America is at Canary Wharf in the Docklands of London, it instantly closes its doors. The future of 5,500 staff is in the balance. Workers are shown on the TV news leaving 25 Bank Street, also clutching their cardboard boxes.

There is panic not only in the City but also in Westminster and Whitehall. Will other US or UK banks fold? Will the entire international financial system collapse? The Royal Bank of Scotland and Lloyds Bank look perilously close to the edge. Gordon Brown's Labour government steps in. Effectively it nationalises both institutions at a price of £37 billion. These private capital corporations, who for three decades have criticised state intervention, are now bailed out by public funds as ministers declare they are 'too big to fail' and have to be rescued in the national interest.[1]

The sight of a Labour government using public money to support private banks and thereby save the jobs of bankers, who'd often been paid excessive wages and bonuses, inflames public opinion. The treatment that was meted out to workers at Baglan, where the state failed to intervene to prevent a closure, is not applied to these Masters of the Universe. The press howls and the Occupy protest movement, born in the streets of New York, springs up in London on the steps of St Paul's.[2] The *Financial Times* produces a special edition nervously entitled 'The Future of Capitalism', and there is a widespread sense in business and political circles that the financialised model of capitalism is in deep trouble – that the era of 'there is no alternative'³ is coming to a close.

Prior to the fall of Lehman Brothers, the neoliberal model had developed a sense of inevitable and irresistible growth. The global economy had been expanding fast for 15 years – not least on the back of China – and with it had come a rise in living standards for some and a boom in commodity prices.

After a period of relatively low oil prices in the 1980s, the global value of crude had risen steadily from February 1994 to June 2008. This pattern was disrupted by a price slump in the late 1990s, but the trend was generally upward over a 14-year period.⁴ In part this had been a reflection of the rapidly increasing levels of global oil consumption, itself a mirror of the vast growth in international trade. The quantity of physical goods moved around the world by ship, plane and truck rose dramatically as, for example, consumer items manufactured in China were transported to supermarkets in London, Aberdeen or Liverpool. Globalisation swept around the world on a tide of oil, and the realm of oil flourished on the back of globalisation. A perfect symbiosis. BP and Shell, which long before had taken the role of Britain's 'national champions', were now among the midwives and champions of globalisation.

Cheap credit flowed as lenders felt secure that this growth would ensure borrowers could pay back loans. Most iconic were mortgages on homes across the US, with families who had only the scantiest ability to keep up repayments being induced to take out those mortgages. On Wall Street these loans were sold on, in theory to spread the risk, but crucially to make a profit on the back of debt in the subprime mortgage market.

When this credit system collapsed the process was thrown into reverse. Unable to keep up repayments, millions defaulted on their mortgages and had their homes repossessed. As recession hit, they lost their jobs. Then, as the citizens of middle America could no longer afford to drive their 4×4s, or even sold them, the demand for gasoline fell and with it the global price of oil. In June 2008 crude was selling at an historic high of $165.48 per barrel, seven months later it had fallen to $51.06. In Britain the downturn impacted the likes of Shell workers at its factory in Wilton, which closed because of a slump in demand for the raw materials of plastics. Employees at the Coryton Refinery were worried by the fall in the sales of petrol and diesel. As the oil price lost $100 in half a year, so too fell the profits of BP, Shell and most other oil and gas companies.[5] Two pillars of the UK economy began to shake.[6]

As the oil price falls BP and Shell have to continually revise their plans. Exploration and development projects planned to stretch out into the 2030s, that seemed viable only six months before, are rapidly put on ice and staff axed. Jobs had been stripped away over the previous decade through the closure of plants such as in Carrington and Birkenhead. Now there are redundancies under review at head and regional offices – in the case of BP and Shell in London and Aberdeen. It seems the financial crisis hits the oil companies, and the oil corporations impact on the wider UK economy, but perhaps the relationship between the two is more complex?

We meet Helen Thompson, professor of political economy at Cambridge University, author of publications such as *Oil and the Western Economic Crisis* and a regular media commentator. The focus of her recent work has been the origins of the 2008 financial crisis. In the café of the campus building she explains,

I was going to write a book about the fallout of the financial crisis and the Eurozone crisis through a set of different things, one of which was energy. At that point, I was trying to introduce oil into things that I knew about. But the more I thought about it, the more I thought it's not about fitting oil into other stories, oil is the story.

She had dug back into the minutes of meetings of the US Federal Reserve, the Bank of England and the European Central Bank. 'I was gobsmacked. Oil was just there, and on like page, after page, after page. Every meeting at a certain point, they're discussing the problem of oil prices.' She came to understand that this commodity had played a huge role in the recession that preceded the financial crisis of 2008, a recession which was then compounded by the credit events such as the collapse of Lehman Brothers.

What puzzled her was that in the business press during these years, despite forensic coverage of the policies of the central banks, there was virtually no discussion about oil:

> I think it's too difficult for people to think about. It's almost like it's a parallel world, because it's so big. It permeates everything. We can't think about our day-to-day life without oil, even though most people don't consciously understand that, with the implications then of actually thinking that we are all complicit in lots of those very bad things that result from this.

She explains that she means not only climate change but also much of the UK's foreign policy decisions. 'Part of my conclusion is that when you try to draw attention to this reality, it has to be pushed away, because if it becomes part of our politics in ways which people really understood them, it makes our politics much harder.'

She explains her shock at seeing that the flag at Westminster Abbey was lowered when the King of Saudi Arabia died. But then recognising that this only revealed the intensity of the UK's dependence on Saudi oil and the wealth that comes from it.

Helen Thompson is clearly frustrated by those involved in the climate debate who demand the summary end of hydrocarbons. She fears this would just hand over power to state-owned oil groups in, say, Venezuela or Russia:

> I'm talking about people who simply do not want to engage with the difficulty of replacing oil as an energy source ... If you look at the world economy since 2008, pretty much for every year that's passed, the world has used an extra million barrels of oil a day. We're not moving away from an oil-based economy. We're going, at a world level, which is the only thing that matters in relation to climate change, in the opposite direction.

'I don't really see any way ultimately in dealing with climate change while the world uses as much energy as it presently does. I know the middle-class affluent people who are flying around everywhere don't want to hear that.' We quiz her on her own attitude to flying. Foreign trips have long been seen as a perk of many academic jobs that often require long hours and relatively modest pay. 'I will only go on one trip a year that involves a flight. That means me turning things down. I'll take one trip, for work or pleasure. It will be there and back.' She hasn't driven for 25 years and hasn't eaten meat since she was about 20. 'I think it's delusional to think that it is possible today for us to carry on living as we are presently living and think that renewables are going to do all the work for us rather than that we have to make any sacrifices.'

The global financial crisis, in part driven by the challenge of oil prices, produced a shock to the system, which led to a collapse in that oil price. This in turn impacted on the oil corporations and they impacted on the British economy. More fundamentally the financial crisis ushered in a Conservative government with its austerity policies and shifted the direction of the nation. But this change in political tempo may provide something of an opportunity in relation to climate change.

We discuss the politics behind personal sacrifices:

Ten years ago, before the financial crisis, it was pretty difficult to see how you would get a politics of any kind of sacrifice. But there's a sense now that things can't carry on as they are, in quite a number of ways, not obviously just about climate change. The fact that there has been so much disruption, and the sense that people have of existential dread, in some other form or another. Since the financial crisis there's a better political space for thinking about the harder questions of climate change and oil than there was previously.

But, she concludes, 'It doesn't mean it's anything like easy at all. Oil is something that people don't really want to think about.'

21 APRIL 2010

The breakfast TV channels carry it as the headline feature to homes and offices across the UK. The news of the *Deepwater Horizon* oil rig exploding the night before in the Gulf of Mexico, 41 miles off the coast of Louisiana. Images of the drilling platform as a ball of flame in the darkness were broadcast globally. The initial death toll is clear: eleven men have been killed in BP's disaster. The events reverberate through the community of North Sea offshore oil workers, reviving the memory of the *Piper Alpha* disaster 22 years previously.

Initially *Deepwater Horizon* fit the story of an industrial accident, but within a few days the human fatalities are lost behind panic over an environmental crisis. The well that the rig had been drilling is gushing millions tonnes of crude into the sea, the geological pressure of rocks driving oil up into the ocean above. TV footage show a brown spill spreading across the Gulf of Mexico and heading for the beaches and towns of Florida and Louisiana. In communities along the Gulf Coast there is deep anxiety about livelihoods in fishing and the tourist industries. Driven by social media, and the voracious appetite of the mainstream media, the news spill out across America.

BP, like its peers in the oil industry, has long been used to being able to exert a high degree of control over media coverage. TV footage of rigs in the UK North Sea, for example, has been almost entirely obtained only at corporate behest. But despite BP apparently trying to restrict access by journalists, the slick spreading across the Gulf of Mexico is visible to camera crews that hired light aircraft, and tools such as Facebook are weapons in the hands of all those determined to drive the state to protect citizens' lives from this corporate disaster. BP exacerbate the crisis. The CEO Tony Hayward speaks to the TV cameras, but he proves to be spectacularly inept.[7] His crisp British accent spills out lines such as 'It is relatively tiny' in comparison with 'the very big ocean', and 'You know, I'd like my life back'.

The media move the crisis for BP to Washington. The Obama administration is pushed by civil anger and the news channels to take an ever stronger line against BP, declaring 'we will make BP pay for the damage their company has caused'.[8] The narrative constantly emphasises the foreignness of the corporation. This is 'British Petroleum' threatening American lives, despite the fact that for a decade the name of the company had been officially just BP PLC, and despite the fact that a far greater amount of shares in the corporation were held by US institutions than UK ones. It is arguably more American than British. Still, Interior Secretary Ken Salazar says: 'Our job basically is to keep the boot on the neck of British Petroleum.'

Then the crisis moves to Wall Street. Within days of the explosion in the Gulf BP's share price begins to fall rapidly. And it keeps on falling. There is increasing concerns that the company might not survive, that the costs of the Herculean efforts to cap the oil well, and the prospects of a massive clean-up and compensation bill, will fatally hole BP's balance sheet. Anxiety in Wall Street spreads to London. The collapse of BP's share value mean a collapse in the value of most of the UK's institutional investors, particularly the pension funds.

So the crisis moves to Westminster. There are protests that the Obama administration is being unnecessarily harsh on a pillar of the British financial sector and poses a threat to UK pensioners. The first politician to speak out is Boris Johnson, then mayor of London, ever trying to get one up on his great rival, Prime Minister David Cameron. Others such as Lord Norman Tebbit also speak in BP's defence. Soon Cameron is on a much publicised phone call with Obama. But in truth the UK establishment exerted little influence on Washington.

By 15 June BP's share price has lost 40 per cent in just eight weeks and the credit ratings agencies Fitch and Moody's downgrade the company's status. After years in which the media has tutored the public on the importance of the credit ratings of states such as Greece, this downgrade seems to presage BP's imminent bankruptcy. News leaks out that contractors were demanding to be paid up front in cash for their services as the creditworthiness of the UK's largest company begin to be in doubt. Haywood and a team of senior executives go to the White House more or less cap in hand. BP agree to set up a $20 billion oil spill fund and to suspend dividend payments to shareholders. In return Obama acts to prevent the company's collapse, declaring that BP is 'a strong and viable company, and it is in all of our interests that it remain so'.[9]

The share price quickly begins to recover and it seems that BP had survived. But the events of spring 2010 have longer-term implications. Shares in BP have for almost a century been considered 'blue chip' in City terminology, meaning they are utterly reliable, a financial instrument that can provide solid foundations for an institution such as a pension fund. Now the very vulnerability of BP's share value – emphasised by the company suspending dividend payments – raises deep doubts over the dependability of an investment in the corporation. And if one company was this vulnerable, could it not be true of Shell and other oil majors as well? Doubts spreading in the investor community begin to reshape the relationship between the UK finance sector and the oil industry, and this in turn force the companies to change.

There were some who criticised David Cameron and other politicians for speaking out on behalf of a private corporation, while within BP itself there was anger that the UK's actions in its defence had been so weak. Perhaps both criticisms were well founded. They illustrated that the British government was more half-hearted in its support of BP than previous administrations, such as that of Tony Blair. And they also showed that the power of London in Washington was far weaker than during the vaunted decades of the 'special relationship'. If Westminster was becoming lukewarm in relation to an oil major, then the reverse was also true. The *Deepwater Horizon* crisis proved that the UK as a source of geopolitical support for the corporations was becoming weaker.

28 JUNE 2010

An evening soirée is held at Tate Britain to celebrate the twentieth anniversary of BP's sponsorship of the gallery. Several hundred guests from the arts, politics, the media and civil service attend. Strolling into the Duveen Galleries they are served drinks in the shadow of Fiona Banner's installation *Harrier and Jaguar*, a Sea Harrier jet fighter suspended by its tail a few feet above the floor. Several tonnes of modern weaponry hung like the carcass of a captured beast in a museum of art. The guests mingle with the staff and trustees of Tate, including the director, Sir Nicholas Serota, and BP senior staff.

Since 1990 BP had used sponsorship of key cultural institutions to bolster its position with the 'special publics' that it targeted, and now those publics went out to bat on behalf of their benefactor. Ten weeks before the Tate reception, on 21 April 2010, the *Deepwater Horizon* disaster hit the Gulf of Mexico and BP's profile in Britain. There was widespread anger in the UK media at the massive oil spill and the company's poor handling of it. There were calls for those institutions sponsored by BP to speak out or revoke their contracts.

It seems that part of the financial world is turning its back on BP, but the cultural elites stand firm. Asked if Tate would break off its long-term relationship with the company, Sir Nicholas Serota declares: 'You don't abandon your friends in times of difficulty.'[10]

The twentieth-anniversary soirée is rich in symbolism. When CEO Robert Horton had begun to sponsor Tate in 1990, BP was still based in the Britannic Tower, among the grandest skyscraper in the City, with several formal function suites. The company had a share price of £1.60 and was a mid-ranking Western publicly quoted oil company, one of 'The Seven Sisters'. Twenty years on its shares were worth almost four times that amount and it was the world's second largest private oil corporation. Yet its head offices in St James's Square, to which BP had moved under Browne's leadership, were relatively modest, without grand state rooms.[11] The corporation had grown in financial scale and yet it had outsourced its formal occasions to the public institutions that it sponsored. It used not only Tate but also the British Museum to host the formal events at which BP entertained its clients and dignitaries. This use of architectural space perfectly illustrated the symbiosis of the private corporation, the state and the elites in UK society, far away from Baglan and Coryton. By coincidence Fiona Banner's discarded fighter plane echoed the reality that the corporation now depended less upon the UK's military firepower and more on its global cultural weight.

Among the throng are two unexpected guests: 'Toni' and 'Bobbie' have arrived in their floral frocks and heels. They sip wine and chat. At a given moment they begin to leak. A black substance runs down their legs, crude oil spills onto the marble floors of the gallery. Other guests look on amazed and back away as Bobbie and Toni get on their hands and knees and try to mop up the growing slick, declaring loudly, 'It's a tiny spill in the ocean … not that much to worry about.' Their lines mock Tony Hayward's infamous phrase. After a surprisingly long while, Tate security guards usher them away and erect screens around the pools of black that disrupt the soirée.

At the same moment, a procession of figures dressed in black, wearing black veils and carrying two-gallon containers emblazoned with the BP logo pour a black liquid down the steps at the entrance of the building. As they scatter feathers upon the slick, demonstrators with banners hand leaflets to the guests who were still attempting to enter Tate.

Through this action Liberate Tate burst into the media and catalysed a new wave of the campaign to drive oil out of sponsorship of the arts that had been running for almost a decade and a half. Over the coming five years this movement, that was made up of a number of other groups including Art Not Oil, Platform, BP or Not BP? and Shell Out Sounds, challenged Tate's contract with BP. Not only this but also the company's sponsorship deals with the British Museum, Royal Opera House, National Portrait Gallery and Royal Shakespeare Company. The scale and intensity of the campaign was radically stepped up and a range of new tactics employed.

Liberate Tate undertook a series of performance actions in Tate Modern and Tate Britain creating pieces such as *Human Cost* and *Birthmark* that were seen as significant artworks in their own right. Powerful aesthetics meant they gained extensive media coverage including on the front page of the *Financial Times*. The arts media began to recognise the quality of the work and the seriousness of the arguments.

Alongside this, Platform pursued an analytical and legal route, launching a judicial review against Tate and requesting that it disclose the exact sum that BP paid in sponsorship to the gallery. It was widely believed that without BP's support several UK cultural institutions would cease to be free to the visiting public. But the long pursuit of Tate by barristers and solicitors revealed the oil company provided only a tiny percentage of the gallery's funding. For a relative pittance the corporation was gaining the free advertising of its brand on all Tate literature and the use of the galleries for corporate events such as the twentieth-anniversary soirée. Campaigners saw the public purse underwriting BP by funding a cheap advertising hoarding, rather than the corporation underpinning the key functions of civic life.

Inch by inch the work of Liberate Tate gained traction. Eventually the crack came. On 11 March 2016 Tate announced that in December later that year it would not renew the sponsorship contract with BP. After 25 years the partnership would cease. Significantly, as was later revealed, the end of the deal coincided with Sir Nicholas Serota stepping down as director of Tate after 27 years, and John Browne leaving Tate's board of trustees after eight years. A break between the corporation and the cultural institution had been forced and the campaign had been successful. Tate had been liberated from BP.[12] The long battle that had raged, and the eventual victory of the campaigners, illustrated the level of unease about oil in British cultural institutions, exactly parallel to shifts taking place in the UK finance sector. The industry, which for decades had helped shape British culture, was now itself being forced to change shape by that same culture.

For 15 years activists had been drawing attention to oil sponsorship of the arts sector.[13] That this criticism had finally gained mainstream attention and support reflected a shift that the wider British public was becoming more doubtful of oil. This doubt was accelerated by *Deepwater Horizon*, a disaster akin to the *Torrey Canyon* oil spill four decades before. The oil companies had utilised sponsorship as a key tool in building their social licence to operate, to gain access to their 'special publics'. Now this tool was slowly being taken from them, a route of access was being blockaded. The dwindling of cultural sponsorship may seem a detail, but here the culture of the nation was shaping the companies.

31 DECEMBER 2012

All of a sudden we see the man we've been searching for. He's there, only yards away, walking along the roadside, white square of dog collar at the neck, slightly stooped, bald head, hands behind his back – the Reverend Sir Philip Watts. We had wondered if we would recognise him after all these years. As our car passes, the priest suddenly looks up and stares straight at us. There is no doubt that this, a decade on, is the disgraced former CEO of Shell.

Watts was at the centre of one of the most humiliating episodes in Shell's history. It was fined over $120 million by the US financial watchdog for misleading the investment community by overstating the amount of oil and gas reserves it had. The Financial Services Authority said it was 'unprecedented misconduct'[14] by Shell. Watts and his two most senior lieutenants resigned but he always protested his innocence. We wanted to interview him about this but needed to track him down first. We found his name linked with a rural parish and decided to investigate.

We had driven around the Warfield and Binfield parishes south of Reading in Berkshire, past jackdaws in the oak trees and beneath buzzards circling in the blue sky. On the narrow lanes we constantly passed Porsches and the occasional electric vehicle or hybrid. Luxury cars. This is asset manager country, the land of the hedge fund owners. Wealth is set back from the road, discreet down the gravel drives. This corner of England is so rich, so seductively serene.

The church of St Michael the Archangel is packed.[15] The congregation, whiter than white, is here for a Sunday communion and a baptism. The Victorian Gothic building looks firmly C of E, but the service is a long way from the Book of Common Prayer. Andy the 'priest' is breathlessly Evangelical. He comperes the hour together with Rupert on the electric keyboard. Towards the end of the first song Andy, in black leather jacket, jeans and trainers, throws his head back and stretches both arms in the air. He seems suddenly quite ecstatic. Three young women in the front pew are singing with gusto as they swing their hips.

The informality of this 'song service' is intense and includes an address by Andy in which he prowls the aisle, mic in hand. It has the feeling of a TV game show. Appropriately the sermon is based on the Parable of the Talents. At the end he advertises a course in money management run by the church for students about to go off to university.

'We don't live perfect lives. We mess up sometimes. Let's say sorry to God. Let's confess our sins to God.' The prayers are projected onto a screen hung from a rafter over the crossing of the church. As the congregants say the words, some do so with outstretched arms and upturned palms.

When the main service comes to a close Andy passes control of the proceedings over to 'Phil'. The Reverend Sir Philip Watts looks strangely formal, owlish in black, heavy-rimmed glasses and grey shirt stretched over a belly that falls over a black belt. With his hand on the baby's back he starts the baptism, speaking into the head mic, in his strong Leicester accent:

Bless the Lord o my song,
I'll worship your holy name,

Every giant will fall
And nothing is impossible with you.

When the service is over everyone is invited to stay for tea and coffee.
We approach Watts and explain we would like to talk with him about
how it is to move from business to theology. He is friendly but wary: 'I
keep a low profile. I would not want to say anything in public', he says.
He agrees to talk on the phone and tells us to get his number from the
parish church office. 'I think we had better end this conversation', he
says. And that is that.

Watts, now 71, left the Anglo-Dutch group in 2004 amid the controversy over the reserves replacement scandal. Unlike John Browne
there was no Lazarus moment for Watts. He disappeared from all
public life, went to study theology at Ripon College, Oxford, and
then was appointed priest in charge of the benefice of Waltham St
Lawrence in Berkshire, the parish next door to Warfield. Here was a
man who once occupied the same role as Ben van Beurden, CEO of
Shell, surveyed the world from the same offices in The Hague and had
the same powers extending around the globe.

Watts seems to have avoided any public statement for nearly a
decade. Many we asked thought that he might have passed away. Only
in 2013 did his words appear in the press when he told the *Maidenhead
Advertiser*: 'It [the scandal] stopped what I was doing short but opened
up the way to what I am doing now. I am grateful for that.'

Before he moves off to the pots of coffee and a throng of people
at the back of the church, we ask what he thinks of Andy's style of
service. 'It [the Evangelical wing] is the only part of the Church of
England that is growing', he says pragmatically, giving a nudge with
his arm as he talks. We do eventually get that phone call in but it lasts
a matter of seconds as the former executive immediately clams up as
soon as the word Shell is mentioned.

The service in Warfield could not have seemed further from life in
the villages and towns of the Niger Delta. Watts' world has changed
utterly in the 13 years since he was ejected from Shell, but life in the
Delta has altered little. However, the reverberations of the CEO's fall
changed Shell's course and a decade and a half on they are still being
felt. The reserves replacement scandal saw a 20 per cent reduction in

the value of Shell's assets and the largest drop in the company's shares in many years. Jeroen van der Veer, who stepped into Watts' shoes, was determined to make good this failure. He called for a change in culture within Shell and a renewed drive to obtain oil reserves above all else.

The fall of Watts had shown who had the whip hand in Shell. Watts, the CEO of the corporation, had hoped to ride out the storm that would inevitably arise from the downgrading of 20 per cent of Shell's reserves, but the institutional shareholders demanded explanations, and when these were not forthcoming the value of Shell's equities began to fall. In order to halt the fall, the directors of Shell threw Watts, the head of Europe's most powerful company, overboard. It is a tale that van Beurden must constantly have in his mind. He got his key career shift, to become a vice president, in the rearranging that van der Veer undertook in the aftermath of Watt's demise. If van Beurden pushes the company in a direction that badly damages shareholder value, the directors of Shell will throw him overboard. It doubtless impacts upon any personal enthusiasm he might have for the corporation to undertake a rapid energy transition.

* * *

By 2005, Jeroen van der Veer was facing a room full of asset managers and analysts in the conference suite of the Shell Centre on the South Bank in central London. This was the first presentation of Shell's development plans since the fall of Watts and investors wanted to know the substance of van der Veer's proposal. A few months before in Houston he had outlined his priorities in two slogans: 'More Upstream and Profitable Downstream' and 'Enterprise First'.[16]

Running through his PowerPoint slides van der Veer outlined his trajectory for Shell, upstream and downstream. Closest attention was paid to the exploration and production prospects, the 'more upstream', with its emphasis on investments in the Albertan oil sands, or tar sands, and the oil and gas projects offshore of Russia's most eastern province, Sakhalin Island in the northern Pacific. Only the briefest of mentions was made of Shell's purchase of drilling licences in the Chukchi Sea in the US Alaskan Arctic.

Since BP's discovery of oil beneath the Alaskan North Slope in 1969 it was widely assumed, in petroleum geology circles, that there would be similar resources offshore. In comparison to the drilling depths in the North Sea and the Gulf of Mexico, the distances between the seabed and the sea surface in the Chukchi Sea is relatively slight at 260 feet. However, drilling in the Arctic Ocean is a challenging prospect for the international oil industry. Between October and March there is almost no daylight. By New Year's Eve the sea can be completely frozen over and temperatures in the Arctic can drop to -30 degrees Celsius. During the summer months drilling rigs would need to be protected from the impact of icebergs and marine waves driven by strong winds.

Although the US Geological Survey estimated in 2008 that 13 per cent of the world's oil and 30 per cent of its gas reserves remained to be found and extracted from beneath the Arctic Ocean, it was these conditions and the extraordinarily high capital costs that had prevented oil companies from exploring the polar region for oil.

The exploration and production divisions at the heart of companies such as Shell and BP have always been dominated by engineers driven by the lure of responding to seemingly impossible technical challenges, such as drilling in the UK sector of the stormy North Sea. The fundamental myth of engineering is that given enough time and capital no problem is unsolvable. Throughout the 1990s, these divisions had been starved of resources. Now, hounded by the demands of asset managers at the institutional investors in London, the board of Shell chose to plough vast sums of capital into exploration. The delighted engineers got to work.

As a first step, licences in the Chukchi Sea were purchased from the US government in 2008 for a record-breaking $2.1 billion. Acreage for exploration was also acquired in the neighbouring Beaufort Sea. So desperate was Shell to be seen by investors to be getting in on the act that the corporation had paid massively over the odds for these licences.

The exploration engineers set to work with the research and design team in Rijswijk, five kilometres south of The Hague in the Netherlands and Shell's office in the Alaskan city of Anchorage was reopened.

A year later the first seismic testing vessels were hired from subcontracted firms. They steamed out of Valdez, passed Dutch Harbour, traversed the Bering Straits and entered the Chukchi Sea. The results of the seismic survey looked promising, the rocks beneath the acres of Arctic Ocean over which Shell had brought rights could hold billions of barrels worth of oil.

From the centrally heated comfort of the board room in The Hague, or the Aviva assets managers' office in London, the Arctic could seem as distant and as 'empty' as it had appeared to the board of BP in 1959 when Maurice Bridgeman agreed to purchase 613,000 acres of the Alaskan North Slope for exploration. But just as then, the Chukchi was *terra nullius* or rather *mare nullius*.

The disaster in the Gulf of Mexico, coupled with the profound convulsions of the financial crisis and its impact on the oil price, had reverberations even beyond the Arctic Circle.

On 31 December 2012 the *Kulluk* drilling rig that Shell had hired to explore in the Chukchi Sea broke its towing hawsers while being dragged south towards Seattle and was driven ashore in a storm. Shell tried to claim that the incident was unrelated to the exploration plans, but the event served to illustrate the treacherous nature of the Arctic seas and the seeming hollowness of Shell's reassurances to investors and the public.

The senior executives of Shell responded to the criticisms with Olympian self-confidence: the Arctic is merely an engineering challenge and, as with the 'insurmountable problems' of the North Sea, would be overcome. Once again, here is a territory for oil that could be utilised in the pursuit of higher, or rational, goals.

In the hushed halls of Tate Modern, Suzanne Dhaliwal, a part-British, part-Canadian campaigner working with indigenous communities against the oil industry, is explaining her decade of activism opposing the strip-mining tar sands operations in Alberta and offshore drilling in the Alaskan Arctic Ocean.

'This is a continuation of the violence of the patriarchy that started 500 years ago. We think that colonialism ended. But the process has just changed pace and become more vicious through the oil industry.' Dhaliwal has been a pivotal person in this struggle, the lynchpin between individuals from First Nations communities in Northern Alaska and civil society movements in Western Europe. 'When I've sat in those shareholder meetings, and listened to the way they play out, the amount of gaslighting, the amount of disrespect to indigenous women, that violence, that perception of the land. The oil corporations are fundamentally the new force of that violence. I don't think that, essentially, they can be transformed.'[17]

Opposition to Shell's drilling plans spread right across Canada and the US. In the mind of metropolitan America, Alaska had long been constructed as an empty wilderness, a place of pristine tundra, of polar bears, caribou and bowhead whales. Alaska was literally a place set apart, physically separate from the Lower 48 US states. None but a tiny fraction of American citizens had ever visited Alaska, let alone seen the Arctic Ocean, but it was vivid in imaginations fed by the pages of *National Geographic* magazines, wildlife documentaries and brochures for 'once in a lifetime' cruises. This vision of the frozen north as fundamentally unpeopled perfectly illustrates Dhaliwal's dissection of colonialism.

Catalysed in part by disasters such as the 1989 *Exxon Valdez* tanker grounding and oil spill in Alaskan waters, a great movement of NGOs and media built up in defence of this image of a place in the American mind. Despite the flaws in this fantasy, the threat of industrialisation mobilised opposition to Shell's exploration plans to drill in the Chukchi Sea. A swathe of international public opinion moved to support the campaign to halt any drilling in the Arctic Ocean, and particularly to stop the plans of Shell. The battle cry to prevent the extraction of fossil fuels was 'Keep it in the ground'.

For a culture and industry built on the inevitability of technological progress and the inherent virtue of territorial expansion, this public support represented a remarkable shift. For here was a campaign built on the underlying assumption that Shell's technology would not be failsafe and that the Arctic Ocean was a frontier to which the lust for energy should not go. The limits to growth were being set.

This quiet revolution in consciousness was fuelled in great part by BP's *Deepwater Horizon* disaster, which to Shell's misfortune came six years into their Alaskan project. The Obama administration's response to the civil society outrage at the BP blowout was to shut down all deep water drilling. Only after five months was this suspension lifted but the cloud from BP's recent and disastrous *Deepwater Horizon* oil spill in the US Gulf hung over Shell's plans.[18] The campaign against

Arctic oil constantly repeated the refrain: can Shell prove that this will not be another *Deepwater* disaster?

Greenpeace, Platform and Share Action utilised the shock that the Gulf of Mexico disaster had produced in the London investment community to drive home the possibility that Shell's Alaskan exploration could destroy the fragile ecosystems of the Arctic and destroy Shell itself. *Deepwater Horizon* had brought BP to the brink of bankruptcy, the Chukchi Sea threatened to do the same to Shell. The message, pushed home through several reports and endless meetings, proved remarkably successful.

The Shell investor relations team came under intense questioning, by asset managers from several major shareholders – including Aviva in London, APB in Amsterdam, CBIS in New York and Natixis in Paris. This not only pressured Shell to justify its investment plans in the Arctic but it also spread doubt about the wisdom of the oil corporations' financial decision making. It began to raise concerns about the long-term reliability of Shell shares as the foundation of any investment portfolio. The difficulties that Shell faced in the Alaskan Arctic threatened to erode the security of its position in the London finance sector. Events in the far-off corners of the empire of oil altered attitudes in the heartlands of the City of London.[19]

21 MAY 2013

The formal part of the AGM itself runs for four hours. As ever, it is a strange ritual process in which all parties act their role, with the directors of the company seated like an array of bishops up on high looking out over the congregation. On the front of the podium is the title, Royal Dutch Shell PLC, on the wall behind is the Shell logo, and above it a video screen that shows a close-up of the face of whichever of the directors is speaking. Most of those seated in the half-full auditorium are obedient adherents – generally pensioners and almost all Dutch. It seems the British presence at these events is inexorably dropping away. The archbishop – company Chairman Jorma Ollila – invites comments on the annual report and accounts. Some of those who have made the journey to this auditorium raise issues concerning the impact of Shell's business on the peoples and environments of the far corners of the Earth, from Canada to the Ukraine, from Nigeria to the US.

Ollila responds a little to each question and then passes it on to a trusted bishop – usually Peter Voser, the chief executive, but occasionally Simon Henry, the chief financial officer,[20] or Hans Wijers, chair of the remuneration committee which decides executive pay levels. Charles 'Chad' Holliday is asked to comment a couple of times as chair of the corporate and social responsibility committee. The remaining nine members of the board, from Sir Nigel Sheinwald GCMG, former UK ambassador to the US and Blair's foreign policy and defence adviser from 2003 to 2007 (the Iraq War years), to Gerrit Zalm, former minister of finance of the Netherlands, sit there in stony silence. They stare straight out at the audience. Impassive. Immovable keepers of the orthodoxy. Patient guardians not only of the shareholders' dividend, but also of a particular way of looking at the world, a way of thinking.

Each of the questions and statements from the floor seem to be carefully honed and considered. They are delivered not in the hope of getting some kind of meaningful answer from the podium, nor in the hope of achieving a miraculous change of direction by the company. No one expects Zalm or Wijers, Holliday or Sheinwald, to have a Damascene moment. Mostly the responses from Ollila, Voser or Henry are equally honed and considered. Answers that have been carefully prepared by the public relations team to ensure they deliver the appropriate amounts of dismissal or denial, emollience or gratitude – ideally without creating any hostages to fortune.

Only occasionally does a statement disrupt the ritual exchange. Louise Rouse from Share Action, a charity promoting responsible investment, questions the company's oversight of contractors. Voser responds that they've been meeting the challenge of overseeing the nearly one million, direct and indirect, employees by the means of 'supplier principles', 'performance monitoring' and 'enterprise framework agreements'. But Rouse claims that these schemes, which Voser says have been in place since 2009, failed to address the fundamental flaws in contractor oversight that was revealed by the fiasco of Shell's attempts to drill in the Alaskan Chukchi Sea in 2012, in particular the grounding of the *Kulluk* drilling rig. Rouse's critique is warmly supported by institutional investors in the audience, who express strong concerns over Shell's plans.

Mae Hank of Point Hope, Alaska, and a representative of Resisting Environmental Destruction on Indigenous Lands, takes the floor. She is here on account of the work of solidarity carried out by Suzanne Dhaliwal. With her solid frame wrapped in a light blue crocheted shawl Hank leans into the microphone, and after a few words in English she begins to speak in her own Inupiat language. Seconds later she has created a ripple across the room – there are murmurs from the rows of Dutch pensioners in the red velvet seats, and there are shrugs and smirks among the men on the podium. Hans Wijers, a non-executive director and former Dutch politician, begins to laugh and exchanges a comment with his neighbour, Michiel Brandjes, the company secretary. It is an uncomfortable moment – as though the far distant provinces of the company had suddenly been brought into this Western European theatre. Ollila's voice booms out over the loud-speaker system: 'What is your question?' Hank continues calmly:

> We all share the same food out of the ocean ... We are terrified that our food is threatened by the offshore drilling ... You are coming into our garden – the ocean is like a garden to us ... The ocean is covered with ice for six to eight months a year and Shell does not have any proof that they can safely contain an oil spill during the winter months amidst the strong currents and powerful storms ... How will Shell compensate us for any spill that kills our food? How will they compensate 20 generations, to keep them going through the winter?

After the formal meeting has finished, what remains of the attendees spill out into the foyer to devour the sandwiches and pastries. It is a chance to try and engage a director in a one-to-one conversation. With Hank we catch Simon Henry. She repeats her questions: 'How will Shell compensate 20 generations?' Henry says: 'You explained that the ocean is your garden, and we can't compensate you for the loss of your garden. But I repeat that Shell is taking a very prudent approach – there will not be any accidents.' Hank is quietly insistent: 'That is what you say – but we saw what happened when the *Kulluk* rig went aground in January and we are frightened – how will you compensate 20 generations?' Gradually Henry becomes testier: 'The *Kulluk* was nothing to do with the drilling, it was purely a maritime incident. As for compensation, we will abide by all the US federal and Alaskan state laws set down on these matters.'

Hank explains that for her a generation is a period of 50 years. We ask Henry, 'What is the furthest date that your plans run to, that your window of financial projections stretches to?' He replies, 'In the oil sands. To 2050.' We try to engage him on the contrast between his sense of time and Hank's, that he is thinking about 50 years, about one generation, and she about 20 generations. Henry responds: 'This is a philosophical question. The world's population is growing and energy demand will grow with it, we have to help meet that demand and that is why we need to explore for resources in the Arctic.' Hank gently insists that this oil project is done not for world development but for the company's profits. Henry looks puzzled: 'But a profitable company is a good company – one that can invest in the future.'

After the exchange has ended we are struck by the sense of two fundamentally contrasting views of the world – both in terms of place and in terms of time. Henry is seven weeks off his 52nd birthday. He had started employment in Shell at 21, as a refinery engineer at Stanlow on the Mersey. In the intervening years he'd worked in a wide range of places around the world, from Egypt to Vietnam. When he was only 17 Hank started fighting against the despoilment of her home by the oil industry. For nearly four decades she's remained in the same place and struggled to defend the land and ocean that has belonged to her community, and to which her community has belonged, for thousands of years.

Henry meets the question of compensation by saying Shell would abide by the laws, though of course Shell has far greater power in framing those US-wide laws than the community of Point Hope, so distant from Washington. Shell sees the possibility of compensation as, at worst, stretching across a decade, as was the experience of BP on the US Gulf Coast. But Hank, standing face to face with him, sees it in different terms. For her 20 generations is 1,000 years. Hank believes her community would either be destroyed by the oil industry or might live with the consequences of its impacts far into the future. Henry is set to retire from Shell within a decade, at which point full-scale oil production in the Chukchi Sea might have only just begun. He will walk away. For him the issue of future generations is a philosophical question. For Hank it is a visceral matter of survival.

There is a saying from Colombia that comes to mind: 'With patience and saliva the ants devour the elephant.' With passion and courage Hank and others are overturning the understanding of the world that lies behind Simon Henry's responses, the orthodoxy that the podium represents, the colonialism that Dhaliwal had described.

27 AUGUST 2015

The rain lashes down. A strong south-westerly drives it along the Thames, across Jubilee Gardens in London and against the railway viaduct that separates the Royal Festival Hall from the Shell Centre. The crowd huddles together on the wide pavement beneath the arch. Across the narrow roadway is the orchestra performing the four pieces of the *Requiem for Arctic Ice*. Beyond the musicians, slightly hidden from view by the crush of damp heads and backs wrapped in anoraks, is Charlotte Church – seated, awaiting her moment.

In early spring 2015 it had seemed that Ben van Beurden, CEO of Shell, might announce a cessation of the company's Arctic plans. There was a sense of wavering. But later he reaffirmed to investors plans to drill again in the Chukchi Sea in July and August 2015. The movement to stop Shell had geared itself up to fever pitch with 'kay-aktivists' trying to block the passage of the drilling rig.

At the height of the global campaign we attend an event in London. We cram under a railway arch, avoiding the pouring August rain, listening to the strains of cellos, violins, viola and double bass. Now and again there's the rumble of a train pulling slowly overhead. At frequent intervals the audience's view of the seated sextet is blocked by a passing single-decker red bus, a white delivery van or a refuse truck. Despite the distractions, the orchestra plays on, echoing what we read on the leaflets that we are handed, the famed musicians on the deck of the *Titanic*.

The front row of the audience is bristling with press photographers and TV camera teams – they're here for Charlotte Church and because of the labours of the Greenpeace press office. Bill Oddie has also come to lend his visible support to the cause. In the distance the head of the Greenpeace media team is chatting with Jon Snow. Channel 4 is clearly creating a package about the event.

The final bars of the fourth of the *Requiem* pieces ends. There's a pause and the cameras ready for action as Charlotte Church rises to sing:

> This bitter earth
> Well, what a fruit it bears
> What good is love
> Mmm, that no one shares.

We stare out of the archway. Sheets of rain are moving across the empty lawns of Jubilee Gardens, in the distance is a line of brightly coloured cagoules, tourists queuing for the London Eye. Nearer too, the chairs of the funfair whirligig spin in the air, flags fluttering. A policewoman pushes back the press to let a motorbike through the arch. It is so far from the imagined Arctic, yet we are moved to tears.

> And if my life is like the dust
> Ooh, that hides the glow of a rose
> What good am I
> Heaven only knows.

At the opening of the event the artistic director, Mel Evans, and campaigner Anna Jones, had spelled out the rationale for the *Requiem for Arctic Ice*.[21] While Shell drills for oil off Alaska, the team of composers, musicians and facilitators have been holding, and will hold, daily performances of the four pieces outside the front doors of the Shell Centre. Only on account of the appalling weather and the size of the crowd has today's event been moved 20 yards east to the shelter of the railway arch.

> Lord, this bitter earth
> Yes, can be so cold
> Today you're young
> Too soon, you're old.

One of the composers, Chris Garrard, has been a stalwart of the Shell Out Sounds group that has used live performances, both outside and inside the auditoria of the South Bank Centre, to draw attention to the fact that Shell sponsored a classical music programme. The campaign has been remarkably successful. Within a short few years the company and the South Bank Centre ceased the sponsorship contract in January 2015. Today's performance of the *Requiem*, and those over the past three weeks, are an evolution of the Shell Out Sounds actions. Here is a professional sextet and a world-renowned singer using music as a weapon of protest against the oil industry. But this is not about driving Shell out of the South Bank Centre and other cultural institutions, but about driving them out of the Arctic.

> But while a voice within me cries
> I'm sure someone may answer my call
> And this bitter earth
> Ooh, may not, oh, be so bitter after all.

After the applause that closes the performance we head out of the arch into the rain, which thankfully has died down a little. Around the other side of the Shell Centre, on Belvedere Road, the building is covered in scaffolding and looking up we can see that on many of the floors the windows in the smooth wall of Portland Stone are without glass. Blind eyes in the cold face of rock. The main tower is under reconstruction. This edifice, and the block that lies beyond the railway viaduct behind the Royal Festival Hall, were once owned by Shell, parts of the largest office complex in Europe.[22] Now the entire eastern wing has been sold off, turned into luxury apartments in the 1990s. The main tower itself was bought by Qatari Diar in 2011: Shell now only uses a small part of the office space that it did in the 1960s. The refurbishment is part of the new owner's efforts to attract tenants.

The company is slowly vanishing from public gaze – first with the closing of Shell-Mex House on the Strand and now with the shrinking of its presence on the South Bank. Perhaps in a few years the Shell flag will cease to fly over the Shell Centre Tower, just as the Shell logo has ceased to emblazon the Southbank Centre's season of classical music.

It seems likely that none among the crowd huddled under the railway arch have ever been to Alaska, let alone seen the vast drilling rig that is right now labouring over the horizon of the Arctic Ocean. Yet the strings of the sextet and the voice of the singer performed a great function of art – they rendered the invisible visible. Music that, from Vince Taylor and the Beatles onwards, helped form an oil culture, is now being turned against that same industry.

28 SEPTEMBER 2015

Marvin Odum, director of Shell Upstream Americas, declares, 'Shell will now cease further exploration activity in offshore Alaska for the foreseeable future.'[23] Eleven years after Jeroen van der Veer fixed upon the Alaskan Arctic as part of Shell's bright future, the $7 billion venture has come to naught.

On this day of the announcement the news ripples through the City of London and the Shell neighbourhood of The Hague. Invariably decisions by the boards of corporations to pursue this project or choose that strategy are presented to staff, investors and any interested public as part of a calmly unfolding plan towards greater profitability. But here is a decision that Shell can not portray as anything other than a *volte-face*. The company had pushed relentlessly for this goal, devoted capital to persuading others that it was a vital investment, and now it is having to admit that it is coming away empty-handed. It brings some satisfaction to those like Terry, who had campaigned for years at the *Guardian* to bring an end to the threat of environmental damage and military confrontation in the Arctic.[24]

A few months later the head of Shell in the US, Marvin Odum, left his job. A little more than a year on the head of group finance, Simon Henry, was gone – although soon after he was awarded a UK government post. And there were wider strategic implications. BP and Shell had long depended upon the geopolitical weight of the UK and the Netherlands to support its investments in the US. Between 1959 and 1966, BP is understood to have been assisted by the Macmillan and the Wilson governments as CEO Maurice Bridgeman invested $6 million in the purchase of 613,000 acres of the Alaskan North Slope. After all, during this time BP was 51 per cent owned by the British government.

This was UK state money. The Cameron government had been called in by BP to defend it against the attacks of the Obama administration in the summer of 2010 during the *Deepwater Horizon* crisis.

Now the combined weight of Shell's Dutch and British support and its influence in Washington had been unable to establish a political climate that was sufficiently supportive of Arctic drilling to ensure the security of the capital investment. The international movement against oil exploration had been strong enough to undermine political support for the project. It emphasised the declining power of the oil companies.

Pressured by the environmental movement and the First Nations, Obama's administration decreed that no further licences for drilling in the Chukchi Sea would be issued and later a total ban on Arctic oil exploration was announced in December 2016.

It was widely interpreted that part of the rationale of Obama's decision was as an act of defiance against the then president elect, Donald Trump. Only six weeks earlier Trump had won the White House with his extraordinary and largely unexpected victory believed to have been partly bankrolled by the oil and gas industry. Trump had declared on his campaign trail that he would overturn any restrictions on offshore drilling, especially in the Arctic. True to his word, on 28 April 2017 an executive order was signed to have the ban lifted.

Despite Trump's rolling back of the Obama administration's presidential decrees, the defeat of Shell in the Arctic seemed to presage a more structural shift in relation to high-capital-cost, high-risk projects. In contrast to the decision of the Trump administration, the French government under Emmanuel Macron, driven by the pressure of civil society, declared a moratorium on oil and gas exploration in all French territory on 24 June 2017.[25] Mainland France has no real oil prospects and only marginal gas potential, but the overseas departments such as French Guiana have existing hydrocarbon production and prospects highly attractive to the industry that could have been exploited by Shell or BP. Far more than this, here was the symbol of the government of a G7 state turning its back on oil and gas.

In parallel the global campaign against climate change took heart from the win in the Alaskan Arctic. Since 2011 ecological activists and aboriginal groups had been fighting against the exploitation of the Great Australian Bight. BP purchased licences there, and for five years progress was slow but solid until it met resistance from the government of the state of South Australia, catalysed by a civil society campaign. Australia's National Offshore Petroleum Safety and Environmental Management Authority called in the plans and the project ground on more slowly. Meanwhile an international campaign, which involved Greenpeace, built up pressure. Eventually BP pulled out of the project and wrote off the costs.[26] This announcement came barely twelve months after Shell stopped drilling in the Arctic Ocean.

Shell's advance into offshore Arctic drilling had been part of then chief executive Jeroen van der Veer's drive for new reserves in 2004. The Alaskan adventure had been mirrored by the corporation making substantial investments in the Canadian tar sands. Between 2004 and 2017[27] Shell invested billions in projects to mine this heavy bituminous substance from beneath the forests of Alberta. Now, in an echo of the Alaskan defeat, van Beurden announced in March 2017 that Shell would sell off $8.5 billion of its tar sands assets. Clearly a withdrawal from high-cost, long-term projects like the tar sands was underway. A wider strategic shift was playing out for Shell and BP likewise.

Shell's abandonment of Alaskan offshore drilling was not only about this one corporation shifting plans, but more importantly about the future of a whole set of regions. For Shell's Chukchi Sea project was widely seen as the beacon that would light the way for oil and gas exploration across the whole Arctic. If Shell could demonstrate profitable extraction of hydrocarbons from the polar seabed then other companies could follow in their wake. The Chukchi Sea, the neighbouring Beaufort Sea, the Arctic waters around Greenland, Norway and all along Russia's northern coast would rapidly be exploited. The massive spaces of ice and tundra would be quartered out with production platforms, pipelines, onshore gas and oil terminals, helicopter bases, airstrips, ports and access roads. The unimaginably vast space of the Northern tenth of the globe would be occupied by industrial capital in a manner similar to the North Sea, and at a rate that would accelerate as the impact of global warming reduced the area of the polar region that is icebound for most of the year.

However, the reverse may also be true. As Shell was unable to proceed, with its mountains of capital, its phenomenal political muscle and its engineering prowess, then shareholders in other private companies could lose heart about such high-risk ventures.

Van der Veer had hailed the Albertan tar sands and the Arctic frontier as central to Shell's future strategy, with production in each of these provinces stretching out to the 2050s. Canada and the US were assumed to be intensely stable political entities highly favourable to capital. Van Beurden's abandonment of the Arctic and the tar sands not only affects the future of those projects and those regions, but also of the corporation itself. If part of that future was relinquished what then was the new future of Shell?[28]

International civil society, by undermining governmental support for Shell's plans, was placing limits on the expansion of the territory of oil and gas. As this expansion has historically been linked to the profitability of the oil corporations, then these limits impacted on the financial pillars of the UK economy.

12 DECEMBER 2015

I wanna see this world, I wanna see it boil
I wanna see this world, I wanna see it boil
It's only four degrees, it's only four degrees
It's only four degrees, it's only four degrees
I wanna hear the dogs crying for water
I wanna see fish go belly-up in the sea
All those lemurs and all those tiny creatures
I wanna see them burn, it's only four degrees.

Anohni, '4 Degrees', 2015

Anohni released her track '4 Degrees' on the eve of the UNFCCC Climate Summit on 12 December 2015 in Paris. The summit had for several years been trailed as the most pivotal convention since that held in Copenhagen six years previously which was widely understood to have failed.

In the lead-up to the Paris summit there was a mobilisation at every level of society. Government departments and ministers honed their proposals, corporate PR teams and CEOs prepared their pitches, and civil society strategised on how to take to the streets and ensure that any agreement was not merely empty rhetoric with targets set far too high. At the heart of the corporate lobbying lay the requirement that although climate change must of course be taken seriously it should not impact on the global production of oil and gas. Civil society, as expressed through NGOs, activist groups and thousands of citizens who joined the marches, demanded that red lines be drawn and that there be a rapid wind down of carbon extraction abiding by the rallying call to 'Keep it in the ground'. In the end the Paris Agreement, though weaker than the hopes of civil society campaigners, was crucially negotiated by 196 states[29] who set a target to keep average global temperatures to a maximum of two degrees above pre-industrial levels, with a commitment to strive to keep it below 1.5 degrees of warming. This changed the landscape within which the oil companies were operating and forced them to at least pay lip service to 'Paris'.

We arrange to meet one of the most dogged campaigners around the 25-year-long UNFCCC process. Jeremy Leggett[30] has a footballer's frame and a shock of grey hair sitting on top of a lined but still youthful face. The 65 year old is dressed in black: dark shirt, jumper and padded jacket. We are at the Royal School of Mines, housed within Imperial College London, the university in South Kensington.

Leggett sweeps past the security desk with the air of someone who belongs: he lectured here 40 years previously after completing a geology DPhil at Oxford University. At the time he won work as a consultant for Shell, BP and other oil companies. His later disenchantment with academic life, plus a realisation about the threat of climate change, meant he left to join Greenpeace International as their chief scientist.

We have come partly to see whether the former Royal School of Mines is moving with the times and has left behind its carbon-heavy past. We have been told that courses on oil and gas elsewhere are struggling to find students. Up the grand stairway, all brass handrails, grey slab treads and white-tiled walls, and into the corridors Leggett reflects on this familiar space. He swears that during his teaching years the lecture rooms were branded with the names of Shell and other corporations, and there was also a model drilling rig on the main landing. Those were the boom years of the UK North Sea. Students came primarily to prepare for work in the new offshore colony.

The famous model oil rig is still there, but it's now shunted into a corner. A young academic we meet says companies such as Texaco are still funding research, but overall budgets are tight. There is a pattern emerging across the UK: local student demand for these training schools for the oil and gas industry is dwindling. Here too it seems the body of British society is turning away from the oil industry, and that in itself is forcing the industry to change shape.

As we mooch around the empty corridors a door suddenly opens. An impeccably dressed elderly man appears clutching a felt homburg and dark mac. Leggett immediately recognises him as Professor Dennis Buchanan and introduces himself.[31] Buchanan smiles broadly: 'I retired 22 years ago at the age of 50, but as you see I am still here.' Leggett probes him about the 'energy transition' and its impact on the Royal School of Mines. 'We are more in demand than ever', says the Emeritus Professor of Mining Geology. 'Electric cars are all about copper. You were ahead of your time', he laughs.

We settle down in a café. Leggett talks of his personal connections with key oil executives such as the former BP CEO, Tony Hayward:

> We were mates when he was at Edinburgh University and I was at Oxford. We were involved in the British Sedimentological Research Group, part of the Geological Society. We sank many pints together, but we grew apart. I followed the *Deepwater Horizon* episode with interest and cringed when I saw Tony's comments: 'I'd like my life back.' Amazing! And then he pops up as chairman of the mining group Glencore, and I hear him arguing the case for coal. That was not good for my blood pressure.[32]

Leggett had been central to Greenpeace's campaigning against climate change. In September 1990 the group published *Global Warming: The Greenpeace Report*, edited by Leggett.[33] This was a foundation stone for seven years of work, much of it centred around the UNFCCC, the body behind the Paris Agreement. The intergovernmental process achieved its initial florescence with the Earth Summit at Rio de Janeiro in June 1992 and a key pivot with the Kyoto Protocol drawn up at the Conference of the Parties (COP) in December 1997. Leggett wrote a powerful eyewitness account of the climate negotiations in the 1990s,

The Carbon War: Global Warming and the End of the Oil Era.[34] It graphically portrayed the way the fossil fuel corporations tried to derail the international talks.

Leggett explains he was sympathetic to John Browne, CEO of BP. 'I used to be an admirer. I think he really did have a go with the "Beyond Petroleum" strategy, but it didn't work.' He explains how in 2015, during the run-up to the Paris climate talks, he was talking with Browne, 'in a hotel bar in Portugal, late at night. We both were drinking from a nice bottle of red wine. I told him that he could write his place in history and achieve remarkable things for the planet if he renounced oil.' He continues, saying that Browne 'could have driven a wedge between the oil companies willing to follow a green agenda and those who weren't. He said my views were "very interesting" and he would go away and think about it. But then nothing happened.' Soon after Browne wrote an op-ed piece in the *Financial Times* defending the industry. 'I lost it with him then. I wrote a blog saying it was bullshit. I got an email from his *chef de cabinet* saying how deeply upset Browne was. He didn't even contact me himself.'[35]

Following Kyoto, Leggett dedicated himself to being a renewable energy entrepreneur, mainly through the company Solar Century. He is a passionate advocate for this transformative technology. When we'd met 18 months prior to Imperial College he'd enthused, 'There really is a possibility of a full decarbonisation of the world by 2030, given the exponential growth curve displayed by solar and batteries. Obviously they would have to be combined with electric vehicles, biofuels and mini grids, but it could be done. In fact it pretty much must be done by 2050 to fit the Paris climate agreement.' He made a point of emphasising, 'That's not just my view but that of others too, such as Ray Kurzweil, director of engineering at Google.' He stressed the need for other technologies. 'All this is being researched as we speak. EasyJet is talking of electric planes flying short haul services within ten years and Södra – a Swedish paper manufacturer – is expecting to be carbon negative by 2030 through planting three trees for every one it cuts down and running its logging trucks and ships without fossil fuels.'

What Leggett describes is a coming world, which sidelines the oil and gas companies. It is a world evolving in large part as a consequence of civil society action driving governments to create the international

Paris Agreement which is effectively placing limits on the expansion of the realm of oil.

14 APRIL 2016

A pale Bob Dudley, the BP chief executive, looks up from his notes and addresses a number of aggressive questions that have been put to the board. He peers out at a sea of grey upholstered chairs and grey-headed pensioners at the three-quarters empty hangar-sized hall of the Excel Centre, in London's Docklands. This is the 2016 AGM and it is a difficult time. BP has run up a $6.5 billion annual loss – partly from paying out for the *Deepwater Horizon* disaster – but has also antagonised some City shareholders by awarding Dudley a 20 per cent pay rise to a salary of nearly £14 million a year.

He concentrates his opening speech on the projects BP is advancing around the world before turning to the North Sea:

> Back here in the UK we remain committed to the country's energy needs. Our Clair Ridge project west of Shetland is part of an ongoing £10 billion investment programme in the UK North Sea. The two Clair Ridge platforms are nearing completion and will start producing the first of an estimated 640 million barrels of oil towards the end of next year. They will all be starting up in the near term. They will go on producing energy for the long term – decades in fact. And that means they will be generating value for shareholders for decades to come.

Dudley carefully outlines BP's responsibility to provide heat, light and transport for the two billion people that will eventually be added to the global population, and the company's measured shift away from oil towards natural gas. He explains that BP can only move as fast as the wider policy environment allows and he speaks of his earnest desire for this to be facilitated by government policymakers.

But the questions keep coming. One by one speakers stand before the mic and address the 14 directors on the distant podium. Louise Rouse and Charlie Kronick of Greenpeace UK on BP's Canadian tar sands projects; Lyndon Schneiders of the Wilderness Society of

Australia and Michael Thornton of Carbon Analytics highlight the threat to the seas of the Great Australian Bight by the company's drilling plans; Jags Walia of APG Investments and Abigail Herron of Aviva press the board to keep to their climate commitments; while Chris Garrard of Art Not Oil quizzes BP's motivations in sponsoring the British Museum.[36]

In his opening CEO speech Dudley refers to the 'historic Paris Agreement' in December 2015, and the way that BP is addressing climate change. James picks up on this when he gets his chance to represent Platform at the microphone.

In the light of the Paris Agreement it's widely recognised that two-thirds of fossil fuels will need to be left in the ground. There is a shift underway, illustrated by the world's largest coal company, Peabody, filing for bankruptcy yesterday. Mr Dudley, you spoke of how BP had a long history of adapting to substantial changes, such as the national-isation of oil in the 1970s by the Arab states. Why are you now not truly adapting to the changes demanded of you? Why do you, Bernard Looney, chief executive, Upstream, Lemar McKay, deputy CEO, Brian Gilvary, chief financial officer and others continue to sanction projects that will bring more carbon into the atmosphere?

He continues:

In your address earlier, you talked about safety. In BP it is empha-sised that safety is a personal responsibility, that individuals are held accountable. After a disaster, such as *Deepwater Horizon*, rightly great efforts are made to determine which particular individuals were to blame for the accident. Now you yourself and your immediate colleagues are making decisions that undermine the safety of the Earth's climate. My question is: what systems are in place to ensure that individuals, not just the collective of the company as a whole, but particular individuals are held personally responsible for the actions that they take in sanctioning new oil fields?

Dudley's reply is very direct: 'You seem to come from the premise that we are guilty already.'

The sea area known as West of Shetland, off the northern coast of Scotland, had been a target area for oil drilling since the late 1970s. Oil was discovered there by BP in the Foinaven Field in 1990 and the Schiehallion Field in 1993. Foinaven started producing oil in 1997 and Schiehallion in 1998. These cold seas that had long been held in the human imagination as a place of birds and fish, ships and stories, were turned into a petroleum asset. They became the subject of presentations such as Dudley's AGM address. But the right to drill into the rocks beneath this seabed was contested by environmentalists.

So deep are the waters in West of Shetland that in order to exploit the oil BP ordered two special floating production, storage and offloading vessels (FPSOs) – the *Schiehallion* and the *Petrojarl Foinaven*. The latter was created by converting a former oil tanker in the shipyards of Ferrol in Galicia, Spain, that, on 21 October 1996, was launched by Paula Browne, the mother of then CEO of BP, John Browne. Greenpeace attempted to block the ship leaving the Spanish port and to harry her progress up the Channel.

In 1997, the following year, the oil companies were undertaking seismic work on exploration blocks in the West of Shetland and the Rockall sea areas. That May, Greenpeace UK showed an advert in cinemas promoting their Atlantic Frontier initiative to prevent drilling in West of Shetland. Meanwhile Greenpeace International joined the campaign by linking climate change to oil exploration at the frontiers in Alaska, Australia and the Atlantic.

On 10 June 1997 two Greenpeace activists, Pete Morris and Robbie Kelman, began a 42-day occupation of Rockall in a survival pod, declaring it the new global nation of Waveland. A month later Greenpeace boats were in action attempting to disrupt the seismic work of *Atlantic Explorer*. In August activists managed to fix the survival pod, recently removed from Rockall, to the anchor chain of the *Stena Dee* drilling rig being hired by BP for the West of Shetland exploration. BP took out a £1.4 million damages claim against Greenpeace on account of its occupation of the rig. This looked like an attempt to bankrupt the organisation, but the corporation backed down following a public outcry.

All of these events took place in the summer before the UN climate talks in Kyoto in Japan. It was this COP that adopted the Kyoto

Protocol on 11 December 1997. At the time that seemed to be a major step forward in the fight against global warming.

The exploitation of West of Shetland continued. In April 2015, BP, in partnership with Shell and Austrian-based oil company OMV, announced that it would begin a seven-year drilling campaign to open up new wells in the Schiehallion and connected Loyal field, with the aim of extracting oil there until 2035 and beyond. In December 2015 a new FPSO, the *Glen Lyon*, built in South Korea, was towed across the Indian Ocean, bound for Norway and then West of Shetland. Meanwhile at the Clair Field, also in West of Shetland, work was well underway on BP's project that aimed to be producing oil up to 2050. It was progress on this that Dudley announced at the 2016 AGM.

The discovery and exploitation of the fields in West of Shetland took place in the full knowledge that the burning of oil, and therefore its extraction, would affect the climate. Indeed on 19 May 1997 John Browne, CEO of BP, had made his groundbreaking speech at Stanford University in California in which he publicly recognised climate change, and was lauded as the first senior executive in the oil industry to do so. Three days later Browne said, 'If we are all to take responsibility for the future of our planet, then it falls to us to begin to take precautionary action now.'

There have been CO_2 emissions from the burning of crude extracted from the West of Shetland fields since 1997. And it is estimated that the carbon from them will remain in the atmosphere for a century, until 2097.[37] So who takes responsibility for the CO_2 emissions from Foinaven, Schiehallion and the rest of the North Sea fields?

Campaigners say the destruction of the Earth's atmosphere through the extraction of fossil fuels for profit is an act of criminality. As in all bureaucratic crimes, there is in the fossil fuel industry a diffusion of responsibility, but this does not lessen the personal responsibility. It seems likely that the criminality of knowingly extracting fossil fuels will at some point be formally established. This illustrates the importance of the work of a number of climate change cases being brought to court around the world. This lies at the heart of the #ExxonKnew campaign[38] and the case against RWE in a court in Germany, as well as in the work of the Scottish barrister Polly Higgins to realise an ecocide law.[39]

If criminality is established, there will be a necessity to determine at what point 'people knew they were involved in an ecological crime', and either actively or passively ignored this reality. Perhaps that point will be 1988 when the World Conference on the Changing Atmosphere was held in Toronto. Or 1990 when the *Global Warming Report*, edited by Leggett, was published. Perhaps it will be 1992 at the Rio Earth Summit. Or 1995 at the first COP in Berlin. Or 1997 at the adoption of the Kyoto Protocol. Or 2015 at the Paris Agreement. Or at the moment when the principle of criminality is established.

* * *

Among the attendees at the AGM in the Excel Centre are a handful of asset manager representatives from key institutional shareholders such as the pension funds. These bodies hold significant percentages of BP's total share issues, and therefore if they can effectively marshal their voices they can pressure the board on issues such as climate change. Their power lies in the importance of the pensions system in the UK's finance sector.

From the Old Age Pensions Act of 1908, through to the National Insurance Act of 1946, to the Occupational Pensions schemes in the 2010s, the structures to ensure provision after retirement have become fundamental elements in the financial systems of modern Britain. Pension funds by their very nature are cautious investors. Their managers have striven to place capital in assets that give reliable, rather than spectacular, returns. Consequently these funds invested heavily in land, government bonds and blue chip stocks. For the pensions funds, the oil companies were seen as reliable sources of revenue.

The century-long lifespan of these institutions has meant they have amassed vast sums of capital. Indeed it was said that by the 1980s, when the Conservative government began its drive to close it down, the National Coal Board had become what some described as a pension fund with a mining subsidiary. For the real value of that state-owned industry lay in its vast superannuation scheme.

Many pension schemes are geared to the individual policyholder, but some are aimed at the institution for which an individual works. Thus a miner employed by the National Coal Board, whose pension was in the Coal Board Superannuation Scheme, saw the cash pile as something that belonged to the labour force.[40] The sense of collectivity came alongside a desire for greater democratic control. Indeed, the right of mine workers to have some influence over their pension scheme became an issue in the lead-up to the 1984–5 miners' strike.

Democratic involvement in pension schemes raised its head as an issue once again 30 years later. In 2012 Bill McKibben, the director of 350.org, an NGO working to prevent climate change, launched an initiative to persuade institutional investors to sell their shareholdings in oil, gas and coal companies.[41] It was a campaign to persuade asset managers to divest from fossil fuel stocks, on the principle that it was both ecologically unsustainable and socially unethical to profit from the destruction of human habitability on Earth. Initially McKibben, and other key activists and writers such as Naomi Klein, hoped that a few private colleges and some churches in the US could be persuaded to divest and provide at least a symbolic representation of the principle that all such institutions should stop financing fossil fuels.

However, the snowball rolled. Within five years a host of universities, faith groups, insurance companies and pension schemes had declared that they would go fossil free. In the US a pivotal moment came on 11 January 2018 when Mayor de Blasio announced that New York would divest its city pension funds, valued at $200 billon, from the oil, gas and coal industries within five years.[42] By September of that year another group of investors said they would divest more than $6 trillion from fossil fuels.[43] Meanwhile the world's largest sovereign wealth fund belonging to Norway announced plans to sell $13 billion worth of oil, gas and coal investments. This had particular symbolism because the $1 trillion fund was created on the back of Norway's oil and gas wealth.[44]

These shifts abroad were mirrored in the UK, where a number of local authorities, under intense pressure from their citizens, chose to divest. In 22 June 2017 UNISON, one of the UK's largest trades unions, representing a substantial proportion of local authority workers, declared their support for the divestment movement.[45]

In a café beneath his office in Bermondsey we meet Mark Campanale. He has spent his life in the City of London, pioneering the field of sustainable finance by designing systems for investors to make money without impacting the planet. He was a co-founder of the Jupiter Ecology Fund in 1989, the first unit trust that strove not to hold shares in companies with environmentally damaging practices. Together with Nick Robins and others he established dedicated funds at Henderson Global Investors where investors could be sure that none of their money would go to oil and gas companies.

In the mid-2000s Campanale and Robins began to develop the idea of the carbon bubble. The spectre of investor bubbles has haunted the world of finance since the development of modern capital markets in the seventeenth century. The threat lies in individuals finding themselves holding shares in companies operating in a sector that is suddenly revealed to be vastly overvalued. The shares become next to worthless and the investor crashes into ruin. The most infamous such crisis was the South Sea bubble of 1720, but far more recent was the dot com bubble. From the mid-1990s speculative money had been pouring into internet based companies, such as pets.com and WorldCom. Throughout 2000 a series of events shook faith in the sector. By November 2002 pets.com had gone bankrupt and $1.75 trillion had been wiped off the value of internet company shares. Most famous of the dot com casualties on the London Stock Exchange was lastminute.com.

The carbon bubble presents the possibility – even probability – that at some point due to climate change there will be a wholesale switch away from fossil fuels. Companies who put such intense effort into acquiring or discovering oil and gas fields to exploit at a future date will then discover that these assets have become untenable, that the crude in the rocks has no market and the rights to extract them cannot be sold at a profit to some other corporation. The oil fields would, in the terminology of finance, become 'stranded assets'. Hydrocarbon leviathans, purchased for billions of dollars, would quite suddenly become like whales dying on the beach as the tide of profit ebbs away. At this juncture shareholders in corporations even as well established as BP and Shell could be holding near-worthless stock and face profound losses. The threat of this possibility, Robins and Campanale argued, should encourage long-term institutional investors, such as pension funds, to wind down their holdings in fossil fuel companies.

In November 2011 the duo formed the group Carbon Tracker Initiative with solar power entrepreneur Jeremy Leggett as chair, and published *Unburnable Carbon: Are the World's Financial Markets*

Carrying a Carbon Bubble? The effect was electric. Their findings quickly gained traction in the business media and the City of London – partly on account of Robins' and Campanale's strong reputations built up over three decades. The report established Carbon Tracker as a group dedicated to developing the research and driving home its message with asset managers in London and internationally.

Eight years on, Campanale has seen the idea behind the carbon bubble become all but accepted in the world's finance markets. But the historic legacy – both financial and cultural – of fossil fuels makes it a challenge for City investors to pull the plug. He explains to us:

> The dependability of BP and Shell shares is becoming weaker, but the difficulties of moving equity portfolios out of fossil fuels is not to be underestimated. In London the inertia is due to the City being historically wedded to the oil companies. The shift is taking place far more rapidly among French, Dutch and Swiss investors.

Campanale explores some contradictions:

> The return on allocated capital in the oil corporations is falling. So they are paying dividends out of debt and this is increasing their levels of indebtedness. But their share price has gone up, which defies logic. We try and persuade asset managers, but they reply, 'What do I say to the companies when they say oil and gas demand will inevitably grow over next 20 to 25 years and with it our share value?'

He explains how the companies are arguing that the price of oil, which at that time was low and making massive projects unprofitable, would inevitably rise. 'We have to show that the change in the fossil fuel world is structural and not cyclical'; that the low oil price will not inevitably be followed by a high price, although this was the trend in past decades. 'Actually the oil corporations do see this. That's why Shell and BP have been pulling out of Arctic projects and the tar sands in Alberta.' He concludes, 'It's a huge game of chicken, the carbon bubble. Investors don't want to believe the worst can happen. It is like those that stayed on in Nazi Germany after *Kristallnacht* in 1938. Recently I did a presentation in Rotterdam. There were 200 people in the audience. Gripped. It was amazing.' Here is the muscle of the finance sector being used to shape the oil industry rather than the other way around, as happened so potently in the 1970s and 1980s.

The logic of stranded assets feeds directly into the divestment movement, providing intellectual backing to the call on public bodies to sell their shares in oil and gas companies. Not only because of climate change but also on account of their fiduciary duty, their legal requirement to protect those that depend upon them, such as future pensioners.[46]

The great parental corporations are being spurned by the younger generation behind the divestment campaign. These actions are a statement of a lack of trust. Trust is key element in capitalism, just as the key to credit is credo. The companies have for so long asked civil society to trust in their ability to handle the future for it. Now this campaign reveals a level of distrust. This is a process of unravelling, a disentangling of our hopes from that future described by the corporations. It is a slow and painful defusing.

12
Rough Trade

An evening walk brings us to the entrance of BP Integrated Supply and Trading at 20 Canada Square, Canary Wharf, in the London Docklands. There are a few lights on in the building and the ubiquitous security guard sits at the front desk. Despite the late hour there's still activity within. However, work is not being undertaken by women and men at their desks but by the computers on the trading floors that continue their incessant labour while staff drink in the bars nearby or go home to sleep. For the London office here is linked constantly with a BP office in Singapore and another in Chicago. The process of oil trading takes place without a break, 24 hours a day, 365 days a year. It is 22.30 as we stare up at the building, but it is 6.30 in Singapore and 16.30 in Chicago. There are still staff working in the US, in a couple of hours human activity will move to Singapore for the day, it will start again here in about ten hours' time.

Volatility is the precondition of profit in this office. Traders, human and machine, are constantly looking out for disruption. Early on 11 December 2017 news spread that the Forties Pipeline System had been shut down. Forty per cent of the UK's crude oil was instantly stopped. The crisis had occurred in the corner of a pony field on the Netherley Road near Stonehaven, Aberdeenshire. The residents of a couple of houses nearby were evacuated. INEOS, which owned the pipeline, announced that it would take weeks to mend the leak. Though JCB diggers were working around the clock in the field, it meant a loss of oil production in the UK and Norwegian North Sea to the value of £20 million per day until the pipe was reopened on New Year's Day 2018.[1] The oil platforms sustained production losses of several million pounds a day, but the traders made a profit on an unexpected volatility in price.

We have come to this office block to explore the world of the oil traders and to consider how the ever more dynamic flow of capital is reshaping Britain's oil system and rendering it less resilient. It is often said that business likes stability; that if large companies following a manufacturing model are to make investments they need to be assured that the near and the medium term will be fairly predictable. For example, that governments will not suddenly raise taxes or change employment legislation. The alarm raised by business over Brexit, claimed by critics to be a self-serving element of Project Fear, was focused on the possibility of sudden changes in access to markets, tariffs, taxes or legislation. This left the future too unstable to safely undertake investments in the present.

However, other businesses follow a trading model where profit is made by buying a commodity on the expectation that you can sell it for more. It can involve speculation, and is the same whether trading in shoes, railway tickets or barrels of crude oil. A period of predictable and stable oil prices is of little benefit to an oil trader whose income derives from astute judgements in a market that is unpredictable and unstable. Volatility is seen as the precondition of profit.

In the autumn of 2008, as the likes of Shell, BP and ExxonMobil scrambled to cope with the collapse of the global oil price, the business media turned their attention to the one sector of the industry that was thriving, the oil traders.[2] After a 40-year evolution this field was now dominated by a small number of large but privately owned trading houses such as Trafigura, Gunvor and Vitol.

Their business model seemed almost entirely different to that of BP and Shell. The latter's philosophy sees the oil corporations endeavouring to embed themselves in the societies, economies and political structures of key states so that they can appear to act as their national champions. This enables them to draw, among other things, upon geopolitical and military support: BP in the UK, ExxonMobil in the US, Shell in the UK and the Netherlands and so on. However, many of the oil traders have offices in London but are headquartered in Switzerland, which as a nation can offer little geopolitical weight and has not taken military action abroad since 1815. What it does offer is a financial system that is barely transparent and a tax system that is extremely beneficial to the corporations. In return the traders make no pretence of acting in Swiss interests on the world stage. They keep a low profile, traditionally spending next to nothing on sponsorship, community programmes or corporate advertising in Switzerland or anywhere else.

These big traders are both parent and child of the oil market. For them return on capital is generated by buying crude or refined product at a shipping terminal, moving it by tanker to some distant port, and selling it on to an industrial customer. Simple though this may seem, by the 1990s it had become immensely complex and could affect the physical flow of oil, especially as traders constantly speculated, meaning that they would buy oil without having agreed the contract to sell it, thus taking on all the risk of potential losses. A tanker might pick up 300,000 barrels of crude, purchased for $18 million, at the Nigerian terminal of Bonny, part-owned by Shell, aiming to sell it to, say, the BP refinery at Texas City in the US, nine days later at a price of $20 million. However, should demand for oil products unexpectedly increase in Western Europe, perhaps due to a cold snap or a refinery fire, then a wholesale company would be willing to pay a few cents more on every barrel of crude above the US price. The ship would then be redirected mid-Atlantic and deliver its load to Europe, say to Shell's Stanlow Refinery on Merseyside.[3]

We are used to living in a densely interconnected digital world, in which information flows around the globe ceaselessly day and night. We understand the notion of 'just-in-time delivery' in which goods are shuttled from distribution depots to supermarkets or vehicle parts to assembly lines on a daily basis. The same reality of constant flow exists in the world of oil. As you read these words there are hundreds of pipelines pumping crude from wellheads to sea terminals, thousands of tankers on the world's oceans and scores of refineries waiting each hour to unload the delivery they have ordered.[4]

The oil-trading corporations have grown spectacularly in the past three decades. Vitol, founded in 1966, had revenues of $225 billion by 2019,[5] not far off BP's revenues of $300 billion.[6] Vitol handles more than seven million barrels of oil and products a day. By comparison the daily consumption of the world's fourth largest oil consumer, Japan, is less than four million barrels per day.[7]

Furthermore, Vitol has been steadily consolidating its power and moving into areas traditionally under the control of the oil corporations. For example, Shell sold some of its African and Australian businesses to Vitol for $3.6 billion.[8] And, as if to emphasise this evolution, Vitol was headed by a former Shell executive, Ian Taylor, from 1995 to 2018. These shifts could be read as Vitol morphing into a Shell or BP, but in fact it appears the reverse is also taking place: the oil corporations are becoming ever more like the oil traders.[9] Highly mobile money – what we call 'dynamic capital' – is reshaping the oil sector.

BP's financial results for the year 2015 showed a $6.5 billion loss, the company's biggest ever annual fall in its century-long history. This exceptionally poor performance, which was the background to CEO Bob Dudley's presentation at the 2016 AGM, was blamed on the low oil price that had continued since 2008 and the exponential costs of the *Deepwater Horizon* disaster. But the overall figure disguised the fact that the quality of performance was not evenly distributed across all the divisions of the company. One division had made spectacular profits. Indeed, its gains had effectively rescued the reputation of the corporation from an even more disastrous set of results. That division was BP Integrated Supply and Trading (IST).[10]

This was a step change in the nature of the organisation, as profound as those orchestrated by former CEOs Walters in the 1980s and Browne in the 2000s. The corporation was thriving on volatility.[11]

Buying and selling in and out of the constant flow of data is now a powerful source of profit for BP or Shell. This profit helps pay the dividends to shareholders around the globe and the salaries of traders and associated staff on at least three continents. In families across London, the south-east and beyond, it funds cars, private schools and health care. Meanwhile many pensioners will find they unknowingly benefit through their pension schemes holding shares in BP and Shell.

Although the refineries at Coryton and Shell Haven have gone, and the chemical works at Baglan and Carrington have been demolished, there are still many families in these regions that live off the profits of Shell and BP. Payments from the corporate pensions still buy shopping and holidays long after the industrial infrastructure has gone.[12]

Ever-increasing numbers of citizens are living off the back of vola-tility. And this volatility in the economy mirrors, and may indeed drive, a volatility in the nation's politics, seen for example in the turbulence around Brexit. Sections of the political class in the UK increasingly seem to reflect the guidance of Mark Zuckerberg, the founder of Facebook, to 'Move fast and break things. Unless you are breaking stuff, you are not moving fast enough.'

4 OCTOBER 2016

Oil traders are even more discreet, privacy-conscious and media-wary than the oil corporations. So we are pleased to have tracked down David Jamison and secured an invitation to his stud farm in West Sussex.

We turn off a B-road onto a narrow driveway, flanked by thick rhododendron bushes, which leads to the stud. We pull up beside a timber-framed house surrounded by black barns. At first sight Jamison[13] reminds us slightly of Conservative Prime Minister Edward Heath, with his thick silver hair and broad frame. He has those red trousers so ubiquitous among men of a certain age and background, a striped blue, green and yellow shirt and a gold ring on his index finger. He is a big man, not one to fistfight with in his younger days. He greets us warmly.

Jamison, a former chief executive of oil trading giant Vitol, leads us through an extensive dining room with a high ceiling and what look like ancient oak beams. He tells us that most of the house is in fact quite modern, built by an architect to his exact specifications which he sent by fax from his office in Singapore. There is a copy of the *Daily Telegraph* and an open bottle of red wine on the dining room table. His study is chaotic, with papers all over a small wooden desk and the faded red carpet. Among the documents is a notice of an AGM for Savannah Petroleum, where he is a non-executive director. By the half-curtained window there are pictures of three young women – his daughters. Nearby are many photos of polo players and horses.

By a strange twist of fate Jamison lives just two miles south of the grave of Sir Maurice Bridgeman and eleven miles east of the home of Sir Peter Walters. All these men of oil wealth settled in this bucolic corner of England, but they spent their lives in different realms. Some of the storms that shook BP and dominated the careers of the two chief executives were storms that made Jamison wealthy.

In May 1967 the Six Day War altered the terms of the oil tanker trade overnight. It weakened the oil corporations and fuelled the rise of the shipping tycoons such as Aristotle Onassis who owned the tankers. It also put power in the hands of the oil traders who bought and sold tanker-loads of oil. This shift catalysed a change of thinking in Walters that, in retrospect, spelled the passing of Bridgeman's social democratic oil company.

Among the new players was Henk Vietor, who with Jacques Detiger established the trading firm Vitol in Rotterdam in 1966. Jamison started in Vitol's London base three years later.[14]

By 1972 Jamison was in charge of the London office, at Crawford Place, Marylebone. 'The Dutch wanted to hire somebody older and I said "Look, I understand the business now. In three months, you phone up, I'll either be here or not here. If I'm not here, it means I can't handle it." I was fairly arrogant at 28. "In the meantime, let me get on with it." The president, Detiger said, "All right, David, you can do it but no speculation, all right?"' Jamison's job was to put the buyer and seller together and not take the financial risk onto Vitol. Speculating was against the trading rules.

The following year, 1973, the oil industry was in turmoil due to the Arab–Israeli War. 'I bought 10,000 tons of oil from Romania, and put it on a ship. I had an idea of what I was going to do, but wasn't sure. I made $100,000. Ten dollars a ton. Which was a lot. Detiger phoned up and said, "You speculated that, didn't you?" I said, "Well ... not really, Jacques. I had a good idea where I was going with it." He said, "That's very good, carry on".'

'After he said that, I bought three more cargos in the afternoon, as the trouble in the Middle East started to increase. I decided to start speculating, and I put 500,000 tons of oil into storage in Rotterdam and Amsterdam. That was one speculation.' This, he explains, meant buying oil without having a client to sell it to and so sticking it in storage in the Dutch ports. It was against the rules.

Life can act in a strange way. I had a big lunch in the Ritz Hotel with the largest competitor of mine, Atlas, owned by the Koch Brothers in America. We had a lot of wine. We were comparing notes, and he had more oil than me. As I was walking back from lunch, it suddenly struck me. If he has that amount and I have slightly less, what has everybody else got? I'm going to start selling. I sold the first cargo to the new boss of Esso at $20 a ton under the market [price]. The guy from Atlas went back to his office and expected to be congratulated but they said, 'You made a mistake. David Jamison has caught you.' The market started collapsing, as I'd been selling heavily. There was just too much oil around, causing oversupply. I think the war was over too. Well, Vitol London was very successful that year.

We listen to his story of 45 years previously and it seems to perfectly illustrate how trading profits from volatility – indeed drives volatility.

The fates of Walters and Jamison were also intertwined around the Iran–Iraq War. This conflict began in the autumn of 1980 and eventually provided a catalyst for the thorough restructuring of BP when Walters became CEO the following year. Jamison recalled:

The Iranians started the war with the Iraqis. I was playing polo in Buenos Aires. My Spanish was terrible, but I could see pictures on the TV of jets attacking a refinery. So I got on a plane. It was over Christmas. I went back to London and started buying heavily. There were certain people who were away skiing, and the rumour went round that Jamison had gone mad. But that was the making of Vitol, that crisis.

Indeed, this was a coming of age for oil traders, just at the birth of Thatcherism.

Two years later, exhausted by differences of opinion with the Dutch office, Jamison threw in his London post and went to be the director of Vitol Singapore. In the East he traded hard and played polo, while at the same time buying this stud farm in Sussex and coordinating the rebuilding of the house. He also financed his own polo team, the Centaurs, based at Cowdray Park.

He explains to us his driving force in life: 'I had no money in my 20s, so the idea was to make money. To me it's a fairly simple equation; if you're married, you have to have money. Then I started playing polo where you needed money.' Making money requires risks, the very thing he thrives on. 'I'm a trader, really. It's addictive. I don't like heroin. I've never had it. Polo, I like. It's definitely addictive.'

As we wind up our conversation Jamison becomes more melancholic and reflects on his 74 years:

> Polo was my love and then age killed it. Golf is too slow. The oil still has a buzz. I'm doing some things with oil at the moment, just to show Vitol I can still do it. To show them that it's still there. I keep in touch with Ian. I got his congratulations because I found the biggest oil find in Alaska in the year.

Ian is Ian Taylor, former chairman of Vitol, whom three decades back Jamison had headhunted from Shell.

That spectacular find on Alaskan land facing the Arctic Ocean had been made on 4 October 2016 by NordAq, a private oil company in which Jamison was a significant investor.[15] It was heralded as one of the largest discoveries in the history of the province. At a time when the big oil corporations were slowly pulling out of Alaska, operations by small private equity-backed companies heralded a flicker of new life for the oil region.[16]

However, the contrast between BP, which had been the driving force on the Alaskan North Slope for 60 years, and NordAq reveals a deeper shift in the industry. Some of NordAq's finds would only be of value if they could utilise the massive infrastructure that had evolved over six decades – not least the Trans-Alaska Pipeline, which was part funded by the public finances of the UK and US. So whereas BP could claim that its exploitation of oil on the rim of the Arctic Ocean – encouraged by the US government – was for wider national development, Jamison's investment in NordAq was simply about return on capital.

Twenty years prior to financing NordAq, Jamison had been one of the first London private capital owners to risk money in the Russian oil fields after the demise of the USSR. Jamison had founded Sibir Energy and made a substantial fortune in the maelstrom of post-Soviet business life. Corruption had been rife in Yeltsin Russia, but it also surfaced in Alaska.

'The problem was the management. One of the guys originally associated with me then behaved very badly. It's all in the press', says

Jamison with a spark of anger. Four months after the spectacular oil find, the CEO and the founder of NordAq were facing criminal charges on misuse of company funds.[17] We enquire whether this is a picture of the coming age of oil and gas, as the large public corporations decline and in their shadows rise the private equity companies, for whom exploitation of hydrocarbons is an extension of the casino?

Jamison's response clarifies the difference:

> Private equity is a new element in the oil investment world. It certainly makes it hard to know where the money comes from. There is a lot of Russian money flying around and the owners don't want their identities revealed because they could get sanctioned, or Putin might be after them. A lot of investors these days want to operate under the cover of darkness. Even if they are as white as white, they would not want to be seen.

23 NOVEMBER 2018

The private equity companies Jamison talks of are a symbol of a shift in the financial structures of the oil industry, which is as profound as the development of trading. To understand part of its impact on the UK economy we journey to north-east Scotland.

The Arnold Clark car showroom in Aberdeen is busy with customers organising hires at the start of the day. We too are renting a diesel engined Volvo in order to track the gas pipeline system running from the St Fergus Terminal away under the fields and forests. With perfect timing, the radio news blares out: 'Today the first oil was pumped from the new giant Clair Ridge field, West of Shetland. Seven years after it was sanctioned, this £5 billion project is coming on stream. BP says the 600 million barrels of oil will keep it producing up to the year 2050 at least.' The bright halogen lights reflect the sheen on the spotless Daimlers, Fords and Vauxhalls. The large room is warm, protected from the raw November day by walls of sheet glass. Despite the dwindling of North Sea oil we continue to be sold the message of its bounty and all the dreams that go with it. Before we leave Aberdeen there are three people we need to meet.

274 · CRUDE BRITANNIA

Trevor Garlick[18] is a man of gentle calm. Slight in build and wiry, he is fit for his age and tells us of how he and his wife are just off to Patagonia on a trekking holiday. For five years he was head of BP North Sea, responsible for the company's assets in both the UK and Norway, and now he is director of ONE, Opportunity North-East, a government agency that encourages investment into the region of Aberdeenshire. In a room set aside for meetings at ONE's sedate offices he sits before us, dark suit, white shirt, hands on the table and kindly eyes. It feels almost as though he's waiting for us to interview him for a job.[19]

Starting work offshore in the early 1980s, after studying geological science and then petroleum engineering at university, Garlick's working life has been intertwined with the North Sea. He has lived through significant phases of this extraordinary colony, from the horror of the *Piper Alpha* accident to the growth of wind power. Soon after Garlick joined BP Exploration to work in the Norwegian sector of the North Sea, John Browne was appointed as head of that division of the corporation. Twenty-two years later, just a few months after Garlick became head of BP Norway, Browne was ejected from the company.

For six years from 2009 Garlick was head of BP throughout the whole North Sea. During his term he oversaw a constant stream of oil projects evolving while he battled with unexpected events – from the aftermath of a helicopter crash on 1 April 2009 when 16 men were killed returning from BP's Miller Field, to escorting Prime Minister David Cameron on an obligatory platform trip in February 2014.[20]

Under Garlick's North Sea leadership BP moved into significant new oil projects. The Clair Ridge field[21] with its 600 million barrels of crude, whose opening we'd heard announced on the radio, had arisen in part as a consequence of Garlick's guiding hand.[22]

We turn to discuss Garlick's view on the future of Aberdeen, gleaned from his new role as head of the government's regional development arm. He explains that the oil and gas industry still dominates. Some estimate that it provided 40 to 50 per cent of employment in this part of Scotland, but following the oil price collapse of 2014 to 2016 that level is far lower. Despite the continued dominance of oil in the economy, one thing has dramatically shifted – the decline of the oil and gas majors and the rise of private equity companies. In 2000 BP and Shell accounted for about 39 per cent of the UK's oil production; by the time we met Garlick in 2019 they were producing just half of this. The vast majority of the difference had been taken up by new, far smaller companies backed by private equity.[23]

The oil majors have concentrated their attention on drilling in new deepwater locations off Brazil and West Africa in the low-oil-price environment. By selling off fields to firms such as Chrysaor, BP and Shell are making the UK's offshore industry inherently more unstable, for smaller firms are far more vulnerable to going under in the event of price shocks. This shift illustrates a new phase in the UK North Sea industry, and very possibly the last phase of its life. UK oil production has slumped by over two-thirds since the late 1990s.

We ask Garlick if the province is about to 'run out'. He defers to the opinion of the old sage of the North Sea, Professor Alex Kemp at Aberdeen University, who had recently asserted there was $13–15 billion worth of oil and gas in the UK sector, and predicted that it would not 'run out' for another 25 years at least, not until 2045.[24]

Kemp, who has written a definitive history of the North Sea,[25] remains optimistic that despite the exit of many oil majors, climate issues and the downturn in activity due to low oil prices, the North Sea can still play an important role in Britain's energy future.

We find the professor of petroleum economics on the campus of Aberdeen University in a small study that is almost completely inaccessible due to a mountain of research papers. Kemp[26] is a tiny man with a serious demeanour, thick square glasses and tweed jacket. He is somehow like an Elfin king.

'The industry is now recovering but it's still quite difficult. Structurally we now still have at least some of the majors with us but they have been selling off mature assets and there's a lot of new entrants, including quite small ones and including quite a lot of private equity-funded companies', he tells us.

'That's a big difference from the 1970s and it's a big difference from the 1980s when it was still greatly dominated by the majors, before and after the oil price collapse in 1986.'

Before we go, we ask Kemp whether oil played a major role in helping the Thatcher government defeat the miners. 'Oh, yes, a big role. If you look at the data, the fuel oil use [in power stations] went right up.' And for bankrolling that government's wider political and economic programme? 'It enabled tax cuts. They reduced the higher rates of income tax so it helped procure that as well.'

Just as Kemp highlighted, BP and Shell are among the raft of large companies continuing to sell on their North Sea oil fields. Total sold almost $1 billion worth of North Sea assets in 2015, Shell sold $3 billion of UK North Sea projects to Chrysaor in January 2017[27] and the sale of all of DONG's (now Ørsted) oil and gas assets to INEOS came in May 2017.[28] BP itself sold $625 million worth of North Sea interests as recently as January 2020. In order to render assets attractive to buyers the large corporations used their political muscle to remove some of the decommissioning overheads that had been company liabilities since the 1960s.

Since the beginning of North Sea exploration it was widely understood that the construction of oil platforms offshore would require vast sums of capital, but also that the removal of these structures at the end of their lives would be tremendously costly. Whereas private corporations had an incentive to make their initial investments – the guaranteed prospect of being able to extract several million barrels of crude – the dismantling of the rigs, after all the most profitable oil has been pumped out, had little to attract it.

As this latest phase in the oil journey got underway, the question of what was to be the extent of clean seabed decommissioning required by law, and who pays for this, once again became live. There were long debates between the oil industry and Whitehall, as represented by a range of government departments. Eventually, in 2010, the industry achieved its goal. Aided by the dramatic fall in the global oil price, the impact on the North Sea industry and the weakness of the Conservative–Liberal Democrat Coalition, industry got an agreement, coordinated by Treasury Minister Nicky Morgan MP, in which effectively the UK state would henceforth take on 50 to 75 per cent of the liabilities, through tax relief on the removal of old oil platforms and pipelines. Critics saw the private corporations leaving the state to half-fund the clean-up of 'the heap of shit' that had been long foreseen.

As Texas Jim had sung in 'The Cheviot, the Stag and the Black, Black Oil', to the fiddle of a hoedown:

All you folks are off your head,
I'm getting rich from your seabed.
I'll go home when I see fit,
All I'll leave is a heap of shit.

8 NOVEMBER 2019

There's a sharp contrast between the offices of ONE and those of the RMT (the National Union of Rail, Maritime and Transport Workers). Only 20 minutes walk away across Aberdeen, the headquarters of the union for North Sea oil workers is distinctly more battered. We are here to meet Jake Molloy,[29] regional organiser for the union.

Molloy's tale is sobering after Garlick's technocratic calm. The union leader explains how the downturn in the global oil industry from 2014 led to a wave of layoffs, with the offshore workforce falling by almost a third, down to around 26,000 employees.[30] Those who remain are being pushed onto far worse terms and conditions. 'We've gone back 35 years!' he thunders, recollecting his experience offshore in the 1980s and 1990s. An increasing majority of the workers on the platforms are now on ad hoc employment, which means they are hired for a two- to three-week trip, then return onshore hoping to be re-employed. This is the 'gig economy', except these are not bike couriers or baristas, but highly trained men working in an extremely dangerous industry, where the consequences of personal error can be catastrophic. As an ex-rigger told us, 'You are always sitting on a bomb of hydrocarbons.'

Molloy is in a dark mood:

I keep telling the other unions that it's all very well talking to nego-tiators from the oil companies. When they're in the room they say one thing, but when they get up and leave the room, they go off and do what the fuck they want. I keep saying to the other unions that the only way to get at these people, to get under their skin, is to take it out into the public. Now if that means going to stand out in the street shouting shit through a megaphone then that's what you do, and you expose them.

In essence this offshore colony, which for half a century has funnelled wealth to mainland Britain while the oil operators exploited its resources, is now fading. As it does so, and private equity companies move in as the major corporations move out, the conditions for those who labour in its oil and gas fields become worse. Molloy worries that life offshore is returning to the wildcat early days, before disasters such as *Piper Alpha* and union action brought significant changes to working life. He says the blacklisting of supposedly troublesome labour that was carried out in the 1980s and 1990s has returned today.

We move on to the decommissioning of rigs. Molloy explains the process as he sees it. 'At BP Miller, say, you have the "Cessation of production phase". You've got to make the wells safe, plug everything, abandon everything, kill it all. That is done by the crews that have serviced the installation during the production phase. They turn out the lights and go home.' The next phase is overseen by firms subcontracted by BP, among them Petrofac and the Italian corporation Saipem, the latter hired for the task of removing the 'topsides' and the legs of the platform.

Saipem used the massive *S7000* crane barge[31] to lift up the redundant accommodation unit that had housed 100 men and transport it across the sea to Stord in Norway where most of the structure was recycled by the Norwegian company Kvaerner.[32] 'In one of the fjords, with gas axes, they chop it up and break it down into razor blades.' It sounds simple, and indeed is made to look so in the video that BP released of the process, with its time-lapse photography and rock anthem soundtrack, in which the human beings are toy figures in dayglo overalls.

But Molloy expresses outrage as he explains the conditions in the furious North Sea, over 100 miles from land. He claims the crews of the *S7000* crane barge are Filipino, flown into Bergen Airport in Norway, from where they are transferred to a seagoing vessel that transports them to the crane barge. Once on board they sail to the Miller platform and set about its dismembering, flensing the beast with gas axes. They do this in all weathers, flat calm and raging sea, freezing cold and sweltering heat. They work 12- to 15-hour-long shifts and are paid $45 a day, he says. This would be less than half the UK minimum wage. 'The companies can get away with this because the minimum wage doesn't apply beyond 12 miles offshore. And because they classify these men as seafarers!' But clearly they are not seafarers, like those that crew the container ships of the world, he adds. Their wages are cut to the bone because the companies are striving to spend as little as possible on this part of the 35-year lifespan of the Miller project, this oil field that began when Sir Peter Walters was head of BP.

Later on we contact Saipem and ask it to comment on this. Through a public relations official in London, we are told that no one working on the Miller platform itself earns less than the UK minimum wage. Those employed on the crane barge are paid an (unspecified) rate agreed with the International Transport Workers' Federation, says Saipem's PR.

Molloy says he has spoken to supervisors who are deeply worried about safety when it comes to decommissioning. He says they come to him and claim 'this is unsafe, we're not compliant with HSE [the UK's Health and Safety Executive] and everything else'. Molloy adds:

> These supervisors, they've been working on platforms in the North Sea for 35 years and they know that a rigger has to go through three OPITO [Offshore Petroleum Industry Training Organisation] courses to become what they believe to be a competent rigger. And they were watching these guys out at Miller, they were untrained and it was as rough as it gets, and it was as unsafe as it gets. So they were very concerned. They were complaining at me, complaining to the HSE, and asking questions.

After pressure from Molloy, HSE went to check what was happening on Miller. To Molloy's dismay the HSE inspectors returned to say that all was in order, partly because the field was no longer 'producing hydrocarbons' and was no longer considered an oil well, so different standards applied.

Clearly there are legal changes that allow this work to be done as cheaply as possible, 'But this is an unsafe operation. This is exploitation of foreign nationals. These Malayan, Indian, Filipino guys, they are nice lads, but this is serious and taking place in the UK sector.' He fumbles with his phone and brings up an image of a man appallingly scarred. He was badly burned while working on decommissioning.

'There is bloody good steel out there. It could be brought to land. Melt them down and fabricate wind turbines. But there's no strategy.' Molloy talks of *Northern Producer*, the oil production vessel used on the MacCulloch field for 17 years. Its owner, Maersk of Denmark, sold it to a St Kitts and Nevis post box company, which falsified papers to the UK authorities and towed the vessel from Teeside to Chittagong. 'They sailed to Bangladesh and dumped it on a fucking beach.' Maersk said it was duped, but certainly someone gave precedence to profit over the people and ecology of Bangladesh while the UK North Sea found itself embroiled in the illegal export of another 'heap of shit'.

24 SEPTEMBER 2018

'What's that? That must be it!' We are 30 miles north of Aberdeen, following the coast as closely as we can. Over the treetops we spy our prey, the tall orange flare of the St Fergus Gas Terminal. We turn off the A90 and follow the slip road into the complex. There is no sign of the arteries of energy flowing underground. About a quarter of the UK's gas is running through steel tubes close to the road, pumped away to the generators, kitchens and central heating systems of the nation. We glide past the grey wire fence. There's a windsock tight above the distant storage tanks and flare stacks. A buzzard stands sentinel on the neatly mown verge, unperturbed by our passing. There are lines of signs – 'National Grid Transmission System', 'North Sea Midstream Partners', 'No Photography', 'Visitors Report to the Gatehouse' – but not a soul about in the terminal car park. The Audis and Hyundais are all reversed into their slots. We follow the instructions and come to a halt, back end in. Clearly it is intended that drivers must be able to ensure the fastest exit in case of an emergency.

We sit in the car. Beyond the windscreen everything is grey. Before us, and either side, run endless mesh fences, ten feet high, topped with razor wire. They are three deep – defences running in parallel around a massive area that has a footprint equivalent to Stanlow or Carrington. Between the wire the ground is white gravel, herbicided to perfection. At regular intervals steel towers stand bristling with CCTV cameras. We've never seen an industrial site that looks quite as much like a prison. The slush, slush, slush of the gas flare in the wind off the sea is the only thing that gives a sense of what is happening here.

The complex is owned by North Sea Midstream Partners (NSMP). It is the reception place for the Frigg UK Association (FUKA) which controls a pipeline that draws gas from the northern North Sea – including the Norway sector. Here too is the landfall for the Shetland Island Regional Gas Export System (SIRGE) pipeline – majority owned by NSMP[33] – bringing in gas from the West of Shetland fields. NSMP also owns the UK southern North Sea gas pipeline that runs to Teeside.

What is this NSMP that seems to have so much control over the nation's gas? It is not a household brand name or even familiar to those of us who like to think we know the industry. Registered in Jersey it was sold for £1.3 billion on 24 September 2018 to Wren House Infrastructure, a subsidiary of the Kuwait Investment Authority, the sovereign wealth fund of Kuwait. And who sold it to them? ArcLight Capital Partners, a private equity firm based in Boston in the US. Clearly the key player in NSMP before the sale was Mike Wagstaff, the man who built the company six years earlier. We've never heard of Wagstaff. Google suggests a man of that name now owns a vineyard in Surrey and likes to go sailing in Croatia with his wife and two children.

It seems that not just offshore assets but the apparently unexciting world of onshore pipelines is shifting out of the hands of corporations and into the hands of private equity companies. The industry is being reshaped by a new style of dynamic capital.

Just as we get out of the car a Ministry of Defence police vehicle cruises by and a face in the passenger seat scrutinises us. We make for the plant's entrance. The gateway is like something from a TV thriller about organised crime. We speak into an intercom. A tiny grille at head height slides back. A section of face is visible. 'We'd like to visit the terminal.' 'Not possible. You need to ring Aberdeen 241300. That's the office of PX Group.' [34] End of conversation. We turn back to the car. There's a constant hum of machinery. Herring gulls hang on the breeze.

We managed to get in touch with Wagstaff[35] and he suggested we come down and meet him at Guildford station in Surrey. 'You will recognise me. I'll be driving a dirty blue Land Cruiser.' He pulls up and offers a friendly handshake before taking us to an Italian restaurant where he is on first-name terms with the proprietor. We settle down in a quiet room beneath the oak beams. He talks enthusiastically and fluently about his life and work. Most men like doing this, but he particularly so. He speaks at considerable speed and remembers every historic date without hesitation.

A vigorous 57 year old, Mike Wagstaff's manner is casual, smiling a lot, but he looks you straight in the eye when speaking. He still enjoys going to rock concerts of bands such as Arcade Fire or The National. Together with his wife Hilary he's a figure in the world of English wines – their vineyard is celebrated on websites such as Local Food Britain.[36]

After private boarding school in Jersey he studied mechanical engineering at Oxford University and did a master's degree in petroleum engineering at Imperial College London. Upon graduating he worked at Shell. He spent 1987 and 1988 in London helping to push the company to develop its southern North Sea gas fields, just at the point when the Conservative government was privatising British Gas and selling off more of the UK's resources to multinationals such as Shell.

But Wagstaff got bored at the corporation. 'Like all big companies, they were too slow and bureaucratic … it's a bit like a civil service.' He left to join the investment bank Schroders in the City. This was the heyday of 'loadsamoney', when Britons were encouraged to build – and be proud of – their own cash piles. His team advised the now-privatised British Gas on how to manage its assets. Wagstaff was posted by Schroders to New York in the mid-1990s. These were crucial years, learning about the rapidly evolving financial system amid the frenzy of Wall Street.

One of his deals was to raise $6 million for a tiny outfit based in Aberdeen called Venture Production. Founded a couple of years previously, the company's model was to buy up oil and gas fields that the larger corporations deemed too small to bother with. Venture focused attention on Trinidad and the UK North Sea. By 1999 Wagstaff had left Schroders and become chief financial officer of Venture. 'Putting in every penny I had', he tells us. Through intense labour, cost cutting and tax efficiencies, plus new ways of working with the contractors who did the drilling, Venture grew rapidly and had over 15 UK offshore gas fields under its control within a decade. Wagstaff wanted Venture to remain independent, but the company had been floated on the London Stock Exchange and in 2009 fell victim to a hostile takeover from Centrica, one of the private corporations that had come out of British Gas.

Wagstaff explains to us at length his woes at Centrica's move, not only because the gas giant had bought Venture for less than he believed it was worth, but also because he claimed it quickly discarded many of Venture's innovative practices. Still, there were compensations from the sale: 'Everybody was a shareholder in Venture, even the girl answering the telephone. Out of a staff of 180 people, we created 19 to 20 paper millionaires, which was probably a higher percentage of the total staff than Microsoft did.' It is public record that he himself made 'about £20 million' but, he says, 'It sounds very blasé, but when you get beyond a certain point, actually, the money is … it's a scorecard, whether it's 20 million or 30 million, obviously, it's not going to make a difference to my personal lifestyle.' He reflects, 'Venture was a fantastic twelve years, we achieved something that I think was great. It was memorable and it was different. We had a great team and we had great fun doing it.'

On leaving Venture, aged 47, he explains he did not want to retire into playing golf, so he brought a Surrey vineyard. His tale echoes that of Jamison, who 25 years earlier had left Vitol aged 42 with a fortune and also settled in the Home Counties, purchasing a stud farm and polo team. Like Jamison, Wagstaff could not resist returning to the world in which he had worked.

Two and a half years later, Wagstaff built on a longstanding relationship with ArcLight Capital, a US private equity firm and one of Venture's largest investors. He went into partnership with them to buy the Teeside Gas Terminal and the associated pipeline system. Alongside ArcLight and the Teeside management team, Wagstaff put in an undisclosed sum of his personal wealth, and NSMP was born. In August 2015 they brought up the FUKA and SIRGE pipelines that pull gas from more than 20 fields in the northern sector of the North Sea and various fields in West of Shetland, where Trevor Garlick had had a pivotal role. This skein of steel tubes runs along the seabed directly to the St Fergus Terminal where we had sat in the car park. Now a key part of the UK's energy infrastructure was in the hands of a US private equity firm and a Surrey-based multimillionaire. Three years on, Wagstaff and ArcLight sold NSMP to an arm of the Kuwait Investment Authority for £1.3 billion. How much Wagstaff came away with is not public, but he says, 'We did well. We all did well financially out of it.' Now the profits generated by St Fergus flow to Kuwait and Wall Street investment bank J.P. Morgan. As Wagstaff speaks to us in this Surrey restaurant there's a sense that it is all about money; the actual infrastructure, the ten-foot-high fences at the terminal and the gas platforms 200 miles offshore, seem impossibly far away.[37]

The St Fergus Terminal was built in 1977 by two state-owned companies, Total of France and British Gas. It was paid for by public bodies. Wagstaff was first involved in British Gas in 1988; he used the knowledge and skills acquired in the intervening three decades to generate profit and sell on this state-built pipeline system.[38]

What was for 38 years a stable piece of state-owned strategic infrastructure is now, in the years of declining gas production, a counter in a financial game. It is a step in the industry's move into volatility, driven by an ever more fluid form of capitalism.[39]

28 SEPTEMBER 2016

Close to the gas pipelines is the Forties Pipeline System carrying oil from its landfall at Cruden Bay. These tubes are sisters on the North Sea shore, arteries of Britain's industrial system, running through the fields and hills of north-east Scotland. Mirroring Wagstaff's ownership of NSMP, the Forties Pipeline is in the hands of the owners of INEOS – Jim Ratcliffe, Andy Currie and John Reece – three of the UK's richest men, with a combined private wealth of over £35 billion.[40]

James – now Sir James, but always referred to as Jim – Ratcliffe became a powerful figure in the UK oil and gas world in 2005 when his company, INEOS, bought out the entirety of BP's global chemicals business known as Innovene for $9 billion. However, he remained unknown to wider British society until he was named as Britain's wealthiest person in the *Sunday Times* Rich List. The closer we explored his biography, the more we understood that his – and Mike Wagstaff's – lives symbolise the UK's future far more potently than BP's John Browne or Shell's Ben van Beurden. As a navigator of the passage of the ship of state, he stands in counterpoint to Jake Molloy of RMT.

Ratcliffe grew up downwind of BP's Saltend chemicals plant near Hull: it was part of the city's future in the 'white heat of technology' described by Labour Prime Minister Harold Wilson. Ratcliffe's father ran a company making laboratory equipment and supplying the plant.[41] During his time studying chemical engineering at Birmingham University, Ratcliffe junior took an internship at Saltend and on graduating became a trainee engineer at the works in August 1973. Peter Walters was the head of BP Chemicals at the time.

After three days Ratcliffe was laid off. He had a history of eczema, which the company doctor feared could be exacerbated by toxic chemicals handled at the site.[42] He switched to the finance side of petrochemicals and in the autumn of 1973, as Britain struggled with the 'Oil Crisis', and became a trainee accountant at Beecham Pharmaceuticals. By 1980 Ratcliffe had been a management accountant at Esso and completed an MBA at London Business School. From the latter he took away the clear understanding that 'it is all about money, and the numbers. Money and generating equity return', as he later explained.[43] Ratcliffe moved to the Midlands working at Courtaulds for seven years as head of his own division of the corporation, Advanced Materials.[44]

In 1987 he was headhunted by a US venture capital firm, Advent International, which lured him from Courtaulds with an increase in salary from £37,000 to £115,000 and the prospect of being plunged into the growing field of private equity. Like Walters at BP, Jamison at Vitol and later Wagstaff at Schroders, Ratcliffe was tutored at the cutting edge of neoliberalism. 'Private equity was a new world to me. I knew absolutely nothing about it. I didn't have the tools in my armoury, but I just had to get on with it. I sweated a lot, but I survived', he later recalled.[45]

Ratcliffe's fate soon intertwined again with that of BP. In early 1991 the corporation, on a wave of cost cutting driven by Chief Executive Robert Horton, determined to close one of its two fine chemicals factories – either Hythe or Carshalton, which had long before been found to be the source of toxins that polluted the River Wandle. Ratcliffe decided to go it alone, leaving Advent and setting up his own venture capital unit to buy up industrial plants. He bid £37 million for the fine chemicals division. 'I managed to persuade Bill Teasdale of PricewaterhouseCoopers to perjure his life away and explain to BP that the money was as good as in the bank.' And he risked the home of his wife and two sons. 'I put all my chips on it. So if it had gone down. I'd have been in a mess.'[46] By September 1992 the lengthy negotiations were complete, Ratcliffe owned Hythe and Carshalton and BP's commercial manager of the latter, Andy Currie, had joined Ratcliffe in his new company.[47,48]

Three years later BP was keen to be rid of its vast works at Antwerp in Belgium.[49] Ratcliffe and Currie bought BP's cast off[50] in May 1995, turned it around as a profitable venture and maintained production. 'Nobody realised just how much money that Antwerp business made, because they didn't do their accounts properly', said Ratcliffe.[51] It remained a hub in the chemicals world of north-western Europe. Each time Ratcliffe made a purchase such as Hythe or Antwerp, he obtained not only hardware but also staff. By the early 2000s around 70 per cent of his employees were ex-BP.[52]

On the back of these purchases Ratcliffe formally founded INEOS together with Andy Currie in 1998. They were joined by John Reece from PricewaterhouseCoopers, on the advice of Reece's mentor Bill Teasdale. INEOS was not listed on the London Stock Exchange. It was a private company without outside shareholders, owned by the trio at the top – Ratcliffe, Currie and Reece. A new type of private equity-backed vehicle; a new form of capital for a new era of oil and gas in Britain.

Whereas the shareholding that van Beurden, the head of Shell, has in that corporation constitutes an infinitesimally small fraction of the world's several billion Shell shares, and John Browne, despite a personal fortune built on his BP shareholdings, has only ever owned an equally tiny percentage of the company, Ratcliffe owns 60 per cent of the shares in INEOS. This is a new model, also arguably a return to the early years of the oil industry, giving Ratcliffe a level of power in INEOS only perhaps equivalent to that Deterding had over Shell up until the 1930s.

In the two decades since its foundation INEOS has grown from a number of plants in Western Europe into a global corporation with sales revenues in 2019 of $85 billion. It has done so largely by borrowing money and buying the factories which corporations such as BP, Shell, Dow and ICI didn't want. INEOS did not follow the model created by the likes of BP and Shell in the early twentieth century, in which capital is raised through traditional bank lending and issuing shares, thus spreading ownership and creating a public limited company, a PLC. Rather it had expanded by raising cash from a wider pool, issuing high-yield 'junk' bonds[53] and cash from hedge funds and other private equity providers who themselves utilise money from a pool of high-net-worth individuals.

This system ensured that ownership of the real assets remained in the hands of Ratcliffe and his two colleagues. It was itself a private equity model and an example of the coming world of oil and gas corporations. Now there is a raft of these smaller companies offshore in the North Sea which the general public rarely hear of. These include Siccar Point Energy, run by a former Centrica man Jonathan Roger and backed by private equity groups Blackstone and Blue Water Energy. Or Chrysaor, backed by Harbour Energy, itself backed by EIG Global Energy Partners. Or Neptune, backed by venture capital firms Carlyle and CVC.

Not satisfied with selling off individual plants such as Antwerp, BP wanted to dispose of its entire petrochemicals arm. The plan of John Browne, CEO, and Byron Grote, chief financial officer, was to spin off the division as its own PLC, to be called Innovene, which would issue shares. Innovene would have been the fifth largest petrochemicals company in the world, with 19 plants in the US and Europe. To their immense surprise there was a bid from INEOS to purchase the whole unit for $9 billion. Ratcliffe planned to raise this from a set of banks.[54] Browne refused even to meet Ratcliffe and told us when we met him that he did not really know who the INEOS man was when he first appeared on the scene as a buyer.[55]

A mighty negotiation ensued. Many times it looked as if Ratcliffe's bid would fail. At the eleventh hour, almost as an aside, BP threw Grangemouth Refinery into the package it wished to sell. On 7 October 2005 the deal was closed and INEOS, with no knowledge of running refineries, took ownership of the plant and town on the banks of the Forth.

Ratcliffe's autobiography relates that he called Jim Dawson, a former Shell executive, and by then an INEOS board member, for advice. 'Dawson, only half tongue in cheek, sent him a link to a website called HowStuffWorks.com. It is a running joke that the company learned nearly everything it knew about refineries from that website.' Bill Reid, a director of INEOS, recalled: 'refining was a 10-billion-dollar business, but our understanding of it was very, very thin ... We were utterly clueless.'[56]

We drive along the shore of the Firth of Forth west out of Edinburgh. To our right we catch glimpses of an oil tanker moored at the Hound Point Terminal. Just past Linlithgow we reach the crest of the hill and see before us the expanse of Grangemouth Refinery on the banks of the Forth. White steam drifts from the cooling towers and the orange flares quiver on the stacks. To our left, occasionally screened by birch trees, is a morass of pipework, towers and steam-swathed buildings. We turn on to the highway towards the heart of the complex. Here are the storage tanks of Kinneil, the Stabilization Terminal at the end of the Forties Pipeline System, where gas is separated from oil. The former is piped into the refinery while most of the latter is pumped east to the tank farm at Dalmeny and from thence to the jetty at Hound Point where tankers ship it around the world.

The highway plunges into Grangemouth. To left and right the same mesmerising array of pipework, stretches of seemingly abandoned land, distant cooling towers and lines of parked cars. There is no human figure in this manmade vastness. We wind down the windows and let in the roar of distant flares and the air's acrid smell. At the point where two highways meet is a bold new blue block of a building. This is the head office of PetroINEOS, its novelty untarnished, its shiny surfaces standing out from the other blocks built before World War II and in the 1960s. Soon we arrive at a neat housing estate, bang up against the sprawling expanse of what seems like steam punk, not sophisticated, industry. This was a town built for those who laboured in the refinery – two-storey, well-tended detached houses. The proximity to the plant is astounding, the stacks and cooling towers rise just above the rooftops. It feels as though the town of Grangemouth was built inside the refinery and petrochemicals works. This hive of humanity is clustered together on an island set apart from the rest of Scotland, separated by the Forth to the north, the Avon to the east, the River Carron to the west and the M9 motorway to the south. A place of its own, with its own high street, its Sparta Fitness centre, its Masons Lodge Zetland 391 and its Cooperative Funeral Care. We pull up by the Lea Park Hotel.

We are here to meet Mark Lyon,[57] an organiser for the Unite trades union branch at the Grangemouth Refinery and the figurehead of a public campaign which eventually saw him sacked by INEOS in February 2014. Following the 2005 purchase of the refinery from BP[58] there had been a decade of intense struggle between INEOS and the workforce, much of it around the company's insistence that its future could only be guaranteed with a wage freeze and cutback in pensions.[59]

Ratcliffe made little attempt to hide his apparent view that unions were just a brake on good management. In the realm of the PLC there was, at least, a notion that all those in the corporation served the corporation – from the CEO to the lowliest apprentice. Now INEOS, despite having a federated organisational structure,[60] was pretty much the domain of one man, Ratcliffe.

Over coffee in the lounge of the Lea Park Hotel, Lyon recalls:

When INEOS bought [the refinery], there was a meeting arranged for us [in the union] to meet him. We were sort of saying, 'All we want really is a seat at the table.' And we wanted to be involved in decision making and hoped there was going to be investment at the plant. He treated us like he didn't want us to be there. He wouldn't talk to us.

Lyon describes a second meeting four years later: 'He sat side-on to us eating grapes. It was almost like "Peel me a grape". He was speaking over his shoulder to us. "I pay your wages and give you bonuses. What more do you want?" So it's not a contractual thing. It's like: "You work for me so you should be grateful." Of course, it didn't go well.' He muses, 'A strange, strange kind of character, almost like a cult (figure).'

The disputes at Grangemouth between INEOS, the Unite union and the workforce eventually escalated into an all-out strike. In the end, however, Ratcliffe got what he wanted and in his biography he boasts: 'The incident has since remade industrial relations on the site and changed the course of British union politics for good.'

To fight the union and the workforce was to fight the place. Grangemouth Refinery had been opened in 1924 by the Anglo-Persian Oil Company, precursor to BP, and had remained under one ownership for just over 80 years. The town around the plant had been built by the corporation to house the workers. Grangemouth was like the Sandfields estate in relation to Baglan or the houses in the villages around the Kent Refinery.

Lyon explains in his own book, *The Battle of Grangemouth*, how his father worked at the chemicals side of the plant. 'We lived in one of the modest yet comfortable four-in-a-block flats that the company had built to accommodate the workforce ... All the homes had huge gardens front and back. If you needed a repair or had any difficulty in your home, you could call the plant [Grangemouth] and they would send out [someone].'

After our coffees Lyon takes us for a tour. As he drives he describes the scene, from the now closed BP Social Club building to the streets of detached homes:

> The houses that are over here are all BP houses. They built us a new town over here and that's why we loved them. My dad worked for BP. It was very paternal. We were a BP family. It brought a lot of loyalty within the town. Another thing it brought was tolerance. This is a noisy town. The flares go up and the place is lit up all night and there is a lot of noise, smells, pollution. But, as just about everybody was dependent on the industry within the town, everybody was tolerant of that.

At the eastern end of the massive complex, near the Kinneil Stabilization Terminal, there is a dark tower that looks newly built. Lyon explains that this holds the 'fracked' gas from Pennsylvania.

Latterly Ratcliffe came to be troubled over the gas that INEOS was using at Grangemouth, all of which came down the pipeline from St Fergus, owned by Wagstaff's NSMP.[61] How long would stocks of gas in the UK North Sea last, and more importantly how much would INEOS have to pay the suppliers? The question became particularly acute as the shale gas boom in the US was making gas, ideally suited for petrochemical production, vastly cheaper than the market price in Europe. This threatened the viability of Grangemouth and INEOS's other European plants, as they competed with US factories in the global market. The shale gas was usually extracted onshore from rocks through the process of fracking, a new method that involved vast quantities of water and chemicals being injected into the wells. It has environmental and social impacts that were being heavily resisted in the US by many communities and activists.[62]

In a move to break away from the stranglehold of the UK North Sea supply, Ratcliffe tried a two-pronged strategy – first to determinedly pursue his own fracked gas onshore in the UK, and second to import fracked gas from the US.

During the battle between the oil and gas industry and communities across Britain over the attempts to frack gas, INEOS was in the vanguard. It stood alongside another private equity-backed company, Cuadrilla Resources, whose chairman for a while was Lord Browne.[63] There were vigorous campaigns against INEOS, especially in the Ryedale area of Yorkshire. Ratcliffe and the company began to get a poor public profile and their PR consultants, Media Zoo, worked hard to defend the petrochemical company's image. But the anti-fracking campaigners were remarkably successful. In England protests broke out almost everywhere there was new onshore drilling. A battle against Cuadrilla Resources[64] in Lancashire was particularly hard fought. By 2016 the Labour Party and the Liberal Democrats were calling for bans on fracking and, in the run-up to the general election of December 2019 the Conservatives also promised one. Meanwhile in Scotland the SNP, with their majority in the Scottish Parliament, brought a ban into law. Ratcliffe's first strategy was looking at least delayed, if not defeated.[65] He hit out at UK government restrictions on fracking as having 'no basis in science' and said Britain would be left dependent on 'unstable' wind and solar power if gas fracking did not go ahead.[66]

In parallel INEOS developed its plan to import fracked gas from Pennsylvania, by extracting ethane and shipping it across the Atlantic. In this they were entirely successful. The first delivery of US gas to Grangemouth came powering up the Firth of Forth on 28 September 2016.

Ratcliffe invested \$2 billion in the huge project which involved the construction of an ethane export plant at Marcus Hook in Pennsylvania, as well as a 300-mile pipeline from the fracking gas fields on the western side of the state. He also commissioned special vessels to transport the frozen liquid across the Atlantic. These were built in China not in the UK, despite INEOS receiving 'crucial' UK government support for the scheme.[67] It was a decision that incenses Lyon.

'They built ships to bring the gas. People were invited to go and see them launching the ships. Flippantly we were saying, "What shipyard on the Clyde is it in, that these ships are being built?" These ships were getting built in China.' He's outraged by the government's action. 'You might have thought that somebody would have said, "If you're building ships, we want them built in the UK. If you're using steel, we want the steel procured in the UK".'[68]

In our attempt to understand Ratcliffe we draw heavily on his autobiography, *The Alchemists: The INEOS Story, an Industrial Giant Comes of Age,* and the few interviews that he did on the back of its publication. Inevitably this limits our understanding, but we do this in part because of the impossibility of getting an interview with Ratcliffe himself or anyone of board-level seniority in INEOS. Emails requesting factual information about the company remain unanswered, including ones asking for confirmation of where the ultimate holding company of INEOS is based. We repeatedly requested an interview with the chief executive over a period of years and were endlessly rebuffed, in marked contrast to almost all others we approached in the industry. As someone who knows Ratcliffe well told us bluntly: 'Jim doesn't do anything [such as interviews] he doesn't want to.'[69]

We did eventually get to speak to his head public relations man, Richard Longden, after INEOS unveiled a \$5 billion deal in June 2020 to buy yet more BP assets, pretty much the last of its chemicals business. Longden is based in Lausanne, Switzerland. He says he will have to come back to us about the question of whether the INEOS holding company is ultimately based in Monaco, London or the Isle of Man. He never does. But Longden does say that INEOS 'is headquartered in the UK and registered for tax in various locations where it operates'. The formal head office issue may seem trivial, but it is a feature of this new business model that you can have INEOS as one of the largest owners of British industry and yet still find it hard to pin down exactly where it is based. This new model of dynamic capital is even more elusive and evasive of public scrutiny than the neoliberalism of the past four decades.

It is also of interest because in 2010 INEOS made great political play out of moving its headquarters from the UK to Rolle, Switzerland in a row over tax with the Labour Prime Minister Gordon Brown. Then in 2016 INEOS returned. There was a lavish ceremony at new offices in Knightsbridge, in the West End of London, while Ratcliffe became a cheerleader for Brexit and all things British. And yet, in August 2018, two months after being knighted, came a blizzard of newspapers reports that Ratcliffe had himself decamped to Monaco for tax reasons.[70,71]

A group with the huge assets, workforce and corporate power that INEOS has, not wanting to grant any interviews – except 100 per cent on its own terms – signifies a new type of institution in the industry, and gives a picture of the UK oil and gas world as it now is and is set to be in the future. It has scant need for journalists, unlike the corporations which used the media to build a positive profile even as they largely lobbied ministers behind the scenes. The likes of INEOS are straightforwardly hidden and largely closed to scrutiny, except via their own public presentations. They are privately owned, often by individuals tax domiciled abroad. The companies may be also registered offshore: NSMP in Jersey or INEOS which appears to be ultimately controlled by Ratcliffe through a company called 'INEOS Ltd' based in the Isle of Man.[72] At the same time the INEOS promotional literature can appear to be open and confessional. No serving CEO of BP or Shell ever produced anything as frank and autobiographical as *The Alchemists*.

* * *

We thank Lyon for his time and the tour of Grangemouth. We drive to the very end of the Forties Pipeline System. It's a struggle to find the place. Eyes glued to an OS map, we go around the back of a housing estate, under a narrow railway bridge, up an almost single-track road. 'That's it.' 'What?' 'That hill is a fake hill. That hill is not a hill at all, it is a manmade downland surrounding a set of vast storage tanks'. We have found the Dalmeny Tank Farm.

We drive on cautiously; it is dusk, there's not a soul about. We stare up at the great mound and notice that there's a tunnel disappearing into its heart. The tunnel, a roadway, is brightly lit. It looks like the set of a Bond movie. We stop in the empty car park and wonder when we'll be accosted.

Nothing happens. We get out and go over to a tiny Portakabin at the tunnel's mouth. There's a woman in a hi-viz jacket. We knock on the window. 'Do you have any information about the storage facility?' 'No, but you could try ringing this number.' She writes it out on a Post-it note. We return to the car.

We wonder what to do next, feeling slightly underwhelmed by this climax of a two-day journey along the Forties Pipeline System. This mountain contains crude oil of untold value and is highly strategic, but in contrast to the St Fergus Gas Terminal there appears to be no 'real' security whatsoever. Then we notice the guard from the hut is coming towards our car. Oh dear. We wind down the window and look innocent. 'How come you are here? You're not supposed to know about this place. This here's a secret', she says. We explain that we could see it on the OS map, and anyway we know all about INEOS and Jim Ratcliffe. She beams. 'I've met Jim. He came here. He's very down to earth. Nice man.'

13
Nexus of Outrage

We drop into the Apple Store at Covent Garden in London's West End. As with every other one of its 38 outlets in the UK, the place is an oasis of calm. The products are discreetly displayed and the assistants avoid any impression of doing the hard sell. The whole experience says 'Relax. You have arrived in the future.' Indeed it seems that Apple, like many of its rival companies, is selling one thing: the future.

The firmament of Big Data – arrayed with the likes of Google, Amazon, Facebook and Apple – asserts itself as our inevitable future. The brilliance of these stars eclipses the other orbs that for so long suggested the way ahead. For much of the twentieth century it was the oil companies that provided the essentials of 'modern life' – to the motorist through petrol, to the farmers through pesticides, to the householders through plastics. BP and Shell delineated and sold to Britain – and elsewhere – one consistent thing: the future. But their power to do so is waning.

In 2005 Apple was 39th on the list of corporations by market capitalisation and ExxonMobil, the biggest private oil corporation, was number one. On 10 August 2011, Exxon, having been top of the pile for four decades, is briefly surpassed by Apple. By 2019 Apple is number one and ExxonMobil is struggling to stay inside the top ten.[1] The former is twice the value of the latter. The supplanting of Exxon is echoed by the similar decline of Shell's and BP's global ranking. And this shift in capital has begun to have its effects on the UK economy.

The business model of Big Data is infamous. For the digital giants the principle is the same: the strength and value of the corporation lies in its ability to extract the data provided by users of its services. In 2010 Instagram was sold for $1 billion; this was not on account of its 13 employees but its 30 million users.[2]

The Big Data firms have built a remarkable mechanism for generating return on capital. This is not achieved through the ownership
and utilisation of fixed assets – such as a gas field – but through the
control of the means of accessing the data of their users via platforms.
This 'assetless capitalism', or 'capitalism without capital', creates some
bewildering economic realities. In 2016 the world's largest provider of
taxi services was Uber, which did not own a single vehicle. The world's
largest provider of accommodation was Airbnb, which didn't own a
single building. The world's largest retailer was Alibaba, which didn't
even have an inventory never mind a warehouse.[3]

By 2017 a phrase was circulating in the media: 'data is the new oil'.
This shift, of course, does not go unnoticed at Shell and BP. Both corporations realised they could face strong competition from Big Data
in the energy markets. Maybe Amazon will finance the construction
of wind farms in the declining oil province of the North Sea? Perhaps
Google will enter the electricity retail market selling power to its
millions of users in the UK? Or maybe Apple will launch an iVehicle,
an electric car that will sync with your iPhone, iPad and iPod? In this
new realm, as in the century-long realm of oil extraction, power and
the ability to generate the highest return on capital will fall to those
that obtain a monopoly position in the market.[4]

Despite their apparent differences, Big Oil and Big Data have one
crucial similarity: they are both extractive industries. BP and Shell
extract oil and gas, Amazon and Facebook extract data. In both cases
control of the infrastructure of extraction is vital.

The difference in the resources being mined is brought starkly
into view in the form of the electric vehicle. For over a century the
petroleum combustion engine has been a key instrument in the
extraction of crude oil. The second half of that century demonstrated
that the part of the oil chain that is most capitally productive lies in
the pumping and transporting of crude, not the sale of petrol to car
owners. But without the car owners there would be far less demand for
crude extraction. The British car driver is useful to BP or Shell only in
as much as she or he burns petrol.

The UK is one of the world's largest petrol markets; if everyone
shifts to electric vehicles it'll have an impact on the oil industry, but
this switch could be used by the oil companies to their own ends. The

value of engaging in electric vehicles does not derive from the driver using electricity but in the data that can be gathered by the company on the driver's habits, and using this data to guide sales and profits. This is the resource that is extracted by the companies who run electric vehicle schemes, as cars have evolved into platforms following the model of Facebook or Google.

16 FEBRUARY 2018

There is something about Jeremy Bentham[5] that exudes a jovial self-confidence. Perhaps it is the theatrical quality of his dark suit, black shirt with red buttons and large black stone set in a gold ring on his finger. His dress stands in contrast to the regulation blue suit, white shirt and polished black shoes of Wendel Broere, the press officer who sits monitoring our interview at the Shell Head Office in The Hague.

Bentham's team at Shell is one of the antennae of this intensely future-focused corporation. Bentham is head of the Scenarios unit, established by Pierre Wack in 1966. This division of 28 staff[6] with a big budget is tasked with comprehending coming trends in society, economics and politics and communicating them in clear and persuasive terms to the executives of Shell and key political figures beyond the company.

The décor in Bentham's office shows how aware he is of Shell's Scenarios' legendary status. In the middle of one wall hangs a picture of a dodo with a roll call of names beneath it. It is a list of the previous heads of the team. At the top comes Wack and the names run down to Bentham at the bottom. The dodo image stands as a warning that Shell itself could go extinct. This notion remains at the forefront of Bentham's mind. In the midst of our hour-long conversation he says: 'I can't guarantee that we don't become a dodo. Maybe the dodo learns to dance? But it won't become extinct because we're asleep at the wheel: Shell is very aware and very engaged.'

A close look at Bentham and his position is revealing.[7] For Wack, who studied under Armenian mystic Gurdjieff and a Hindu swami from whom he learned the art of sensing the future, the aim of the Scenarios team was not to give Shell executives data by which to guide the ship of the company, but rather to train them to sense the seas into which they were sailing. Trying to encourage them to think outside given frames of thought, to look outside their orthodoxies. Unlike Wack, Bentham himself does not hold any particular status in the world of futurologists. Indeed it looks like his role is less to interpret the direction of society for Shell executives, than to present the direction of Shell itself to journalists and politicians. Critics tell us this is a weakness in the corporation and reflects an anxiety in the company about its future role.

Bentham is seated in front of a whiteboard that shows the names of his team and a handwritten note with a quote from the science fiction writer William Gibson that reads: 'The future is already here, it is just that it is not yet evenly distributed.'

'Did Shell foresee Brexit and Trump?' we ask. He replies:

> Yes. If you were to look back at our 'Mountains and Oceans' scenario, published in 2013, you will see that those forces were being described. The Mountains was really about the concentration of power in particular elites. The Oceans was about the distribution of power. It was also about the way you can get competing elites. In a sense you've had examples of competing elites in both the UK and US, each legitimising their position through the shaping of a broad constituency. Nobody could ever imagine that Donald Trump or Boris Johnson are anything other than elite figures,[8] very powerful elite figures. In our leadership team we have a great geopolitical analyst who was very clear. He said before the referendum: 'I wouldn't say Brexit is the most likely outcome, but believe me this could really happen.' So we had given some thought to what that could mean. So we had Brexit scenarios.

Bentham's analysis of the UK's future strikes us as pretty clear:

> We saw that for at least a decade UK government attention will be dominated by one issue, which meant that policy in other areas will be neglected. We saw the potential that Brexit could have for the breakup of the UK and hence all kinds of question marks raised about the North Sea. We're still a substantial player in the North Sea and so that could be a very complicated set of issues.[9]

He explains how the scenario plans he oversees are put together:

> My role is to speak truth to power in our company, so I have to be as objective as I can. The reason that we have such large networks of external advisers is to compensate for the fact that people like me have naturally grown up in a particular environment and so I need to be challenged all of the time. Typically a scenario plan, like the one going out next month, *A Better Life with a Healthy Planet*,[10] is 'touched' by 150 Shell subject matter experts and 300 external subject matter specialists.

'Are they from Chatham House and the like?' we ask. 'Yes, from the think tanks – such as Adam Posen, head of Peterson, a Washington-based economic institute – and from people in government. I liked to talk with Malcolm Turnbull, former prime minister of Australia, about the way he saw things.'

Bentham is remarkably blasé about Shell's access to global political figures. However, the field of advisers, though broad, seems to echo the elites that he describes in the Mountains metaphor. Later, when quizzed about the role Shell played in the deindustrialisation of Britain as a root cause of Brexit, he accepts that areas of Britain were neglected: 'This was not Shell in isolation but the leadership in general – business leadership, government leadership, academic leadership – took many things for granted that ultimately needed attending to', and this reflected 'a narrow-minded perspective on reality that got absorbed in the minds of those who had influence to shape the reality for others'.

He is keen to move on from Brexit and explain how Shell is leading the way in the energy transition. 'Is that your phrase?' we ask. 'In a sense yes, I obviously absorbed it from somewhere.' Key to this change in the global energy system will be the crowning of electricity as the fuel of the world. As he explains, this is going to be a challenge, a shift from 'less than 20 per cent electrification globally to more than 50 per cent. That's big. How that happens is not clear. But Shell is not asleep at the wheel, we're already North America's second biggest power trader.[11] And you will have seen some of the purchases we've made recently that are in the electric power domain'.

Bentham explains that through shifting to electricity generated from renewables Shell can help the world 'get to zero carbon by 2070'. He describes how the company has the ability to transform itself entirely in a decade, rapidly selling on the oil and gas assets that currently make up almost all of Shell's value. This is a wildly ambitious target and Bentham does not say that Shell 'will' but rather that it 'has the ability to' make this change. We listen to him, amazed at the boldness, but knowing that zero carbon by 2070 is almost 40 years behind the demands of many climate scientists and activists. And it is 20 years after the UK government's own target.[12]

Bentham's declaration comes out of the scenarios that his team created in 2016 and 2017. They explained how the energy world is going through a sustained phase of radical uncertainty, an industrial cycle that neatly mirrors the social and political turmoil seen in Brexit and Trump. In this uncertainty they foresee four worlds and encouraged the board to take up a strategy of energy transition, which would change Shell from an oil company into an energy transition company working in a climate-constrained world. A world in which demand for oil is likely to peak in the late 2020s,[13] meaning that an ill-judged oil company could be left with stranded assets such as crude reserves that are difficult to sell. We see that the ideas generated by groups such as Carbon Tracker are being absorbed by the world's largest corporations. Already this strategy had seen Shell begin to sell off projects that require a high oil price in order to make a profit, such as fields in the UK North Sea and the investments in the Canadian tar sands largely made since 2004.

Bentham has been explicit about the necessity for Shell to pull out of high-cost projects before oil demand peaks. Three weeks before our meeting his comments had been published in *Fortune* magazine. He explained that if Shell held on to the Alberta tar sands projects too long 'you were – gosh, forgive me – fucked'. He repeats the line to us with a happy grin. And we recall the dodo. There had been a decades-long global campaign to prevent the exploitation of the tar sands, on the grounds that it could undermine the lives of Canadian First Nations and have a catastrophic impact on the climate. But Bentham's job is to avoid the extinction of Shell.[14]

It is on the basis of the four worlds, and the subsequent scenario report *A Better Life with a Healthy Planet*, that Shell is advising governments on the ways to meet the challenge of climate change as posed by the 2015 Paris Agreement. For as Bentham explains, they want to ensure that these issues, central to Shell's existence, are handled in the right way:

> one of the concerns is that pressures build up in particular areas, decarbonising being one of them, and then at a certain point, and it may be as much as ten years' time, they have to be addressed and you get knee-jerk and sometimes not very sensible policy changes and policy shifts. So you'd much rather have a steady rational debate that's going on and not being subject to the potential that suddenly somebody is seizing the steering wheel.

So here is Shell desiring to manage the change that they recognise has to happen, in a calm collegiate manner that meshes with their business strategy and does not impact on their capital value. They see that if they can make a plan to transition in a way that does not lose them profits, and can encourage governments to make the changes at a pace that fits this, all will go well for Shell. However, this 80-year model of liberal corporate paternalism is open to disruption, as political events of the last decade have revealed.

Bentham is eager to show how his chief executive Ben van Beurden supports these ideas: 'I was very pleasantly surprised that on coming into post he really did embrace the reality of energy transitions and, beyond that, in his heart, and this is a phrase he's used, he believes that. I believe it as well, that at our best we Shell can be a force for good, because nothing happens in the world without energy.' He shows his engineering roots in his closing words: 'What we're talking about is rewiring the global economy, and it's an awe-inspiring task and the fact that its awe-inspiring is inspiring. I'm still inspired.'

As we leave we ask Bentham where he's from, as he has an accent that is somehow unplaceable. Is he from the West Country? 'Yeah. People tell me that. It's 'cos I've moved around. But when I'm back with my old school friends, and after a few pints, my accent comes back. I was born in Blackpool, but we were from Wigan.' He did not

follow family tradition into the coal mines but joined Shell doing
design work for the Stanlow Refinery in the early 1980s, at the same
time as Kenny Cunningham and Simon Henry were working there.
And when Andy McCluskey wrote the OMD hit.

> The distillation tower was one of my designs. It was the biggest
> single thing moved on the UK roads at the time. I've got some great
> photographs of it somewhere. When I'm landing at John Lennon
> Airport on a visit to Liverpool, which I do fairly frequently as I'm an
> Everton season ticket holder – I know they are awful at the moment,
> and they truly *are* awful – when I'm flying in, I still see some of the
> results of my handiwork at Stanlow.

Shell has left Merseyside behind, and as Bentham explains with a surprising dispassion: 'Shell refining capacity in the UK didn't make any
sense at all, and so that's why we exited, basically, from that area.' Merseyside is part of Shell's past and not part of its future scenario. The
future is unevenly distributed.

17 FEBRUARY 2018

Up the stairway, through the revolving glass doors and we're back in
the lobby of Shell's head office in The Hague. The foyer, decked in
white marble, suits the scale of this building. Behind a long front desk
sit a line of three women, reminiscent of stewardesses at an airline
check-in, they take down details and give out security passes. We sit
waiting for the arrival of our appointed guide, watching the busy to and
fro of staff. Most are between their twenties and forties and are dressed
informally, in jeans, open-necked shirts and jerseys. These people are
not civil servants or investment bankers, nor are they software designers
or entrepreneurs. There's a sense of purpose and calm.

The three receptionists talk between themselves in Dutch. All
around this neighbourhood of the city is an array of Shell offices,
including a company hotel for staff.[15] Despite the notion that Shell
is an Anglo-Dutch multinational, in the capital of the Netherlands
there's nothing to suggest this is a British company. Through the plate
glass windows, down the avenue of Carel van Bylandtlaan, we can look

NEXUS OF OUTRAGE · 303

towards a grid of streets whose names – Balistraat, Javastraat, Suma-trastraat – all echo the Dutch imperial roots of Shell. The company was a key engine of that trading empire.

A sharp-suited communications director arrives. He greets us briskly, organises our fob cards and ushers us through another set of revolving doors into the next chamber of the corporation. In the lift, in impeccable English, he wants to know about Brexit. He says he has a special interest as he was formerly a speechwriter in the Dutch Foreign Ministry. At the sixth floor we step out into the carpeted space of the CEO's suite.

Ben van Beurden's[16] office is on the corner of the building looking out onto the green of Oostduin Park. He sits behind a large L-shaped desk surrounded by executive gifts. There is as a wooden dragon in a glass case clearly from China. The rest of this modestly sized room is laid out with comfortable chairs. Partly drawn blinds let in the late winter sun that catches white orchids in a vase. A well-built Dutchman with floppy greying hair, van Beurden holds out a hand and manages a smile but he seems a bit tired. We have met before on a number of occasions in public places, he has always been friendly and at ease. Curiously he seems more uncomfortable today, despite being in his own executive suite.

Though, just as Bentham predicted, van Beurden quickly gets fired up about the energy transition and plunges into a long monologue:

> Ultimately what society needs to do is to decarbonise energy con-sumption, to the point that we get to zero carbon emissions per megajoule of energy consumed, and we will probably get there in the second half of the century, to be in line with Paris [climate agreement]. At the moment we are at about 74 grams per megajoule of energy consumed and we have to work our way back to zero. Which means that by 2050 we need to be at 40, and what we have said at Shell is that, 'If I want to be a company that is relevant by 2050, and we want to supply the products that society needs on its way to a net zero energy system, I have to be at 40 as well.'

He explains that Shell will get there by saying: 'I'm going to get to 40 by changing the mix of energy products that we provide, and doing all sorts of other things on top of it.' That includes selling more renewable power, which is probably going to be the equivalent to doing five to seven world-scale wind farms a year by 2050.' He nods at the scale of his assertion and reinforces it with a 'Yup' before continuing. 'We will have to sell a massive amount of renewables on forecourts to cars, probably the equivalent of three times the power production in the Netherlands at the moment.' Beyond this, 'We will help society on its way to zero through large-scale nature-based offsets. To offset the remaining emissions from the hydrocarbons that we will still be selling by 2050, think of reforesting something like Spain, that sort of order of magnitude. We have plans to do all this.'

Pretty quickly our heads are spinning: seven world-scale wind farms a year! Reforesting Spain! And that's only to deal with the emissions from this one company's products. Where are we going to plant enough trees to soak up the carbon from everything else? What does this mean for Britain? A return to the past of endless forests?[17] The North Sea crammed with wind farms? We are struck by the power of this middle-aged man, on the sixth floor of his corporate palace, to imagine he can make all of these changes.

Then he picks up on the turn to electrification and explains the logic:

This energy transition going on is undeniable and is fundamentally driven by a number of factors, one is growing demand for energy which is somewhat unstoppable. That will be driven by demographics, by prosperity. The other trend is of innovation, new technology, new ways of doing things, new attitudes of customers, the way that we consume electricity, organise our lives. These are two big trends that are important for the energy system. Now of course what you want is an outcome that respects the Paris Agreement of two degrees C or less. So how do we position ourselves? Or what do we want to believe?

Van Beurden seems unstoppable as he outlines two belief sets:

There is a first nexus where people have ignorance of the energy system, some complacency, maybe the belief that it can all be solved at one minute to midnight, the belief that somebody else needs to move first before I can move. And a second nexus of growing public awareness, growing outrage, calling for better collaboration between different segments in society, business, government, NGOs, general public, driving to really shape the system in a completely different way and forcing us onto that two degrees C pathway.

He calls the latter the 'nexus of outrage' and says, 'We like to believe in Shell that the nexus of outrage, which is very uncomfortable, is also the better nexus.'

So Shell places itself in a nexus of outrage as it tries to stay relevant in the unfolding future.

He reinforces this point: 'I hope I will have painted a picture that this is not easy, that there is radical uncertainty, and that we have to navigate this with a high degree of conviction, but also bearing in mind that we can get this seriously wrong on all sorts of counts'. He explains, 'the energy system will transition to one that is much more electrical. In the second half of this century, the share of electricity in final energy consumption will grow from less than 20 per cent to more than 50 per cent.'

For Shell this will be an opportunity to maximise profits:

It will change the complete dynamic of the electricity system from being a relatively boring, predictable, centrally planned way of doing things to a highly dynamic, unpredictable and flexible way. Whether it is on the generating side or on the consuming side. That will open up a very significant economic opportunity for us. So that is one of the driving forces for Shell to go into the power value chain, as an integrated player because we believe that there is economic rent to be had, and if you are an energy transition company, your objective is to capture that.

Van Beurden is in the flow and we realise that he is running through a well-rehearsed script. Like others we'd met, he was apparently delighting in the unpredictable and volatile as a means of generating profit. The sentences, which contain many phrases from the Scenarios team, feel as though they have been crafted for an internal audience of Shell staff. Only when we interrupt and tackle him more obliquely do we gain a wider view of the man.

We ask: 'You sat in an office at the Shell Centre opposite the Houses of Parliament for many years. That was a period of enormous political stability, which we all got used to. Now we're pitched into something quite different. Could you ever have foreseen that was going to happen? Brexit?' Such is the apparent ongoing sensitivity of this he asks for his reply to be off the record.

Van Beurden's career in Shell, in the manufacturing division and the chemicals division for 35 years, coincided with the era of plant closures. We ask him: 'Among the reasons that people give for Brexit and Trump are globalisation, deindustrialisation, the "left behind". The oil industry was a large part of the UK economy in the 1960s, 1970s, 1980s. Then plants like Carrington closed, Thornton closed ...'

There's a long silence. 'Yup.'

We continue: 'You can see a picture in which Shell has effectively left the UK behind: Thornton closed, Carrington closed, Wilton closed, Sittingbourne closed ...'

He responds: 'Yeah ... that's true ... and I think that has happened, and there is no way I can position it any differently.'

But the head of Shell insists the company was just reacting as much as acting:

There had been major and massive forces at work in society and in the global economy that have moved these things about, true globalisation and the removal of trade barriers and everything else. Some of our assets that had been built to serve a national economy became less competitive when they had to compete internationally, and err ... and that is a major driving force, for instance, on how our petrochemicals footprint has changed, from having something like 150 manufacturing sites worldwide and now having 60.

We move from the past back to the future, onto the ground that he's keen to cover, and the threat posed by the Big Data giants such as Google. 'I think you have to treat it like a threat, but I would also like to think that we have arrows in our quiver that they don't have.' He explains:

Think of the power chain, think of it as discrete components. If you want to succeed in this chain you have to have access to electrons, otherwise you can't play. So you have to have exposure to generating assets. You need to have access to customers. And if you want to play you have to have the piece in the middle, which is the trading, taking advantage of arbitrage opportunities.

We see clearly how Shell is aiming to extract profit from the coming low-carbon world. He continues:

Now a very important component in all of this, if you really want to harvest what is there, is digitalisation. Taking advantage of massive data systems on customer behaviours. Being able to see trends, being able to, maybe … you know, switch off demand in someone's house which they may not notice but which may give you a tremendous trading opportunity. Say in the bathroom if you switch off power for half an hour nobody will notice. But if you do it in two million houses, all of a sudden it becomes a major trading chip for you in a trading play.

So the 27 million households[18] of Britain are seen in a new light, as a chip in an electricity poker game. As van Beurden talks of the trading opportunities we are reminded of the world of David Jamison and of Vitol established in Rotterdam in 1968, 60 miles north of where eight-year-old van Beurden was growing up.
Of Big Data he says:

You know, data companies are very well-positioned to attract data science technicians, so they really hold that part, but what they don't hold is the customer proposition, what they don't hold is the generating capability, what they don't know is the trading play. And they might not be the right people to sit down with governments to talk about market design to enable growth in this sector. So I think we have all the parts in the puzzle.

We see again the way in which one of the tools that Shell has to extract profit is through moulding the energy market of the UK, for example, to its own design. This may be a polite way of describing lobbying of government departments.

We say, 'But what Google and Amazon have in abundance is incredible detail about customers, an alarming amount of private material about your life. Whereas you probably don't know who your customers are in terms of the motorists who pass through the service stations.'

Van Beurden replies, 'I think we do. Increasingly we tap into that, and we do that in a very careful way, because of privacy concerns and all that, but we interact with Big Data, through things like Facebook. If we are in the power sector we will have to do that through the contact points with end customers.' He explains, 'the contract that we strike with those customers is, "Give us access to this information and we will manage that for you".'

So the households have become a chip in the energy markets to trade with, and that depends on gathering vast amounts of data, knowing when we use the bathroom or not.

The Dutchman speaks again of the wider political and social outlook through his 'nexus of outrage' phrase:

I think this is inevitably where we will be drifting as a society, a nexus which is much more assertive, in which we, Shell, have to play a more assertive role as well, where Shell insists that government has to get it right, where Shell shows up with broader coalitions, not just us as the energy sector, but with a broader societal 'coalition of the credible'.

He drums his fingers on the table as he continues:

'You need to do this, and if you don't do it, it's not going to happen, and actually if you don't we're going to make a point of pointing that out to your electorate.' So it's not necessarily naming and shaming governments into doing the right thing, but it is just helping, you know, the forces line up to do the right thing. We cannot demand it. We cannot dictate governments to do it. We don't have the mandate.

He talks of the frustrations of working with the unpredictability of a
Western democratic system:

> In this particular country, the Netherlands, we've been pretty effective
> in sitting down with government, explaining what the energy tran-
> sition should look like, pointing out that. We'd love to act, we will
> put our money where our mouth is, but these are the parameters that
> need to be put in place for us to be able to do that. One of the things
> is predictability in government policy, so don't say 'Yo! This partic-
> ular cabinet will do the right thing, but we can't vouch for the next
> one', because by the time we have lined up our investment proposals,
> there will be a next cabinet. So you know, how am I going to have
> certainty of the way forward?

We have a flash of understanding of a future conflict between
democracy and tackling climate change. And remember that in its
century-long history Shell has worked in hundreds of countries and
political systems, many of which were effectively one-party states. The
company's task is always the same, return on capital, the nature of
the state it works in comes second. This was the driving logic of van
Beurden's predecessor who worked in these offices 80 years before,
Henri Deterding.

The hour is winding up. Before we leave the room, we explain a little
of the book we're engaged in, and our understanding of the connection between music and the oil industry. We slip in our final question
in the hope of hearing something unexpected: 'What's your favourite
band?' After a moment's hesitation he confides – Led Zeppelin and
the Rolling Stones. Jeremy Bentham had named his as Deep Purple,
Uriah Heep and Led Zeppelin. Old rockers in the Shell tower.

We walk down Carel van Bylandtlaan towards the Shell Expatriate
Archive in Surinamestraat. Despite the scale of Shell's plans in wind
or reforestation, and declaring it wants to 'rewire the world', these
projects are dwarfed by the size of their current and future oil and
gas projects. Van Beurden has allocated $2 billion a year to the energy
transition initiative, or 'new energies', but this is a fraction of the $23
billion that was allocated to buying and opening up new oil and gas
fields in 2018. It seems the company is hedging its bets or heading

in two different directions at once. Building a future around the core means of generating profit through carbon, plus a small amount of renewables investment on the side. Partly because of this, Shell's target of zero carbon by 2070 is far below the demands of many institutions in Britain of zero carbon by 2030. Perhaps that means that UK local authorities, myriad activists and NGOs are placing themselves outside the coalition of the credible? Or could this be another false start, like Shell's buying and then shutting down a major solar business four years later in 2006 or BP's 'Beyond Petroleum' initiative, long since abandoned?

In the coming fight over the future there will be those who put faith in Shell and BP. There will also be those who are clear that while working to develop the electricity market, van Beurden should stop exploring for and developing new oil and gas fields and focus the company's attention on shutting down its existing fossil fuel projects. Without doing this, Shell's talk of wanting to abide by the Paris Agreement is not considered by many critics to be credible. The board admits it has, like so many others, known of the threat of fossil fuels to the climate for over 35 years, and during that time has – with the connivance of us the customers – only exacerbated the problem.

In a comment to the press shortly before we met, van Beurden had said, 'What is a challenge at the moment, is that we don't know anymore where the future will go.' But Shell is clearly attempting to map out that future and thereby avoid being usurped from it by the likes of Big Data. There's something about the massive office block, the scenario reports, Bentham and van Beurden that says, 'On the matter of climate change, we are well aware and very engaged. Leave it to us to handle, we will manage that for you.' We emerge from spending two days in and out of the head offices feeling amazed but disturbed by the aura of corporate confidence.

7 SEPTEMBER 2018

The gift shop at the Trump International Golf Links on the Menie Estate, Aberdeenshire, is full of merchandise bearing the brand of the course's owner. There are golf shirts, golf caps, teddy bears, umbrellas, all the paraphernalia of the marketing arm of the 45th President of the United States. As we pay £3.00 for a golf ball marker emblazoned with the Trump logo, the helpful guy at the counter tells us that a good third of the clientele on the course are Americans, flying in from across the US to visit Scotland and play a round on the president's links. A slice of Americana, something of imagined Florida or Texas, all manicured putting greens and this clubhouse, has been dropped onto the sand dunes on the outskirts of Aberdeen. The city was reborn in the 1970s as a colony of Houston, and now, as the oil depletes, this golf course has been bestowed upon it by an American father. Perhaps an unintended memento of better days.

Beyond the counter lies the restaurant. There's a busy gaggle of customers taking lunch amid the piped music. Over the grey heads focused on crockery and cutlery, through the French windows, we can see the bright green fairways, tufts of marram grass, the horizon of the cold grey sea, and a long line of eleven massive pillars. These are the turbine columns of the Aberdeen Bay Offshore Wind Farm, their grey blades turning silently against the grey sky. There is a workboat moored at the base of one of the towers.

It is astounding how close the array appears, how it dominates the seascape. It seems entirely unsurprising that this should be the most heavily written about wind farm in the UK. Donald Trump tried for nearly a decade to block its construction. Again and again his company staff lobbied the Scottish government. They appealed the decision of the Aberdeen city fathers to grant planning permission and the SNP-dominated Holyrood to uphold it. They even took the matter to the Scottish Court of Sessions and then to the UK Supreme Court.[19] Trump threatened to withdraw his investment and close the course. But to no avail. The turbines were erected and opened by Nicola Sturgeon, first minister of Scotland, on 7 September 2018.

We had come to Menie to reflect on the future of North Sea wind energy. We had come feeling entirely sceptical of Trump's motives and the use of his financial and presidential firepower. But the sight is more dramatic than we'd expected. The turbines are indeed a challenge to Trump's domain and the view of the sea from the clubhouse. But more than that, they can be seen as an affront to what Menie stands for. This course is an outgrowth of the oil and gas industry. It is oil that provides some of the citizens of Aberdeen with the wealth to become members

BRITAIN'S WIND

WIND FARMS

1 Aberdeen Bay	18 Lincs
2 Barrow	19 London
3 Beatrice	Array
4 Blythe	20 Lynn
Offshore	21 North Hoyle
5 Burbo Bank	22 Ormonde
6 Dudgeon	23 Race Bank
7 East Anglia	24 Rampion
8 Galloper	25 Rhyl Flats
9 Greater	26 Robin Rigg
Gabbard	27 Scroby Sands
10 Gunfleet	28 Sheringham
Sands I & II	Shoal
11 Gwynt y Mor	29 Teesside
12 Hornsea	30 Thanet
13 Humber	31 Walney I & II
Gateway	32 West of
14 Hywind	Duddon Sands
15 Inner Dowsing	33 Westernmost
16 Kentish Flats	Rough
17 Levenmouth	

Offshore wind farms as of Autumn 2017

Orkney Islands

North Minch

Moray Firth

SCOTLAND

Aberdeen Bay
Aberdeen ● 1

14

Glasgow ●   Edinburgh ●
17

NORTHERN IRELAND

Belfast ●

Isle of Man

Newcastle-upon-Tyne ●
4

26   29

Irish Sea

Ormonde
22
31   2
32

Middlesbrough ●

North Sea

Kingston upon Hull ●

33
13

12

N
W ─●─ E
S

Burbo Bank
11   ● Liverpool
25 21   5

ENGLAND

23 6
28
15
18
20

Cardigan Bay

WALES

Birmingham ●

Norwich ●   27

7

St. George's Channel

Cardiff ●

Bristol ●

Bristol Channel

London Array

London ●
10
19
Kentish Flats   16
30
9   8

Dover ●

Portsmouth ●

Plymouth ●

24

English Channel

0	20	40	60	80	100 miles

0	40	80	120	160 kilometres

of the club. It is oil that enables this place to function, from the maintenance machinery to the plastic golf ball markers. And it is oil that
makes it possible for a third of its clientele to fly in from America. The
turbines stand for another way of being. For the coming world of wind
and solar will provide many delights, but it will not provide anything
like enough energy to keep afloat the likes of this course, a leisure
pursuit built on the deck of an oil tanker.

The next morning is cold as we make our way across Aberdeen
harbour, walking past empty gravel lots and long, low warehouses of
steel. Trucks thunder by carrying containers to distant industrial units.
Herring gulls slide back and forth in the north-easterly wind. At the
door of Unit 6, Natalie Ghazi[20] is waiting. Bubbling with conversation
she leads us upstairs to a meeting room. Kevin Jones, head of Aberdeen
Bay Wind Farm and Ormonde Wind Farm off Barrow-in-Furness,
south of the border, shakes hands warmly and bids us sit. Around
Formica tables and on plastic chairs in this room of bland function-
ality, the conversation unfolds. We are in Vattenfall's operations and
maintenance shore base.

Vattenfall is a Swedish multinational based in Solna,[21] 100 per cent
state-owned and founded in 1909, the same year as BP. It is one of the
largest corporations in the Nordic country[22] with annual sales of £13
billion. Vattenfall – meaning 'waterfall' in English – evolved out of a
hydropower and nuclear electricity company, but in the 1990s branched
into coal-based generating with power stations in several countries. It
also owned coalmines. Only in 2016 did it sell its infamous opencast
mine in Lausitz just south of Berlin. It effectively passed on its carbon
liabilities to others.

The issue of liabilities is not unimportant. In November 2017 a
Peruvian farmer successfully won the right in a court in Hamm in
Germany to bring a case against the German corporation RWE for
'climate damages'.[23] Saul Luciano Lliuya argued that RWE, a coal
mining and coal-fired electricity business, had contributed to the
change in the global climate leading to the growth of a glacial lake that
threatened his hometown of Huaraz. He demanded compensation for
the flood defences he had to build and a contribution to the cost of
the community's defence measures. A study published in 2013 had
concluded that RWE was responsible for around 0.5 per cent of global

CO_2 emissions since 'the beginning of industrialisation'.[24] If this case succeeds then all fossil fuel companies are potentially liable for historic damages – including Shell and BP. Indeed, both of these featured in a list of the 20 largest contributors of CO_2 emissions in a new study by the respected US-based Climate Accountability Institute.[25] Shell was number six and held responsible for 2.12 per cent of historic emissions, BP was fourth, responsible for 2.47 per cent. Vattenfall would make it into a much longer list.

As we settle in for our meeting we are handed company literature. The opening double-page spread in one brochure features a shot of Vattenfall's Kentish Flats Wind Farm emblazoned with the words: 'Fossil free in one generation'.[26] The text below reads:

> Our goal is to be free from fossil fuel within one generation. To achieve this, we need to invest in energy solutions fit for the future. In the UK, electricity needs are expected to double by 2050, as we electrify industries and transport. From our renewable energy parks, plans for more than 4GW of wind energy developments, heat, grid and electric charging networks – we're investing now in delivering solutions designed to offer climate smarter possibilities.

The initiative to build a wind farm just outside Aberdeen's harbour, the logistical epicentre of the UK's North Sea oil and gas colony, was begun by the Aberdeen Renewable Energy Group – the AREG – in 2003, led by the city council. Progress was slow. Eventually the group called in the corporate capital and organisational muscle of Vattenfall and a €40 million research grant from the European Union.

The principle behind the European Offshore Wind Deployment Centre, to give the project its proper title, is that it should not only be a demonstration of Aberdeen's desire to go beyond oil and gas, or at least to embrace renewable energy, but that it should also serve as research into the challenges of offshore wind. The southern North Sea is comparatively shallow and erecting wind turbines in 100 feet of water, building on the experience of 50 years of constructing gas platforms, is a relatively standard engineering task. But the deeper northern North Sea is far more challenging, and Vattenfall wants to know how ever more powerful wind turbines can be erected in the most economic way to cut costs of supply.

Jones is breezy and warm, happy to answer any questions and full of enthusiasm for his work. In a gentle manner he explains the wind farm in outline. 'Aberdeen City Council started the AREG. They were the real drivers, they had the vision.' Construction began onshore in 2016, when Jones started in the project. The wind array will operate for 20 years and produce enough electricity to power 70 per cent of the homes in Aberdeen. Jones is at pains to explain the research innovations, including a novel form of foundations for the turbine towers, a suction bucket jacket that allows faster and virtually noiseless installation. There is also the uniquely high-voltage cabling to shore, making landfall at Blackdog just south of the Menie golf course, and running inland for five miles until it links up to the National Grid. This experiment will reduce power losses between wind farm and grid, he explains.

The Vattenfall man omits to mention there was local resistance to the onshore developments at Blackdog, but makes much of the community liaison carried out around the works. Those that objected to the wind farm ranged from Donald Trump to the Royal Society for the Protection of Birds. The wildlife group only gave its consent following a reduction in the number of turbines and layout. Ghazi pitches into the conversation highlighting the environmental impact assessment, which will continue over the next three years and address concerns over its effects on bottle-nosed dolphins, guillemots and sea trout.

The suction bucket foundations were built from units manufactured in Poland, the Netherlands and Belgium, assembled in Newcastle, shipped north, transported offshore from Peterhead and Dundee harbours by the Dutch company Heerema's vessel *Aegir*, and erected by the massive crane ship *Asian Hercules III*, of Singapore. The generating systems to which the blades are attached were manufactured by Vestas at Lindoe on Fyn, Denmark. The steel of the entire structure came from a hidden array of furnaces and iron ore mines across the globe.

This system to embrace the wind was made from materials wrenched from the Earth. Machines that give citizens of Britain green energy are mostly manufactured elsewhere. These early days of the development of North Sea wind stand in contrast to the local construction of refineries, chemical plants, oil tankers and offshore oil platforms between the 1940s and 1970s, but in line with what happened later when contracts fled abroad. Jones agrees with our assessment of the lack of involvement of British firms. 'It's disappointing', he says.

These shadows are dispelled by Jones' radiant enthusiasm for the fossil-free electric world that he is part of creating. He lives in the Lake District and travels to Barrow and Aberdeen in his electric car. When heading to London for a meeting at Vattenfall's UK head office he takes the train on the newly electrified West Coast Main Line. We quiz him and Ghazi about the place of all these turbines and the electrification of Britain in the wider culture, explaining our delight in tracks such as 'Stanlow' and 'Drive My Car'. They cannot think of any equivalent song about wind farms, although Jones excitedly remembers that his nephew's favourite video game, *Call of Duty*, is set in a wind array.[27]

After our interview with Jones, Ghazi takes us along a corridor festooned with an artwork in blown glass, all bright blues and greens. She says: 'Commissioned from a local artist, Shelagh Swanson who worked with school children to make these pieces for our offices.' Ghazi talks of how Vattenfall celebrated the opening of the project with family events and 'some very young kids illustrating how the power comes into their houses'. This is a world away from the launch of the Forties oil field nearly 45 years previously, with the Queen, 1,000 guests mostly from London and a tent the size of two football pitches carpeted in red.

Meanwhile, in the Vattenfall control room there are four well-built guys in work overalls seated behind computer terminals. They are all Nordic, including Kristian from Denmark. It begins to feel like the coming UK offshore wind industry will be dominated by the culture of a different land, just as in its early days the UK offshore oil and gas industry was disciplined to the norms of Texas and Louisiana.

This is only a maintenance centre but it contrasts with our sense of military culture and danger that we have seen on offshore oil rigs. The cleanliness, the atmosphere of calm and the lack of machismo seems self-evident here. And offshore wind farms tend to have few if any people stationed on them. Terry recalls having stood on top of a wind turbine out at sea, and the stomach-churning sway to and fro of the massive structure. But there is no fear of a blowout, fire or explosion.

One Vattenfall wind engineer in his fifties explains that he used to work on semi-submersibles and jack-up rigs for Halliburton, but he quit and joined Vattenfall on the Aberdeen Wind Farm. 'After the downturn in 2014, working on the rigs became extra tense. This place is a lot less stressful. Of course, safety on the platforms was tightened after *Piper Alpha*, but you are always sitting on a bomb of hydrocarbons, just as *Deepwater Horizon* shows. I do some work on the turbines offshore, but in this job I can always get home every night', he tells us.

Shortly before our visit to Vattenfall we had asked Jake Molloy, the oil trade unionist, about the forthcoming world of wind. Do the wind farm owners recognise trades unions? 'The industry trade body, Renewables UK, talked of a partnership deal, but it is very weak and going nowhere.' But surely, we ask, the Norwegians had achieved strong worker rights in their sector of North Sea oil, might this not be replicated in the Nordic wind industry? Are there unions in Danish renewables? 'Not in any significant way.' There was a meeting in Copenhagen to create a partnership between Stavanger, Esbjerg and Aberdeen to push for common conditions, but little has come from it.

Much of the construction and maintenance work around a wind farm such as Aberdeen Bay is contracted out from Vattenfall to foreign firms.[28,29] Operators of wind farms say British companies are not always active in these markets but there is nothing – apart from perhaps initial cost – to stop fabrication work being done here. Concerns have been increased by revelations that Russian and other migrant workers found to be building the Beatrice Wind Farm for SSE were being paid £5 per hour.[30] The company insists this was a mistake that was being rectified, but for Molloy it's just a symbol of the wind industry following an oil path chasing cheap costs over societal benefit. Ultimately the British state needs to step in to ensure the wind industry develops for the benefit of the nation, just as BNOC attempted to do in the 1970s prior to being sidelined and privatised in the 1980s, Molloy believes: 'I can't get my head around the idea that you'd not take ownership.'

Some day, not too far away, the UK will be ringed with wind farms, the shipping lanes running between them like paths through fields of wheat. They will form a core part of the future British economy, just as oil and gas once did. It matters in this future what the culture is of these massive industrial enterprises, who owns them, the conditions for workers, the respect for ecology. For these values will heavily impact on the culture of the UK as a whole. The conditions in which this new era of North Sea wind is being born are those in which the oil and gas industry is in rapid decline. Just at the UK's North Sea oil and gas inherited some of the repressive work practices of the US southern states and became a test bed for neoliberal relations between capital and labour. Molloy wants to make sure the offshore wind industry 'decontaminates' itself from this inheritance.

* * *

A few weeks later we meet Danielle Lane,[31] the head of Vattenfall UK at the company's London head office in Blackfriars. She's been in post for only ten months, but has spent all 20 years of her working life around offshore wind and speaks passionately on the question of women in the industry. She says there is a far better gender balance than when she started. There is a feminisation, and Vattenfall has a company policy that a minimum of 35 per cent of its workforce should be women, from engineers in the wind farms to the board.[32]

We push Lane on the lack of British involvement in the North Sea wind industry and she heartily agrees that there is a 'supply chain problem' and a poor level of state support, which in itself reflects the ideological bias in the UK towards very simple ideas of the market economy. Have the North Sea oil and gas equipment suppliers got involved? 'No, they were very slow to move over, they were used to high margins in their industry, and so renewables was not profitable enough for them. Until the oil price crashed, then they all complained that we'd been shutting them out.' As a result we can see the UK offshore wind supply chain is provided for by factories in Poland, Denmark, the Netherlands and elsewhere. Lane says: 'There was too much scepticism in the UK about the prospects of North Sea wind.'

She reminds us of a remark by Mark Campanale. We ask, is that because fossil fuels are so embedded in British culture?

Yeah, that's a really interesting way of looking at it. I was at an event in Norfolk we were running with school children to help them understand the opportunity that can come from working in the wind industry. We asked them, 'Why would you be interested in working in this industry?' and one of them stood up at the end and said, 'Well why wouldn't we?' Wind farms are part of our culture here'. I just felt that was the best thing I'd heard 'cos when I first started coming to East Anglia it was, 'What are you thinking of doing? Why would you even do that?' Over 20 years there has been that change. I mean she wasn't even born when I started going to the region. That was one of the most inspiring moments I've had in my whole career, actually. I think it's a shame we didn't start out with that attitude because we could have done so much more.[33]

Still it is intriguing that Lane herself came from the fossil fuel world. She grew up in Chester, her mother was a teacher and her father worked at ICI Runcorn, on the Mersey overlooking Stanlow. The ICI chemicals plant was fed raw materials by the Shell refinery. ICI is now defunct, but the factory still exists, owned by Ratcliffe's chemical giant INEOS.

We ask Lane what she makes of the recent wave of school student climate strikes, criticised by many in the current government:

I think it's amazing and I'm absolutely in admiration of school children who strike for climate change. This goes back to the point about it being part of culture. I think my generation have been a bit complacent. I remember people going to the Body Shop, you buy your fair trade goods and worry about organic food, but you fly away on holiday. So I think, good for the strikers. The willingness of them to stand up and say, 'We need to do something different and we need make some really hard decisions', and I hope that we all do that.

* * *

The wind farm at Aberdeen stands just outside the harbour, in full view of the city's households. Its presence is in stark contrast to the remoteness of every single one of the dozens of oil and gas platforms that, although often served by Aberdeen, exist far beyond the horizon.

These massive machines that steal energy from the wind are present in the daily life of the city, but despite the scheme being initially developed by the city council they do not belong to local people in any way. They are the property of a private capital organisation as much as the Forties oil platforms. And they are subject to similar external forces. BP owned Forties and barely invested in it for a decade until they sold it on to the Houston-based Apache Corporation. Vattenfall could do likewise with the Aberdeen Bay Wind Farm. The Swedish multinational has around 50 wind farms on and offshore in five different countries. If, as BP began to do, it chooses to concentrate investment elsewhere, then the scheme at Aberdeen could be deprived of capital, the offices in the harbour not maintained, staff not replaced and working conditions left to wither. The company has already threatened to quit the UK electricity supply market which it had entered a short time previously.

These uncertain conditions apply not only in Aberdeen, but all around the UK. On the day of visiting Vattenfall's offices there were over 30 offshore wind farms operating, from Barrow to the Moray Firth and from the Wash to Thanet in Kent. Britain is the world's largest offshore wind market, but over 90 per cent of its wind farms are owned by capital concerns based outside the UK.[34] Many of these are state bodies such as Stadtwerke München owned by the City of Munich or Danish public pension funds.[35]

BP, which has historically run some wind farms in the US, has none in the UK. Shell used to have a foothold in the British industry, holding 33 per cent of what was then the world's biggest offshore wind farm, the London Array in the Thames Estuary, between 2003 and 2008.[36] But Shell sold out, with most of its stake going to Masdar, the Abu Dhabi investment fund.[37] In the face of criticism from Environment Minister Hilary Benn MP, Shell explained in 2008, 'We constantly review our projects and investment choices in all of our businesses, focusing on capital discipline and efficiency.' The company said the US government incentives in wind energy offered more competitive returns.[38] Now, a decade later, after co-financing the Blauwwind Wind Farm off the coast of Holland, van Beurden has made clear his plans to buy rights to build in the UK sector of the North Sea.[39] This shuffling and reshuffling of Shell's pack over two decades perfectly illustrates what it means for this massive resource to be subject to the vicissitudes of private capital concerns.[40]

The exploitation of the oil and gas that lay beneath the UK sector of the North Sea was impelled by a set of drivers. From the point of view of the British state, decisions were pushed initially by its hunger to address the spiralling balance of payments deficit, and after 1973 by its anxiety over access to energy supplies. The rocks hidden beneath the grey ocean promised security. From the point of view of the private corporations, anxious about access to oil reserves, the UK North Sea offered the lure of enormous profits. The two parties fought over the speed at which those resources were to be extracted.[41] Both parties tussled but they recognised that this was a finite resource, and that at some point the oil would run out.

The fluid wind over the surface of the sea is not like the fluid rocks beneath the seabed. The wind is infinite. There may be still days, there may be storm days, but it is inconceivable that the wind itself will cease. Maybe the fundamental difference between finite rocks and infinite wind can help us to determine a different approach to the latter's use by capital and the state, an approach not of exploitation but of embrace. Perhaps this will provide the rhythm of the economy of the coming century, just as oil provided the rhythm of the past century. Could there be a rhythm beyond dynamic capital?

14

Heading for Extinction

15 APRIL 2019

We are in the glass conservatory of a neat former council house on the outskirts of Stroud in Gloucestershire. A room of bright blue rugs and comfortable chairs, a wood-burning stove and Nigella Lawson cookbooks on the shelves. A boy in a black football shirt and trackies enters and exits the room. His mum is Gail Bradbrook, a doctor of molecular physics and daughter of a Yorkshire coal miner.

This family home is the birthplace of the environmental movement Extinction Rebellion. Here twelve people on a short video in October 2018 declared: 'We demand to be heard, to apply informed solutions to these ecological crises. We refuse to bequeath a dying planet to future generations by failing to act now.'

Bradbrook seems vaguely amused, if not bemused, that Extinction Rebellion, or XR, has grown from those dozen activists to a global movement. The circled hourglass logo is possibly as well known in Britain as the yellow on red scallop shell of the Anglo-Dutch oil giant. 'We're now in 50 countries, 150 groups in the UK, at least 300 across the world. It's a dream come true but it's also a nightmare, in terms of coordination. And without sounding messianic, I felt sure this was going to happen at some moment. So it feels familiar in a really strange way.'

One of the key differences between XR and many other environmental campaigns is that it encourages civil disobedience on a mass scale and relishes arrests and court appearances as PR opportunities. Bradbrook, who herself faces legal charges for allegedly damaging a Department of Transport building in London, explains why she thinks it is important:

I've spray chalked banks. People can't get their heads around it, 'this is the High Street and there's this woman doing graffiti?' But those tactics have a limited shelf life. However, if they arrest you, there's a story. Or if they don't arrest you and you stand there for the day, there's also a story. You don't lose. If you're just stood there with a placard for an hour or two, nobody gives a fuck, do they?

She laughs to herself. 'I am just chuckling thinking about me doing that "heading for extinction" talk. Just twelve people squashed in here.' Bradbrook,[1] blond hair, blue eyes and pierced nose is seated comfortably on her settee, her legs crossed. She tends to talk into the distance looking at the wall rather than us, but is a warm if edgy presence. She has ended up doing more backroom work, often looking after the finances, while her extravert and quixotic collaborator Roger Hallam tends to act as 'frontman'.[2]

How does it feel to look after children, to run a household, and to take responsibility for a worldwide movement, albeit a self-organising and partly autonomous one? 'Well the kids are mildly irritated by the amount of time it takes from me and a bit amused. I got fan mail off [TV wildlife presenter] Chris Packham the other day, and they were like: You haven't heard of Chris Packham!? They're just ordinary kids. They like football. They'll probably become bankers and have fast cars', she laughs. Bradbrook feels a huge sense of responsibility. 'I do have a spiritual practice and I am a scientist by training so I am not into wishy-washy stuff for the sake of it. But what else are we supposed to do with our lives right now?'

The strategies behind XR did not come out of nowhere: 'I'd been trying for ages to figure out tactics, but didn't realise there was this body of literature around social change. Gene Sharp has been translated into 80 languages, he's the father of civil resistance. He says when the state represses peaceful protesters who've got a just cause, it just backfires on them.'

Forty-seven-year-old Bradbrook traces some of her political convictions back to the miners strikes of 1984–5 and her life in South Elmsall, Yorkshire. 'My dad put his feet up for the year and drove my mum mad. She was hoovering around him. He was a traditional guy but he wasn't out on the picket lines or anything. We wrote essays

about the miners strike in school. It made me want to be in politics.'
And XR itself came out of earlier campaign groups that Bradbrook
was involved with trying to organise mass tax disobedience and
actions against fracking for gas, local waste incinerators and finally the
expansion of Heathrow.[3]

Just a few days prior to our meeting with Bradbrook, XR activists
had used the tactics she had described to attack the Shell head offices
on the South Bank,[4] spray painting the sheet glass frontage and pad-
locking themselves across the doorway for several hours before the
police removed them and took them into custody.

Soon after we attended a presentation by Ben van Beurden, CEO
of Shell, to City analysts. It took place not at the Shell Centre but the
County Hall building next door for reasons unspoken, but probably to
make security easier after the offices had been blockaded by the XR
protesters.

Van Beurden's team has invited asset managers to give them an
upbeat message about the golden future the company is building.
There is plenty of emphasis on what Shell will do in the green arena
– hydrogen ships, electric vehicle charging stations and electricity generation for domestic use. Van Beurden reassures them, however, that
we are far from the end of the line when it comes to Shell's involvement in fossil fuels.

He promises there will be lower net carbon emissions in future but
oil and gas exploration and production will go on because, 'as long as
there is sustained demand for oil and gas there will be sustained commitments from Shell and that means sustained investments'.

We collar him afterwards briefly to ask how he feels about those
activists who graffitied his London headquarters. He looks earnestly at
us: 'I welcome the fact that Extinction Rebellion and school students
are raising the profile of the climate debate. We, Shell, can only move
as fast on the energy transition as society moves and creates demand
for new products.'

But is he not angry that his company, which has been much more
positive about the transition than most other rivals, is still being singled
out as one of the bad guys? 'Of course it's frustrating that Shell is
attacked but it is perhaps not surprising as the Shell brand is so iconic.
Obviously that is good but it can make things difficult as well. Most

importantly we want this debate.' He can partly thank Bradbrook for having it.

23 APRIL 2019

'It's not fair that they're messing up my future. They're messing up my future and it's not my fault.' Elsie Luna[5] has just delivered an electrifying speech in Parliament Square on the first morning of Extinction Rebellion's April 2019 attempt to blockade London. The ten year old spoke passionately to a growing crowd of 500 demonstrators gathered on the thin grass, wielding pink, blue and green XR flags. In the background was the bronze hunched figure of Winston Churchill and behind him Big Ben wrapped in scaffolding. The tiny schoolgirl from Scunthorpe declared: 'Adults should really be taking responsibility, but they are not. So why don't we protect the climate if no one else is?' She told the rebels, some of them seven times her age, 'I decided to go and find the leaders of the companies. I said "Please, please, declare a climate emergency and keep the fossil fuels in the ground".' There were loud cheers.

Minutes later Luna is sitting cross-legged beside her mother squinting up at the sun and throwing clumps of grass around. The bespectacled girl who was a pioneer of the school students' climate strikes in Britain seems playful and slightly distracted. But when she speaks to us about tackling climate change, she becomes instantly incisive and serious:

> I realised this is an emergency and I just got more and more involved. I saw the list of the oil companies, then I thought, 'Why not go up to them and ask them why they're doing this to my future?' Mainly I didn't even get to speak to anyone, it was just to the security guards who would say, 'No, sorry, he is in the other country.' I'd say 'Can I speak to whoever is in the office today?' They'd say, 'No, you need an appointment.' The end.

'Mostly it was like that, but two times I got to speak to someone. The first was Sinead Lynch,[6] Shell country chairwoman for the UK. Sinead Lynch basically just said, "We don't need to go fossil fuel free to save your future." Just all these pretty much lies trying to greenwash me. Basically, we both didn't get very far. We both didn't change, nothing much happened there.' She laughs. 'Nothing much to say about that. I was with her 30 minutes.' She continues: 'The next day I got to speak to Shannon Wiseman, the media person for BP. I decided to just cut it short, and like avoid all the arguments and say, "When will you go fossil fuel free?" as like the only question. She said, "We don't know the answer yet." She dodged the question obviously.' Luna laughs.

However the following day, 21 November 2018, her actions were covered in *The Times*, which reported that 'Elsie reminded Ms Lynch that Shell is responsible for 1.67 per cent of global emissions and asked her "on behalf of the children in the world please, please declare a Climate Emergency, have a change of heart and keep your fossil fuels in the ground".'[7]

There is something magical about the protest, as if it was lifted out of a Hans Christian Andersen fairy tale. We're intrigued that Lynch gave 30 minutes in her hectic schedule to an unannounced visit from a ten year old. Why did Lynch pose for a selfie together with Luna, the photo taken by her mum and then passed on to *The Times* to publish? The companies have always held an iron grip over the images of senior staff, ensuring they are taken by photographers on the strictest of contracts in order to prevent pictures of CEOs with unguarded expressions. In corporations that have around 100,000 staff, not more than 20 ever have their photos in the press. Was Shell unable to resist the magical powers of Luna?

Lynch may be the UK country chair of Shell, but her role is more ceremonial than it sounds, and in her 26-year career she has only spent four years in Exploration and Production, the true heart of the fossil fuel machine. She has long experience in facing the public as a commercial manager, and vice president of sustainability. Perhaps her openness to Luna's visit was to demonstrate Shell's willingness to be part of the 'nexus of outrage' that her CEO van Beurden had described to us ten months before? The school strikes had begun with Greta Thunberg sitting down in protest outside the Swedish parliament on 20 August 2018,[8] half a year after we'd met van Beurden and three months before Lynch crossed paths with Luna. The movement had spread like wildfire around the globe in a matter of weeks. Perhaps the school strikers brought a message that the world was thirsting to hear? It seems that Bentham's scenario about the nexus of outrage was peculiarly insightful.

Certainly Luna's actions caught the imagination of the wider public. She was covered in her local newspaper, the *Grimsby Telegraph*, and a little over three months later won a national award for 'individual inspiration'. Caroline Lucas MP was there to give her the plaque at the Climate Coalition Green Heart Hero ceremony in the Houses of Parliament on 11 March 2019,[9] with her local MP Nic Dakin proudly in attendance. Luna and all the other school strikers were commanding media and political attention.

The following month the catalyst of the movement, Greta Thunberg, was also in Westminster, to address a constellation of MPs. There was standing room only at the most packed meeting that the Attlee Suite had seen for a decade. In front of Michael Gove, the environment secretary, she spoke almost inaudibly into the mic. 'I am 16 years old. I come from Sweden. And I speak on behalf of future generations.' She warmed to her theme:

I was fortunate to be born in a time and a place where everyone told us to dream big. I could become whatever I wanted to. Things our grandparents could not even dream of. We had everything we could ever wish for and yet now we may have nothing. Now we probably don't even have a future anymore. Because that future was sold, so that a small number of people could make unimaginable amounts of money. It was stolen from us every time you said that the sky was the limit, and that you only live once. You lied to us. You gave us false hope. You told us that the future was something to look forward to.

With her grave face Thunberg demanded, 'The most dangerous misconception about the climate crisis is that we have to "lower" our emissions. Because that is far from enough. Our emissions have to stop if we are to stay below 1.5 to 2 C of warming. And by "stop" I mean net zero – and then quickly on to negative figures. That rules out most of today's politics.'

She turned her attention to Britain. 'The UK's active, current support of new exploitation of fossil fuels – for example, the shale gas fracking industry, the expansion of North Sea oil and gas fields, the expansion of airports – is beyond absurd.' She acknowledged the challenge. 'The climate crisis is both the easiest and the hardest issue we have ever faced. The hardest because our current economics are still totally dependent on burning fossil fuels, and thereby destroying ecosystems in order to create everlasting economic growth.' She emphasised the necessity to act *now*, chastising the inadequacy of the gathered men and women, all of them the age of her parents or grandparents. 'You don't listen to the science because you are only interested in solutions that will enable you to carry on like before. And those answers don't exist anymore. Because you did not act in time.' She drew to her conclusion: 'We children ... have not taken to the streets for you to take selfies with us, and to tell us that you really admire what we do. We children are doing this to wake the adults up. We children are doing this because we want our hopes and dreams back.'[10]

Thunberg sat down and in silence listened to the MP's responses. Gove, as she'd predicted, told her, 'When I listened to you, I felt great admiration. Your voice – still, calm and clear – is like the voice of our conscience.' There was a note of contrition in his response. He felt 'responsibility and guilt. I am of your parents' generation, and I recognise that we haven't done nearly enough to address the climate change ... that we helped to create.'[11] There is a tremendous power that comes from Thunberg, from Luna, from all the school strikers. It is the disruptive power of moral authority. Just as the Angry Mums of Canvey proved so disruptive in their occupation of the offices of Occidental Oil in 1973. It seems that Lynch was challenged to navigate this and resorted to a selfie.

We find ourselves reflecting on the school strikes. For most of the last 70 or so plastic, pesticide and petrol years things have improved from one generation to the next with a growth in domestic comfort and private wealth. But now the situation has been thrown into reverse by the political response to the financial crisis, and at the same time what is revealed is the decline in the UK's public realm and ecological richness.

In July 2019 Mohammed Barkindo, the secretary general of OPEC, observed there was a growing mass mobilisation of world opinion against oil, which was 'beginning to ... dictate policies and corporate decisions, including investment in the industry'. He said the pressure was also being felt within the families of OPEC officials because their own children 'are asking us about their future because ... they see their peers on the streets campaigning against this industry'. He accused the campaigners of misleading people with unscientific arguments.[12] The school strikers were delighted by his comments that highlighted the sense that civil society was shaping the oil industry, rather than the other way around.

Terrified by the destruction it sees, the country's youngest generation has turned against their elders. Michael Gove talked of his feeling of guilt. The same could be said for us. Gove was born in 1967. His generation is our generation. We are children of the oil revolution. We have all been weaned and sheltered by the petrochemical world against which these children and teenagers are in revolt.

This rebellion poses a radical threat to the oil and gas industry. Bentham and van Beurden had explained that Shell could get to net zero by 2070, and this is the timetable behind Lynch's thinking. But the school strikers demand that the companies abide by the climate science, leave the fossil fuels in the ground and achieve net zero by 2030. For corporations the size of Shell or BP with the depth of their oil and gas assets, this is very difficult without a destruction of their profitability. Both companies have said that they will be looking to abide by the Paris climate agreement, scale back fossil fuel investments and move faster into low-carbon production. Bernard Looney, the new CEO of BP, said the group would be net zero by 2050 and 'reinvent' itself as a new, greener business.[13]

Experts say one option is to follow the example of DONG (now Ørsted) and sell off all fossil fuel assets and give the financial proceeds back to shareholders in the way of a huge last dividend.[14] However, while they might exit oil and gas the assets would still exist. The majority of DONG's fields in the UK North Sea were purchased by INEOS.

Planners such as Jeremy Bentham's team in Shell have endeavoured to plot out a path to net zero. But the target date they aim for is not 2030. This, they assert, cannot be done. Thunberg declares that this will not stand, and a different strategy must be applied. 'Avoiding climate breakdown will require cathedral thinking. We must lay the

330 · CRUDE BRITANNIA

foundation while we may not know exactly how to build the ceiling.' This runs counter to all the principles of the oil and gas industry. No project – even one with the complexity and 60-year lifespan of the West of Shetland fields – is given sanction without understanding how it should unfold unto its very end. Capital demands it is so. To build without knowing how the ceiling is to be completed would be financial insanity.

So the heads of these corporations pursue the familiar role of great paternal institutions, the fathers who are looking out for their children. We had left the head office in The Hague with Shell whispering reassuringly, 'On the matter of climate change, we are aware and very engaged. Leave it to us to handle, we will manage that for you.' This echoed Lynch's lines to Luna: 'We don't need to go fossil fuel free to save your future.'

John Browne, in an interview about his fifth book, *Make, Think, Imagine: Engineering the Future of Civilisation*,[15] made a surprisingly unguarded comment. Asked, 'What would you say to Greta Thunberg?' he replied:

I would say that I have been at this for longer than you've been on the planet and that [decarbonisation] will take time. And so my proposal is this: remember that energy is a very big system and there is not one solution. We can't have one magic bullet that will make the solution work for us. We need to take all the things people are doing [to reduce emissions] because it will be very difficult to persuade them to change.[16]

In his mode of the kindly patriarch he continued:

It's going to take us a long time to take oil and coal out of the energy system and even longer for natural gas. We have the tools to take a lot of the carbon out of hydrocarbons; what we don't have is the right cost. The more you do of something, normally, the cheaper it gets, with the exception of nuclear power [where] the more you make, the more expensive it becomes.[17]

These sound like the words of an engineer in the service of capital. Asked if we can reduce emission fast enough to avoid catastrophe he concluded:

> If we get on with it now, we stand a chance. If we don't start now, the chances get lower and lower. I said 22 years ago that we should start then. James Lovelock reminds me that we're not trying to save this planet, but simply the humans on this planet. The planet will look after itself. It will have its course of life. What we are doing is adjusting its path. We will go. It will come back. We have to do something to keep it ready for human beings. Because if we don't do things, we are simply consigning a portion, if not all, of humanity, to its death. We are stewards because humanity must live on this planet.[18]

What if the youngest generation is not satisfied with avuncular lines such as these from a 71 year old? Thunberg has declared, 'Expansion of North Sea oil and gas fields is beyond absurd.'This is a condemnation of the life work of the likes of Browne, or Trevor Garlick or Jake Molloy. The children, or the grandchildren, of these men condemn their life's endeavours. As the poet W.B. Yeats wrote in *The Wind Among the Reeds*, 'And our children's children's children will say that we have lied.'

22 MAY 2019

It is a small crowd but the message is insistent. Lena Šimić, councillor for the ward of Anfield in north Liverpool, mic in hand, is speaking on the steps of the Town Hall. Beside her a friend holds up a speaker system. On this street in the centre of the city are gathered activists from Labour organisations across Merseyside, school strikers and members of the local Extinction Rebellion group. It is a novel fusion. Šimić, used to speaking before left-wing audiences, is wary of the brooding resentment of the XR crowd and deeply sceptical of the policies of the Liverpool mayor, Joe Anderson. He is known jokingly as 'Concrete Joe' for his allegedly close relationship with a number of private property developers. However, of late he has campaigned hard for government cash to fund a Liverpool Green Deal.

Šimić's[19] lines are compelling and full of conviction:

In the opening paragraphs of her book *Staying with the Trouble* Donna Haraway writes: 'We – all of us on Terra – live in disturbing times, mixed up times, troubling and turbid times. The task is to become capable, with each other in all our bumptious kinds of response.'[20] And here we are gathered calling on our politicians, our elected representatives, for a response, but also enacting that response by being here. We all know we are in trouble, we have no option but to stay with it, to work through it.

She continues:

Merseyside helped build the oil world. We have here one of the country's largest refineries, Stanlow, and a shipyard that constructed oil tankers, Cammell Laird. That is our history. We also have one of Britain's great wind resources, we have Burbo Bank Wind Farm in our Liverpool Bay. We have the tide of the River Mersey, and the possibility of tidal power. We sometimes even have the sun. We are blessed with natural resources.

We need to DECLARE A CLIMATE EMERGENCY NOW! Let's make Liverpool and Merseyside one of the places that builds the post-carbon world. We need a just transition, we need to build our new world for people not profit. The wind, the tide, the sun belong to us. We need to take the wind farms into public ownership – the people will possess the wind!

This is a unique event, a rare coming together. Šimić is part of the new wave of Labour members who flooded into the party around the election of Jeremy Corbyn as Leader. Across Britain this cohort is uniformly of the left and often from younger generations, full of zeal that change can come and a desired future can be created. The activists from XR are from a different political tradition, but they too are here to chart out the future. The climate emergency declaration, already made by 59 local authorities across the UK, is a policy statement about the future, that Liverpool should go carbon neutral by 2030, in a mere eleven years time.[21]

'We need to live and imagine this change already and you are our guiding lights. You and the young people on school strikes. You teach us living otherwise. And I'll not forget this as I walk into the council chambers. As Haraway says: We become-with each other or not at all.'

A couple of months later, the City Council, led by Mayor Anderson, voted unanimously to declare a climate emergency and to make Liverpool net zero by 2030.[22] There are myriad views as to what a net zero, or carbon neutral, Britain would look or feel like, and hundreds of paths to this goal are still being suggested. However, most interpretations agree that oil – and probably gas too – can have only a very limited role in the future. These hydrocarbon commodities are part of the past and the corporations that extract and market them must transform quickly or die out. British civil society is forcing government, local and national, to declare war on the oil and gas industry or at the very least to do battle over its future.

Given that these commodities and corporations have been part of the foundations of Britain for the past century it is remarkable that councils, borough, cities and indeed parliament itself should have made these emergency declarations. These are statements of intent with extremely dramatic consequences. These are indicators of a shift in the shape of this nation as profound as Abercrombie's Greater London Plan that modelled London as a petrocity. They are heralds of the electric cities of the future.

* * *

Ever since her father fixed tiny electric lights to her doll's house when she was a child, Bridgit Hartland-Johnson wanted to be an electrical engineer. Now, 40 years on, that girl turned woman is using a couple of innovative battery storage projects to help keep the lights on in Merseyside.

'I love these projects', says Hartland-Johnson when we meet her at the Ørsted[23] head office in London. 'People have moaned about the financial cost of storage but it's a game changer', she proclaims with a wide smile.

Wind and solar power has historically been dogged by the intermittent nature of production: energy is only produced when the weather permits. Batteries offer the chance to hold on to that generated power until it is needed by businesses or domestic homes. But the technology is still in its infancy and these kinds of storage schemes have previously been deemed too costly to develop and run. The biggest offshore wind developer in the world, Ørsted, has been determined to break through the logjam by setting up two trial schemes in Liverpool.

Hartland-Johnson, who has sat on a European Commission energy advisory group, has played a leading role in both. The first is a two megawatt scheme in Wallasey, at the mouth of the River Mersey and connected to the Burbo Bank Wind Farm out in Liverpool Bay. The second is the much larger 20 megawatt project at Carnegie Road in Liverpool which handles power coming from the several wind farms in Scotland.

'Our electricity consumption pattern is changing and is becoming less predictable as we use more electronic devices and electrify our transport system', she explains. 'The way we generate electricity is also changing as we add more low-carbon sources from wind and solar to the grid. The combination of storage and renewable energy means we can now deliver infrastructure that enhances grid operations and ultimately delivers much better value to us as consumers.'[24]

The Ørsted battery initiative comes amid the wider strategy to turn the Merseyside region from an oil to a low-carbon economy. In the summer of 2019, on submitting Liverpool's bid to the national government for a £230 million Green City Deal, Mayor Joe Anderson explained, 'We need to be bold, radical and ambitious if we are to meet our target of becoming a Net Zero Carbon City by 2030',[25] adding, 'There are huge opportunities to improve the lives of all residents across the city … with better and more energy efficient housing, use of smart technology and making sure young people have the right skills to take advantage of jobs in growth areas.'

Anderson promised his scheme, if it won government funding, would create 10,000 new jobs, build or retrofit 6,000 homes, set up a Liverpool Mutual Bank to help people onto the housing ladder and support new green businesses. The Liverpool city region is also gearing up to put the first of 25 hydrogen-fuelled buses on the road.

The hydrogen fuel cells for the buses are being designed by Arcola Energy, led by Liverpool-born Ben Todd. The vehicles will start with 'brown' hydrogen provided by BOC and derived from natural gas, but will eventually move on to hydrogen made entirely from renewable biogas. Todd told us: 'We are in the process of setting up our production facility in Liverpool and are looking forward to seeing both fleets of zero-emission buses and new high-value manufacturing being established in my home town.'

* * *

There has been a long-running debate over the possibility of constructing a scheme on the Mersey to generate power from the tide.[26] However, in the Severn Estuary schemes are far more advanced. The former BP petrochemicals works at Baglan has been designated as the site for a substation to receive electricity from the Swansea Tidal Lagoon Power project. In this scheme water would be forced through the blades of 16 subsea turbines, each more than seven metres in diameter, at high and low tides. The force, in different directions four times a day, would generate 400 gigawatts of net zero energy a year; enough for more than 150,000 homes. The developer, Tidal Lagoon Power, has similar plans to roll out a scheme in Cardiff, also in the Severn Estuary.

Ironically the brains behind the Swansea Bay Tidal Lagoon is a man who tried but failed to get on a BP graduate scheme in his youth. Mark Shorrock remembers his early brush with BP:

I had finished a degree in Chinese and Spanish at Leeds University and applied for this fast-track graduate scheme with 50 other kids. We had four days of tests meant to show whether we were the kind of high flyers they wanted. At the end of the programme this woman coordinating the scheme said: 'You will undoubtedly do something important with your life but you are too much of a free spirit for us.'

We speak with Shorrock[27] via Zoom from Saigon, where he's trying to set up a solar business. He looks like an old punk rocker with his rueful

336 · CRUDE BRITANNIA

smile, spikey grey hair and black T-shirt. He slips easily from thought-ful reflection to the excited marketing spiel of a business entrepreneur.

Swansea Bay won planning permission in 2015 and had the support of the Labour Party and the Welsh government, but Westminster baulked at the level of subsidies needed to make it economic. After earlier enthusiasm, Conservative ministers formally refused to provide funding for the £1.3 billion project in the summer of 2018. They claimed it was too expensive, but Shorrock is determined to raise the necessary finance in other ways:

> I made some mistakes early on. I was naïve. I had never really dealt with politicians before. All the wind and solar projects I had been involved with just went ahead without objections. Big iconic schemes such as Swansea Bay would make a real difference to Britain but the very size of it seemed to attract the wrong kind of attention at times.

The 49 year old is convinced the fossil fuel lobby in and outside of government played an important role scuppering the hydro scheme's progress:

> Two meetings I remember vividly because they were the hardest I ever had when trying to win support for Swansea Bay. And that includes others with David Cameron and various right-wing think tanks which were trouble-free. One was with Steven Fries,[28] an ex-Shell executive then with the Department of Business, Energy and Industrial Strategy. The negative body language was so extreme when I talked about the positive elements of renewables.

He continues:

> He listened to my pitch for financial support for Swansea Bay in which I outlined the timetable for payback on this first prototype, but just said words to the effect of 'you don't work on a ten-year payback model so its useless'. And yet clearly making such demands on a new technology was unreasonable. And I knew the fossil fuel sector was getting a very sympathetic hearing for the massive

lobbying campaign they were doing around carbon, capture and storage which would allow them to keep on burning oil and gas.

The second difficult meeting was a gathering organised by a PR friend with John Manzoni, the ex-head of BP Refining and Marketing. 'Every time we got onto the subject of Swansea Bay and renewable power more generally, John Manzoni looked incredibly awkward. He was very polite, very professional but his body language suggested he was viscerally uncomfortable with the whole idea of renewables.'[29] Ultimately it was then Business and Energy Secretary Greg Clark who said 'no' to Swansea Bay, claiming it was not the best value for money.

Shorrock says government never liked talking to small entrepreneurial enterprises like his and he had huge difficulty getting meetings with anyone of any seniority:

Ministers like dealing with corporate power: the EDFs[30] of this world. I did do a five-minute briefing about Swansea Bay to Boris [Johnson] in 2016. He said it was a 'fantastic idea' and we should build them all round the British coast. Boris likes entrepreneurs but it remains to be seen whether he will support it now. And I also spoke to Jeremy Corbyn at a Labour dinner last year. He got it. Once he realised the importance of the British supply chain, the fact that the turbines for instance would be made by GE at a factory in Rugby, he was very enthusiastic.

Corbyn promised a Labour government would back the scheme if it won political power.[31]

Although Swansea Bay is billed as revolutionary, there is nothing new about tidal power. There has been a major scheme running in Western Europe on the River Rance, Brittany, since 1966. It generates 500 GW of electricity a year from a 750-metre-long, 13-metre-high barrage. Shorrock saw La Rance as an eight-year-old boy. His father, a Devon-based commercial adviser to the farming community, stopped the car on a bridge overlooking the power station. 'It made a great impression on me. I was captivated by the water gushing into the 24 turbines.'

8 NOVEMBER 2019

When Princess Diana opened a new lubricant centre at the Stanlow Refinery on Merseyside in 1988, she was given a bouquet of flowers by an eight-year-old girl, a daughter of a worker at the plant. Over 30 years later that child, now a woman, described the event: 'She was lovely. She didn't speak to me for very long, but she said, "Now that we've met we will be friends forever".'[32] That daughter went on to become a member of parliament, shadow energy secretary and a potential undertaker for oil. 'I think there is growing recognition now politically and economically that fossil fuels are on their way out', Rebecca Long-Bailey tells us. 'I think the industry knows that. If they don't adapt to that change and start to diversify their operations, then it's going to be very difficult for them going forward. Particularly if they have governments in power like the Labour Party who would be looking at phasing out fossil fuels in the future as quickly as possible.'

We are talking at her constituency office in Salford, Greater Manchester, a shop front on a busy road just up from the Top Trough sandwich bar and Wizard Tattoos and Body Piercing. The walls of the office are decorated with a historic print for the United Machine Workers' Association, a list of the 'Socialist Commandments' and black and white photos of local musical heroes, Joy Division and the Smiths – the iconic image of the band in front of Salford Lads' Club. Long-Bailey[33] admits that in her teens she went clubbing four nights a week in Manchester, had rainbow-dyed hair and a lower lip piercing, wore baggy trousers with a vest and was a 'bit of a skater girl'.[34]

Nowadays in her low-key but stylish office, Long-Bailey is straightforward, smiley and neatly dressed. The shadow business and energy secretary[35] has just turned 40; she reminds us of the young professional lawyer she once was before winning the safe parliamentary seat of Salford and Eccles. Long-Bailey is talking with us in November 2019 just before a general election which is characterised by a high level of debate around climate change issues for the first time. However, the poll did not go the way she expected or wanted. Long-Bailey retained her seat but the overall victory for Boris Johnson and the Conservatives delivered a huge blow to her agenda.

Despite this, the ideas behind a green industrial revolution – municipally owned wind farms, insulating millions of homes and manufacturing electric vehicles – live on. They featured heavily in the promises made by candidates – including her – in the 2020 Labour leadership elections.

Both Labour and the trades unions have been wary of the environmental movement seen by many in the past as too idealistic, too middle class and too careless around the impact of job losses in the shutting of coal mines, nuclear plants or oil facilities. But Long-Bailey says the school strikers and Extinction Rebellion have played important roles shifting awareness.

She is also clear that a switch from coal and oil to renewables must be a fair process that creates new jobs as quickly as it moves people out of old polluting jobs. There can be no reprise of the closing of the mines and the massive impact on communities that followed. 'To be able to make a transition and to take workers with us and people who aren't even involved in those fossil fuel industries, along with the huge shifts that we are going to see to their homes, their cars and their technology, we've got to show that it's good for them. To do that you need an industrial revolution.'

There is a growing field of female economists such as Kate Raworth, Anne Pettifor and Mariana Mazzucato whose academic work supports this drive. Raworth, Oxford University lecturer and author of the high-profile book *Doughnut Economics*,[36] told us that the big oil companies may not necessarily be welcome in the new green economy. 'Yes, we want out of fossil fuels, but we don't want top-down companies that are most interested in the needs of their shareholders completely capturing new sectors such as solar.'

Is the message getting through to top trades unionists like Len McCluskey at Unite and people on the streets of Salford? we ask Long-Bailey. 'Yes, he's a climate activist now isn't he?' she laughs. 'Even my mum, whose never really talked about climate change before, was ranting at me ahead of the 2019 Labour Party conference about setting a zero-carbon target for 2030. "2030 Rebecca! 2030!" And I was like "All right scientist mum, thanks." It's starting to resonate with people.'

Long-Bailey was born not far away and her dad, Jimmy Long, initially worked for Shell unloading tankers at Salford Docks on the Manchester Ship Canal until it was closed down. He was transferred to Stanlow, where he worked in a lubricants plant and Rebecca won a raffle to give flowers to Princess Di. 'When my dad joined Shell in the seventies without an education, it was a massive deal. It was a highly unionised workforce and everybody got paid very well. He always talked about being the luckiest man in the world the day he got that job.' He eventually left and there is some irony that his daughter wants to see the winding down of that industry, albeit in a just way that preserves jobs and communities.

Long-Bailey was an icon of a political movement, which though it appears to have run out of parliamentary steam still courses through the British body politic. At its heart is an understanding that the struggle to deal with climate change means that the days of oil and gas in the UK are numbered, that the end of Crude Britannia is coming and that the next challenge will be how the world of wind, tide and solar is to be controlled. How will the ownership of this common wealth shape the nation?

* * *

The *Discovery* slips through the lock gates of Liverpool Marina and Gary Flint steers her out into the broad brown stream of the Mersey. The tide has just come on to the ebb and the current helps this fibre-glass catamaran out into Liverpool Bay.

This is a strange crew of artists, writers and activist politicians eager to imagine how another energy world could look. It includes Gary Anderson, a drama lecturer, and Lena Šimić, a city councillor. There is Tim Jeeves from Liverpool Walton Constituency Labour Party, and Zoë Svendsen, an artist making theatre about climate change. And ourselves. We have come not to pull cod, whiting or ray from the murky sea but to take possession of the Burbo Bank Wind Farm.[37]

It is not long before we catch a glimpse of our prey, the array of turbines standing pale in the western sea. Behind us the terraces and tower blocks of the city stretch away to the north and south.

There is a steady Force 2 wind from the north-east. The breeze is cold in this morning hour, but the wind with tide makes for a calm passage. We leave Burbo cardinal buoy to port, the boat alters its bearing and settles in for the 40 minutes it will take to reach the base of the turbines.

Šimić calls us together by unfurling a white mainsail made into a banner with scarlet lettering that reads: 'The People Will Possess the Wind.'

In full voice Anderson reads from Norman MacCaig's 'A Man in Assynt':

Who owns this landscape?
Has owning anything to do with love?
For it and I have a love-affair so nearly human
we even have quarrels. –
When I intrude too confidently
it rebuffs me with a wind like a hand
or puts in my way
a quaking bog or a loch
where no loch should be ...
Who owns this landscape? –
The millionaire who bought it or
the poacher staggering downhill in the early morning
with a deer on his back?
Who possesses this landscape? –
The man who brought it or
I who am possessed by it?

Battling the noise of the boat, we stand close as they loudly declaim their lines to the audience of the grey ocean.

Šimić: 'Who owns this seascape? The millionaire who brought it or we who are possessed by it? What does it mean to be possessed by the sea? What does it mean to take possession of the wind turbines that dominate this western horizon?'

Jeeves: We can see before us Burbo Bank Wind Farm and Burbo Bank Extension Wind Farm. The first array of 90 turbines had been erected by the Danish state-owned company DONG by 2007. The second array, of 258 generators, was commissioned by the same corporation in 2017. Together the blades of these machines generate on average 256 megawatts of electricity. Burbo Bank has the capacity to power 230,000 households. When every home and office, every school and shop, is properly insulated and efficiently lit and heated, then these turbines will provide enough power for all citizens of Liverpool and beyond.

Svendsen: How did DONG, which has since changed its name to Ørsted, seize this resource in the first two decades of the century? The bed of Liverpool Bay belongs to the Crown Estates. Ultimately it is the Queen, of course guided by the government, who grants the right for Ørsted to erect its turbines here. This wind that blows from the coast of Lancashire across the sea towards Ireland, belongs to everyone and no one.

Jeeves: A platoon of accountants marshalled the loan from a general staff of international banks who financed Ørsted to construct these machines. Now Ørsted sells power to the Big Six electricity corporations who go house-to-house collecting their profits on the bills of millions of families and businesses across the UK.
We are getting close now. The extraordinary towers rise above us. Their monstrous blades slice though the air.

Anderson: Every part of each turbine was loaded on barges at the Cammell Laird dockyards in Birkenhead and pulled by tug on the same route that we are following. With the aid of cranes of outlandish size the towers were lifted into place and fitted onto concrete foundations sunk deep in the mud bed of the Bay. At the pinnacle of each pillar was fixed the generating unit onto which three blades were attached.

Šimić: How can we have not noticed the building of these massive structures? Once the western horizon was a grey line, now it is dotted with an army of machines that suck money from the movement of air. That same breeze was the power that filled the sails of the trading ships that made the fortunes of the merchants of this city. How are we to harness the common resource of the wind that rocks our boat on the bay? How are we to harness it for the common good? How can we ensure that the money gathered from the bills of families and companies is turned to repair buildings and seal them from the winter cold and summer heat?

Jeeves: What if the land and sea shift from being spaces under corporate control, utilised in generating return on capital, to places under common control, common ownership? The common-land and the common-sea. What is now the property of Ørsted will have to become the property of the people of Liverpool. The rights that the Crown and government hold over the seabed of the bay will have to become the common rights of those that live along this coastline.

Svendsen: What is the benefit of 'control' over a resource, over a place, without a sense of ownership through the heart? A sense of possession. Before ownership comes possession. And we have come here today to help possess these wind farms. To grab them in our imaginations, to let them seep into our dreams, and fill our daily thoughts.

Šimić: Thirty years back the Merseyside band, Orchestral Manoeuvres in the Dark, released their song 'Stanlow'. They used a wild track recording of a pump at the Stanlow Refinery to create the base rhythm of the piece. To the beat of the machine Andy McCluskey sang:

We set you down
To care for us
Stanlow

Now we will record the blades of turbine number seven on Burbo Bank Wind Farm to provide the base rhythm for a track of some future sound.

The captain cuts the engine and the boat begins to gently rock. We stand silent.

A short while passes with the blades swooping above us.

Epilogue
The Commonwealth of Wind

1 OCTOBER 2020

We had planned another trip to the Thames Estuary to meet with refinery workers at the Pegasus Club. But a notice dated 4 December 2019 in the *Gazette*, the government's official public record, reported that liquidators had been called in. The Coryton Refinery recreational club and sporting grounds that was opened in 1961 by Mobil, passed on to BP and then Petroplus, had shut. More ruins.

There was worse on its way during 2020 for the UK employees of the oil industry: Covid-19. With daily life in Britain utterly transformed by the pandemic from mid-April, car travel stalled, home working increased and airports were closed. Demand for petrol from refineries like Stanlow plummeted. For the first time in decades pipelines such as that pumping aviation fuel from Coryton to Gatwick were thrown into jeopardy. Global demand for oil products fell and, coupled with a price war between Saudi Arabia and Russia, this forced down the value of crude.

The International Energy Agency declared the oil industry was facing the worst situation for a quarter of a century and on 21 April 2020, crude prices in America dropped to zero. Contracts for future deliveries of oil reached minus $40.32 per barrel at one point.[1] This meant producers were paying customers to take oil off their hands rather than go through the costly business of shutting down their wells. Up and down the UK North Sea there were oil platforms now producing crude at a loss, and fears that 40 per cent of the offshore workforce would be laid off.

In the Thames Estuary, in Liverpool Bay off Anglesey[2] and at the mouth of the Firth of Forth, the hulks of oil tankers dotted the horizon. Every available ocean-going tanker was hired to store oil and anchored off coasts around the world. The cost of leasing a VLCC doubled within a week as demand exceeded supply.[3]

Oil companies, traders and others presumed the pandemic would blow over and crude prices would recover. If they could just keep oil stored on land or at sea, in fact anywhere off the market, they could profit from the upturn when it came.

In the face of falling profits and plunging share prices, the major oil companies began to cut back spending, axe jobs and cut payouts to shareholders. By May the 2020 *Sunday Times* Rich List was reporting that Jim Ratcliffe of INEOS had seen £6 billion wiped off his personal wealth.[4]

By mid-summer Shell was writing off $22 billion worth of oil and gas holdings[5] and was slashing its payouts to shareholders for the first time since World War II. In an unusual admission, CEO Ben van Beurden said the group's rarely mentioned trading arm had saved it from heavier losses.

Similarly, BP wrote off $17 billion of assets, leading *Forbes*, the American business magazine, to question whether this was the beginning of the end for the oil industry.[6] The arguments made to us by Jeremy Leggett, Carbon Tracker and climate campaigners about the fear of 'stranded assets' were now being realised.

On 4 August 2020 the new BP chief executive, Bernard Looney, announced the company would cut its fossil fuel production by 40 per cent in the next decade. He declared: 'BP today introduces a new strategy that will reshape its business as it pivots from being an international oil company focused on producing resources to an integrated energy company focused on delivering solutions for customers.'[7]

In addition, the corporation promised to invest upwards of $5 billion a year in low-carbon power sources such as wind. BP wanted to be known in the future as an energy company, not an oil company, Looney insisted.

The 'Beyond Petroleum' strategy started by John Browne two decades ago, but abandoned by his successor, Tony Hayward, was back on, this time because BP's survival was at stake. Browne praised Looney's 'very clear' statement of direction, while Nick Butler admitted BP was facing a highly complex and fast-moving situation. Butler added: 'BP has made the boldest step yet taken by any of the major oil [and] gas companies and this is a sign that we are moving into a new era. What remains uncertain is the pace of change.'

However, would BP follow through on these words, and would other oil companies follow? And even if BP did reduce its oil and gas output by 40 per cent over the next decade, would that be enough to make the company compliant with the Paris Agreement? Many thought not. Greenpeace, impressed by BP's change of heart, noted that the company's investments in third-party oil companies such as Rosneft of Russia were not mentioned in the new game plan.

Other former BP executives had their own mixed views on the future. One said Looney had little choice but to make major changes in corporate direction given British and international civil society and financial and political pressure to tackle climate change.

'Can they deliver on it? It's a big question. The rates of return on capital in the renewable sector are much lower than in oil and gas. One big oil find can keep you in profits for the next five years. Renewables are still largely dependent on government subsidies', he said. 'Quite honestly if I was running the show I would sell up. You are trying to enter an electric economy, which you don't really intuitively understand. I think a smooth transition from oil to renewables is all but impossible. Look at oil companies' [difficult] past experiences with diversification.' He saw there was a high likelihood that a company under less powerful civil society pressure – such as ExxonMobil – could come in and buy up BP.

Oil companies had helped shape civil society in the past, now the boot was on the other foot. Civil society was shaping the crude producers, but the pressure was much more intense in Britain and in the rest of Europe than the US.[8]

However, the world's largest private oil corporation, ExxonMobil of Texas, was certainly not impervious to the impact of Covid, lockdown and the collapse in demand for oil. ExxonMobil was thrown out of the S&P Dow Jones Industrial Index of top US stocks for the first time since 1928.[9]

The oil industry's decline paralleled Big Data's rise, as across Britain teleconferencing, home working and shopping became the norm. That meant a massive boost to the sales and share price of Amazon and Facebook, plus newer companies such as Zoom. The electric economy was on fire. Zoom shares rose from $68 to $457 in the first eight months of 2020. Electric vehicle manufacturer Tesla saw its shares rise from $84 on 1 January 2020 to nearly $500 by 1 September 2020.

For the first time since the 1940s it was possible to raise questions over the future of the oil pipeline systems that had been the skeleton of the country. The sinews of electrical power, running from homes through the National Grid to the offshore wind farms, became ever more vital. It seemed that a symptom of Covid was to accelerate a change in the body of Britain from oil to electricity. A shift that both Jeremy Bentham and Ben van Beurden had alluded to but that neither would have imagined coming about so rapidly. Shell and BP were still banking on natural gas being used as a 'transition' fuel to a full electric economy and they hoped it could be used to produce fossil fuel-based hydrogen to fuel buses and possibly even ships.

We set out on our journey four years ago to highlight how oil and its industry had shaped Britain, past and present. We ended with a sense that the people of this island were beginning to both reshape the oil majors and the nation's future.

Tony Wade, one of the former Coryton Refinery workers who we'd met at the Pegasus Club, was still upbeat about Brexit and still believed in a good future for oil. 'We [the UK] are not going to be self-sufficient with wind power and anyway the Chinese have a stranglehold on the minerals needed for batteries and the like. We still have oil and with Brexit we have our nation back.' Crude Britannia? 'Yes it's still alive', Wade insisted. But despite his loyal commitment to fossil fuels it was with some irony that his local Thurrock Council had just invested £600 million in solar power schemes.[10]

Meanwhile the Welsh government had identified eleven 'priority' areas to develop large-scale onshore wind and solar.[11] A consultation document says: 'Welsh coal, steel and iron drove the industrial revolution, and our wind, solar and tidal resources point forward to a clean, sustainable future.'

And in Aberdeen the city's most celebrated oil services company, Wood Group, was starting work on a decarbonisation plan for the north-east of Scotland,[12] amid a debate on whether the North Sea might have to shut down to meet UK climate targets.

Britain's world of oil was changing fast and popular culture reflected that. Dave Randall,[13] the former Faithless guitarist, who like Wilko was brought up in Essex overlooking oil refineries on the banks of the Thames, released climate songs with his Slovo music collective.

About the *Bread and Butterflies* album released in the summer of 2020, he said: 'This feels like such a strange and pivotal time to be alive. We've tried to explore the issues affecting us all, and to capture some of the feelings they invoke. But above all this is an optimistic album, born of the belief that a better world is possible.'

Slovo vocalist Barbarella urges us on the 'Deliver Us' track to:

Rise up for the people
And rise up for the trees
Rise for the birds and rise for the bees
Rise for the elders and rise for the young
Rise for the songs yet to be sung

And the young are still rising.[14] Elsie Luna, the ten year old we met in Parliament Square and who went on to found XR Kids was still deeply frustrated with oil companies, governments and even Extinction Rebellion. If civil society was changing Britain and the oil companies, it was not nearly fast enough for her.

'I think XR is wrong to prioritise climate over justice issues: neither is more or less important than the other', she says. 'The oil companies are continuing to exploit land and indigenous people just to make themselves rich. All this talk by oil companies of decarbonisation is just greenwash and current systems [of government] are just part of the problem. They will not take the right decisions needed.'

So has she given up? 'No, definitely not. But I think we have to change the politics and that starts with doing work in local communities. So that's what I am going to do now.'

In the ruins of an oil world, the new is being built.

Notes

PROLOGUE

1. All quotes from interview with Terry Macalister and James Marriott, 30 March 2017.
2. The lay offs at Coryton were resisted by the workforce, with pickets and outbreaks of violence with the police on 25 June 2012, www.dailymail.co.uk/news/article-2164473/Running-battles-break-police-union-picket-line-British-fuel-refinery-threatened-massive-job-cuts.html.
3. See Arnold Wesker's play *Beorthel's Hill* (1988). Massey Ferguson opened a tractor factory at Basildon in May 1964.
4. MFG (Motor Fuel Group), www.motorfuelgroup.com, is a subsidiary of Scimitar Topco Ltd, itself 33.04 per cent owned by Alasdair Locke, chairman of MFG, who began as an oil trader in Citibank, later relocating to Singapore. His career echoes that of David Jamison. In January 2018 Scimitar Topco was purchased by US private equity firm Clayton, Dubilier and Rice, www.motorfuelgroup.com/wp-content/uploads/2019/04/MFG-Group-Structure.pdf. MFG has more than 900 sites in the UK, operating franchises of BP, Shell, Esso, Jet, Murco, Texaco, Budgens, Costa Coffee, Greggs, Spar, Subway and Hursts.
5. XPO Logistics controls a substantial section of the UK tanker market following its 2015 purchase of Norbert Dentressangle, a French multinational, https://theloadstar.co.uk/xpo-logistics-swoops-in-3–5bn-takeover-of-nd-famous-red-brand-will-disappear/. XPO Logistics is part owned by US billionaire Brad Jacobs, director of Jacobs Private Equity LLC. Jacobs began his career as an oil trader in 1976.
6. Greenergy, a private equity company based in London, is itself owned by Brookfield Business Partners another private company based in Toronto, Canada. Brookfield is reputedly linked to the interests of the Saudi Royal Family.
7. CLH – Compania Logistica de Hidrocarburos – is a Spanish pipeline corporation largely owned by the giant private equity company CVC Capital Partners, a firm spun out of US banking group Citicorp but now formally headquartered in Luxembourg.

CHAPTER 1

1. Music by Kurt Weill, original text by Felix Gasbarra from *Konjunktur* (1928) by Léo Lania. The song tells the story of a seaside village destroyed by the oil industry. First, oil is discovered and the beautiful waterfront is ruined by oil tanks. Then the local economy is taken over by oil cartels and workers are

forced to work on the oil rigs. Finally, as the thirst for oil grows, and more rigs are erected, burning oil sets the world on fire. It premiered in Berlin in 1928, directed by Erwin Piscator, www.kwf.org/pages/ww-muschel-von-margate. html.

2. Daniel Yergin, *The Prize*, Simon & Schuster, London, 1991, pp. 316–18 and p. 335.

3. Harold Nockolds, *Shell War Achievements, Vol. 4: The Engineers*, The Shell Petroleum Company, London, 1949, p. 15.

4. An intense increase in oil production was required to keep Allied and Axis forces in the field and air and on the seas – see Daniel Yergin, *The Prize*, Simon & Schuster, London, 1991, p. 379.

5. Harold Nockolds, *Shell War Achievements, Vol. 4: The Engineers*, The Shell Petroleum Company, London, 1949, p. 15.

6. James Bamberg, *The History of the British Petroleum Company, Vol. 2: The Anglo-Iranian Years, 1928–1954*, Cambridge University Press, Cambridge, 1994, p. 243.

7. W.E. Stanton Hope, *Shell War Achievements, Vol. 1: Tanker Fleet*, The Shell Petroleum Company, London, 1948, pp. 18–20. The *San Calisto* was bound for Southend-on-Sea, aiming to rendezvous with other ships to form a convoy to head south through the Channel and out across the Atlantic. She sank 2.5 miles south west of Tongue Lightship.

8. The six men were: Arthur Andrew Boswell (42, Hull), Harry Craig (31, Ellesmere Port), Christopher John Holmes (20, North Shields), Evan Leonard (27, Lampter), Ernest Wheartey (61, unknown), John Wheeldon (22, Hull), www.wrecksite.eu/wreck.aspx?73908.

9. During the war Grangemouth Refinery, near Edinburgh, was closed and Llandarcy by Swansea gradually ceased production.

10. James Bamberg, *The History of the British Petroleum Company, Vol. 2: The Anglo-Iranian Years, 1928–1954*, Cambridge University Press, Cambridge, 1994, p. 210.

11. W.E. Stanton Hope, *Shell War Achievements, Vol. 1: Tanker Fleet*, The Shell Petroleum Company, London, 1948, p. 123.

12. James Bamberg, *History of the British Petroleum company, Vol. 2: Anglo-Iranian Years*, Cambridge University Press, Cambridge, 1994, p. 216.

13. Ibid., p. 210.

14. Shell and Anglo-Iranian had been deeply engaged with the European fascist states prior to the war, although Shell undoubtedly had the closer relationship. Sir Henri Deterding, a Dutch-born British resident, was head of Shell from 1900 and for 36 years held his position in an infamously autocratic manner, but nevertheless operated as part of a board with a number of executive directors. Deterding was outspokenly anti-communist and after the Soviet Union expropriated Shell oilfields in the Caucasus in 1921 he tried to coordinate with other companies an embargo on all oil trade with the Bolsheviks. Ultimately the cartel between these corporations failed to hold, but Deterding saw much hope in the possibility of Mussolini and Hitler coming to power. He personally helped finance the Nazi Party in the late 1920s.

In Germany, as in the UK, the boards of Anglo-Iranian and Shell engaged in a constant dialogue with senior government figures. In the summer of 1935 a trade delegation of British businesses, went on a tour of Germany and in September of that year the Anglo-German Fellowship was founded in London, intended to develop closer alliances between British companies and the Nazi government. Sir Frank Tiarks and Deterding were founder members. Tiarks was a director of the Bank of England and a senior director of Anglo-Iranian, having by that time been on the company's board for 18 years. Other members included Sir Montague Norman, governor of the Bank of England, Geoffrey Dawson, editor of the *Times* and Sir Andrew Agnew, a director of Shell and later head of the Petroleum Board. These men were among many British executives in the 1930s that saw much they admired in fascism.

Deterding retired from the Shell board on December 31 1936 after a protracted internal struggle, partly due to his age of 70 and partly due to growing unease within the British government over his proximity to the Nazis. By September 1935 Sir Andrew Agnew had raised his concerns over Deterding's discussions with German officials and asked the Foreign Office to investigate through the Berlin Embassy. An FCO staff member commented, 'The British members of the board are keen that the board should not do anything contrary to the views of H.M. Government.' Deterding retired to his hunting villa in Mecklenburg and became an outspoken advocate for Hitler, writing several pieces in the German press declaring his support. The former chairman of Shell died in February 1939. He was buried with high-ranking Nazis at his funeral and a wreath sent by Hitler inscribed 'Heinrich Deterding, Ein truer Freund den Deutschen Volk'.

15. Dan Gretton, *I, You, We, Them, Vol. 1*, William Heinemann, London, 2019, p. 174–5.
16. Ibid., p. 175.
17. There were reports in the *New York Times* that Deterding agreed to give Germany one year's oil supply on credit and that the company would build a network of petrol stations designed to be protected from air attack along the new German Autobahns.
18. Now ExxonMobil.
19. James Bamberg, *History of the British Petroleum company, Vol. 2: Anglo-Iranian Years*, Cambridge University Press, Cambridge, 1994, p. 210.
20. Stephen Howarth and Joost Jonker, *Powering the Hydrocarbon Revolution, 1939–1973: A History of Royal Dutch Shell, Vol. 2*, Oxford University Press, Oxford, 2007, pp. 469–70.
21. The official history of Shell notes the attempts to frustrate anti-Jewish measures imposed by German authorities in the Netherlands after October 1941. Forty Shell staff in the Netherlands were 'classified as Jewish under German laws'. They were dismissed in April 1942, and 'At least twenty of them did not survive the war.' The history does not detail what happened to Shell staff in other states under Nazi control – Germany, Belgium, Austria, Czechoslovakia, Italy and France. See Stephen Howarth and Joost Jonker,

   Powering the Hydrocarbon Revolution 1939–1973: A History of Royal Dutch Shell, Vol. 2, Oxford University Press, Oxford, 2007, pp. 83–4.

22. Joost Jonker and Jan Luiten van Zanden, *From Challenger to Joint Industry Leader, 1890–1939: A History of Royal Dutch Shell*, Vol. 1, Oxford University Press, Oxford, 2007, p. 469.

23. As the official company history states: 'The far-reaching changes to the Rhenania-Ossag board could not have taken place without the full consent of central offices ... Moving managers and employees to other jobs was probably seen as a cosmetic exercise of the kind occasionally required to placate particular regimes, and no more. No questions of principle or moral judgements about the Hitler regime appear to have arisen and it bears pointing out that, whereas correspondence shows Group managers quick to identify and condemn Bolshevism, they appear not to have had the same sensitivity to Fascism or Nazism. We do not know the Group's treatment of the staff members concerned, nor their fates.' See Ibid., p. 469.

24. Except for private ownership of capital, state socialism and communism offered these conditions as well.

25. Anglo-Iranian's petrol stations in Germany in 1930s were branded with the 'BP' logo.

26. At this time Anglo-Iranian was 51 per cent owned by the British government.

27. Stephen Howarth and Joost Jonker, *Powering the Hydrocarbon Revolution, 1939–1973: A History of Royal Dutch Shell*, Vol. 2, Oxford University Press, Oxford, 2007, p. 78.

28. Ibid., p. 31.

29. Ibid., p. 48.

30. Ibid., p. 86.

31. Ibid., p. 81.

32. A Shell company, Astra Romana, 'was brought into the German fold' when von Klass appointed a prominent Dutch Nazi, J.H.W. Rost van Tonningen, to be its general manager. See Ibid., p. 31.

33. Ibid., p. 69.

34. There is also substantive evidence that Rhenania-Ossag used forced labour to maintain the running of their Hamburg Harburg Refinery which was heavily impacted by Allied bombing.

35. See a receipt for aviation fuel provided by Shell in Delft from Der Kommandant des Flugplatzes in Luftwaffe, 28 May 1940, reproduced in Stephen Howarth and Joost Jonker, *Powering the Hydrocarbon Revolution 1939–1973: A History of Royal Dutch Shell*, Vol. 2, Oxford University Press, Oxford, 2007, p. 81.

36. The exact division of finances between the two parts of Shell during the war is obscure, but at the end of the conflict many of the assets of the Axis body were absorbed back to Shell Group. The head offices in The Hague are a resonant example of this.

37. Philip Ziegler, *London at War*, Sinclair-Stevenson, London, 1995, pp. 122 and 144.

38. On 8 September 1940 Winston Churchill and Chief of Staff General Hastings Ismay inspected the bomb damage in the East End. Ibid., p. 115.

39. See W.G. Sebald, *On The Natural History of Destruction*, Hamish Hamilton, London, 2003.

40. Centre for the Study of War, State and Society, University of Exeter, https://humanities.exeter.ac.uk/history/research/centres/warstateandsociety/projects/bombing/britain/.

41. Nicholas Crane, *The Making of the British Landscape*, Weidenfeld & Nicolson, London, 2016, pp. 492–6.

42. Ibid., pp. 492–6.

43. Abercrombie's intention is echoed in twenty-first-century attempts to find that balance again in relation to the new environment of climate change. See Ibid., p. 493.

44. Peter Ackroyd, *London: The Biography*, Vintage, London, 2001, p. 764.

45. See Lewis Mumford's concept of Paleolithic City in Lewis Mumford, *The City in History*, Penguin, London, 1973.

46. Patrick Abercrombie, *Greater London Plan*, HMSO, London, 1944.

47. The Greater London Plan designated an area of southern Essex on the banks of the Thames as one of the largest industrial zones in the region. It became the refineries of Coryton and Shell Haven.

48. See Edward Platt, *Leadville: A Biography of the A40*, Picador, London 2001.

49. The Autobahns had long been admired. As Stephen Howarth notes, 'in the summer of 1937 one Shell employee, de Maat, went on a 1,800-mile tour … In Germany, the last leg of his tour, de Maat was impressed by "the new *Reichsautobahnen*, the great new motor highways, which aroused our admiration for modern German enterprise".' Stephen Howarth, *A Century in Oil: The 'Shell' Transport and Trading Company 1897–1997*, Weidenfeld & Nicolson, London, 1997, p. 191.

50. Aneurin Bevan MP, minister of health, 3 August 1945–17 January 1951.

51. There were two distinct but overlapping plans, the County of London Plan 1943 and the Greater London Plan 1944.

52. The phrase 'commanding heights' is attributed to Lenin by Daniel Yergin. See Daniel Yergin and Joseph Stanislaw, *The Commanding Heights: The Battle for the World Economy*, Simon & Schuster, New York, 2002.

53. The hydro schemes in Highland Scotland were initiated by the North of Scotland Hydro-Electric Board, established in 1943 and nationalized in 1948. It was chaired by 'Red Clydesider' Tom Johnston from 1945 to 1959.

54. The matter of Shell, BP and other British oil companies not being nationalised is a case of 'the dog that did not bark in the night'. It was an act that did not take place, yet the absence of nationalisation and the impact of this non-act had profound implications in British life. Private British coal companies such as Powell Duffryn and Baldwin were taken into public ownership in 1947 and the coal industry remained a pillar of social democracy until the 1980s. The private provision of health care was largely replaced by the NHS, which today remains a cornerstone of communitarian and civic life in the UK. Could the same thing have happened to the oil industry? In 1945 the UK government held 51 per cent of BP's shares. Meanwhile the board of Shell

had courted the UK government between 1940 and 1942 hoping that Britain might take a substantial stake in the company while the British directors battled with the Dutch directors. Any discussion over public ownership of the oil industry would probably have taken place between Emanuel Shinwell MP and Hugh Gaitskell MP (ministers of fuel and power) on the state side, and AIOC Chairman William Fraser and the chairs of Shell Transport and Trading and two London operating companies, Walter Samuel and Lord Fred Godber on the side of private capital.

55. Both BP and Shell had laid the groundwork for this identity as 'national champions' through their advertising in the 1920s and 1930s which utilized British symbolism to sell products. Shell in particular ran three campaigns: 'See Britain First on Shell' in the 1920s, and 'Everywhere You Go You Can Be Sure of Shell' and 'To Visit Britain's Landmarks' in the 1930s. They employed artists such as Charles D. Fouqueray (1925), Hall Woolf (1931), Vanessa Bell (1931), Frank Dobson (1931), Duncan Grant (1932), Harold Steggles (1934), Edward Bawden (1936) and Denis Constanduros (1937). See the Shell Collection at the National Motor Museum Trust. In parallel Shell published the *Shell Guides* from 1934, covering Cornwall, Buckinghamshire, Derbyshire, Devon, Dorset, Hampshire, Gloucestershire, Kent, Northumberland and Durham, Oxfordshire, Somerset, Wiltshire and the west coast of Scotland.

56. Recent research suggests that Churchill's view of the festival was more pragmatic, both opposing and supporting it at different moments. See Iain Wilton, *Winston Churchill and the 1951 Festival of Britain*, Autumn 2017, https://winstonchurchill.org/publications/finest-hour/finest-hour-174/1951-festival-of-britain/.

57. The 'Beeching Axe' was named after Dr Richard Beeching, a chairman of the British Railways Board who wrote a report advocating the closure of many branch lines.

58. See http://news.bbc.co.uk/onthisday/hi/dates/stories/december/16/newsid_4035000/4035801.stm.

59. See R.E.G. Davies and Philip J. Birtles, *Comet: The World's First Jet Airliner*, Paladwr Press, Virginia, 1999.

60. See Timothy Walker, *The First Jet Airliner: The Story of the de Havilland Comet*, Scoval Publishing Ltd, Newcastle upon Tyne, 2000.

61. See Peter Lane, *The Queen Mother*, Hale, London, 1979.

62. See Wired, 4 October 1958, 'Comets Debut Trans-Atlantic Jet Age', www.wired.com/2010/10/1004first-transatlantic-jet-service-boac/.

63. See R.E.G. Davies and Philip J. Birtles, *Comet: The World's First Jet Airliner*, Paladwr Press, Virginia, 1999. The journey time in 2020 from Heathrow to Tokyo is 11 hours 25 minutes.

64. James Bamberg, *The History of the British Petroleum Company, Vol. 2: The Anglo-Iranian Years, 1928–1954*, Cambridge University Press, Cambridge, p. 295.

65. Stephen Howarth and Joost Jonker, *Powering the Hydrocarbon Revolution, 1939–1973: A History of Royal Dutch Shell, Vol. 2*, Oxford University Press, Oxford, 2007, p. 302.

66. James Bamberg, *The History of the British Petroleum Company, Vol. 2: The Anglo-Iranian Years, 1928–1954*, Cambridge University Press, Cambridge, p. 295.

67. Heathrow catered for 796,000 passengers in 1951 the last year before jet airliners. In 2018, 80 million people passed through Heathrow per year, a daily average of 219,458. Thus in 2018 the passenger load of all of 1951 was carried in just 4.6 days. See Heathrow Our Company, www.heathrow.com/company/about-heathrow/company-information/facts-and-figures.

68. 'The period of Fraser's chairmanship was ... one of unusual strain in the Company's relations with the Government. To a large degree Government criticism of the company was focused on the exceedingly strong personality of Fraser, whose uncompromising temperament was to raise many hackles not only in Westminster and Whitehall, but also in Washington.' James Bamberg, *The History of the British Petroleum Company, Vol. 2: The Anglo-Iranian Years, 1928–1954*, Cambridge University Press, Cambridge, p. 213.

69. These projects depended on the state financing infrastructure around the new plants – for example council housing for workers in villages near the Kent Refinery and close to the Carrington chemical works.

70. W.J. Harvey and R.J. Solly, *BP Tankers: A Group Fleet History*, Chatham Publishing, London, 2006, pp. 108–61.

71. Cammell Laird continued to construct tankers after 19 other yards had ceased to do so.

72. Stephen Howarth, *Sea Shell: The Story of Shell's British Tanker Fleets 1892–1992*, Thomas Reed Publications, London, 1992.

73. W.J. Harvey and R.J. Solly, *BP Tankers: A Group Fleet History*, Chatham Publishing, London, 2006, pp. 108–61.

74. Indeed they were the last tankers for Shell outside those built at Harland and Wolff in Belfast.

75. David Roberts, *Cammell Lairds: Life at Lairds, Memories of Working Shipyard Men*, Avid Publication, Gwespyr, 2008.

76. James Bamberg, *The History of the British Petroleum Company, Vol. 2: The Anglo-Iranian Years, 1928–1954*, Cambridge University Press, Cambridge, pp. 290–2 – the British Tanker Company changed its name to the BP Tanker Company in 1955 and subsequently to BP Shipping.

77. 'To help Britain, *British Admiral*' and British industry deserve a salute from us all' ended the narration on two Pathé News features covering the launching of the BP tanker MV *British Admiral* built at Vickers Armstrong. Launched by the Queen, the tanker was the largest ship being built in Europe at the time, a 100,000 tonne vessel, the heaviest ship built in the UK since World War II. See www.britishpathe.com/video/queen-names-giant-tanker.

CHAPTER 2

1. Viguen's 'Moonlight' was a hugely popular Iranian jazz-pop song released in 1954, it is considered a turning point in Western-influenced Iranian music. The lyrics translate as 'Moonlight! | a lover's companion | Brightness of the heavens | moonlight! The sky lights | luminous of the world | where is my

moon? | with you at nights | we were with her next to you | regardless of the world | we were 'lips on the lips'.

2. Kathy Evans and Douglas Marsh, *Who's Who: A Century of Memories*, Running Dog Press, UK, 2008, pp. 37–45.

3. The Anglo-Iranian Oil Company was the name given to the previously titled Anglo Persian Oil Company, which became British Petroleum and finally BP.

4. James Bamberg, *The History of the British Petroleum Company, Vol. 2: The Anglo-Iranian Years, 1928–1954*, Cambridge University Press, Cambridge, p. 293.

5. The Anglo Persian Oil Company, precursor of AIOC, had from its foundation an institutional link with the Royal Navy. In 1913 First Sea Lord Winston Churchill MP drove the British government's purchase of 51 per cent of APOC's shares with the company in return becoming the key supplier of oil to British warships.

6. Kathy Evans and Douglas Marsh, *Who's Who: A Century of Memories*, Running Dog Press, UK, 2008, pp. 100–1.

7. Ibid., pp. 37–45.

8. Interview, James Marriott with Peter Pickering, 30 November 2016.

9. Both Hogarth and Searle worked for Shell as illustrators of *Shell War Achievements*, vols 1, 2 and 4, published by the Shell Petroleum Company, London, 1948 and 1949.

10. All quotes from the personal diaries of Peter Pickering, Alan Hubbard, *A Post Normal Life*, 2015.

11. James Piers Taylor, 'The Island', in *Shadows of Progress: Documentary Film in Post-War Britain 1951–1977*, BFI, London, 2010, pp. 15–16.

12. Am Dally, 'Dylan Thomas in Iran: New Discoveries by John Goodby', *Dylan Thomas News*, https://dylanthomasnews.com/2017/06/15/dylan-thomas-in-iran-new-discoveries-by-john-goodby/.

13. At the time of Dylan Thomas' journey to Iran, another young Swansea man, Michael Heseltine, was at Shrewsbury School, and preparing to take his exams to allow him to go to Oxford University, a vital step away from Swansea.

14. Am Dally, 'Dylan Thomas in Iran: New Discoveries by John Goodby', *Dylan Thomas News*, https://dylanthomasnews.com/2017/06/15/dylan-thomas-in-iran-new-discoveries-by-john-goodby/.

15. Ibid.

16. James Bamberg, *The History of the British Petroleum Company, Vol. 2: The Anglo-Iranian Years, 1928–1954*, Cambridge University Press, Cambridge, p. 352.

17. Ibid., p. 350.

18. As a consequence of AIOC being 51 per cent owned by the British state, income to the UK Exchequer via dividends and tax revenues was substantial. It is notable that the Labour government who had made nationalization a central plank of their program in the UK resisted attempts by Iran to do likewise, whilst Mossadegh saw the symbolic importance of the nationalization of British coal industry.

19. The spellings of the prime minister's name is variously given as Mosaddegh or Mossadegh. We have settled on the latter.
20. James Bamberg, *The History of the British Petroleum Company, Vol. 2: The Anglo-Iranian Years, 1928–1954*, Cambridge University Press, Cambridge, p. 425.
21. Mark Curtis, *Web of Deceit*, Vintage, London, 2003, p. 307. On 26 January 1952 British forces used tanks and heavy weapons against Egyptian police in Ismaila, Egypt. At least 50 police were killed and 73 injured.
22. For further reading see James Bamberg, *The History of the British Petroleum Company, Vol. 2: The Anglo-Iranian Years, 1928–1954*, Cambridge University Press, Cambridge and Mark Curtis, *Web of Deceit*, Vintage, London, 2003.
23. John Browne, *Beyond Business: An Inspirational Memoir from a Visionary Leader*, Weidenfeld & Nicolson, London, 2010, p. 16.
24. Ibid., p. 17.
25. Ibid., p. 17.
26. Mark Curtis, *Web of Deceit*, Vintage, London, 2003, p. 314.
27. Remarkably the pipeline co-existed for some time with oil distribution from Grain also by sailing barge.
28. The anxiety over Iran fuelled a shift in BP's strategy to diversify its places of oil production which underpinned the decision to buy Alaskan concessions on 3 April 1959. The BP board that took the decision on North Slope acreage included Neville Gass (chair), Maurice Bridgeman, Eric Drake, Harold Snow, John Pattinson and Bryan Dummett. All had lived through the Iran crises, especially Eric Drake. There was a powerful symbolism in this shift from Iran to the US, from the desert to the Arctic, from British hegemony to US hegemony, and in the name, from AIOC to BP.
29. From 1950 the lawyer in the BP Concessions Department liaising with the Kuwait Oil Co. was David Steel, future chair of BP.
30. Story gathered by James Marriott 2018 from Andrew Mitchell, father of Chris Mitchell, who was a Socony-Vacuum manager at Coryton in 1953.
31. The single was originally a B-side release, but swiftly became the most famous of Taylor's tracks. It was covered by The Clash on *London Calling* in 1979, who described it as 'one of the first British rock 'n' roll records'.

CHAPTER 3

1. In the constituency of Right Honourable Selwyn Lloyd MP lay Shell's Stanlow Refinery processing oil from Ogoniland and elsewhere.
2. See entry on Ogoni on website of Unrepresented Nations and Peoples Organization, 11 September 2017, https://unpo.org/members/7901.
3. All quotes from interview by Terry Macalister with Lazarus Tamana, London, 23 October 2019.
4. The Royal Niger Company was formed in 1879 and was absorbed into Unilever in the 1930s.
5. Exploration for oil in the Delta was first undertaken by a German company, the Niger Bitumen Company, from 1907 to 1914.

6. For AIOC the move into Nigeria was partly in response to a loss of concession terms in Iran and for Shell partly in response to the threat of nationalization in Mexico – see Andy Rowell, James Marriott and Lorne Stockman, *The Next Gulf: London, Washington and the Oil Conflict in Nigeria*, Constable & Robinson, London, 2005, p. 56.

7. Following the discovery of oil at Olobiri, Ijawland in January 1956, subsequent discoveries were made in Ogoniland, such as at Dere in February 1958. In many areas of the eastern delta there were the sounds of explosions from seismic testing.

8. Andy Rowell, James Marriott and Lorne Stockman, *The Next Gulf: London, Washington and the Oil Conflict in Nigeria*, Constable & Robinson, London, 2005, p. 66.

9. Ibid., p. 77.

10. The BP executive filed a report describing a visit to these offices in March 1963, as noted by Bamberg: 'In Nigeria, where Shell provided the staff and management for the jointly owned Shell-BP Petroleum Development Company of Nigeria, there were by 1963 about 300 expatriates at Shell-BP's main administrative centre. This expatriate community was … "extremely conspicuous both by its numbers, its relative affluence and the quality of its housing".The Shell-BP general office in Port Harcourt was "by far the largest and most conspicuous building in that place." It was flanked by workshops, a hospital "which to lay observers compares favourably with the most expensive private clinics in the United States", and an attractive Shell-BP housing estate. "The contrast between all this and the town of Port Harcourt remains vivid."' See James Bamberg, *The History of The British Petroleum Company, Vol. 3: British Petroleum and Global Oil, 1950–1975*, Cambridge University Press, Cambridge, 2000, p. 73.

11. The first length of pipeline ran 70 miles from Oloibiri to Port Harcourt, built between April 1957 and June 1958. It was later extended to the Bonny Terminal, completed in summer 1961. See Stephen Howarth and Joost Jonker, *Powering the Hydrocarbon Revolution, 1939–1973: A History of Royal Dutch Shell, Vol. 2*, Oxford University Press, Oxford, 2007, p. 295.

12. See launching of MV *British Admiral* in Chapter 2. See W.J. Harvey and R.J. Solly, *BP Tankers: A Group Fleet History*, Chatham Publishing, London, 2005, p. 149.

13. Crude from Nigeria was shipped not only to Milford Haven, but also other refineries in Britain such as Kent and Grangemouth, as well as to other countries.

14. Crude had been shipped to south Wales with hardly a break since the Llandarcy Refinery was opened in 1922. Only during the war was production disrupted by German bombing. Fifteen-year-old Richard Jenkins in Port Talbot, just west of the plant, noted in his diary on 3 September 1940, a day after the worst raids: 'Oil works still burning and there is a black pall of smoke over the sky. There are now huge barrage balloons within sight.' Two days later he writes: 'We had films in school describing the production of oil. In parts interesting. In other parts it was dry.' However, the oil industry went through a radical transformation between the war and the 1960s. The

ability of private corporations to generate immense profit from the extraction of crude in the Middle East and the sale of petroleum products in the West meant that the transfer of geology from the shores of the Persian Gulf and the coasts of Africa to the engines of Britain became ever more intense. The packhorses of this transfer were the crude tankers and the size of these vessels leapt in the mid-1960s. Queen's Dock and the waters of Swansea Bay were too shallow to handle the VLCCs.

15. Don Strawbridge and Peter Thomas, *Baglan Bay: Past, Present and Future*, BP Chemicals Ltd, Baglan Bay, 2001, p. 32.
16. Ibid., pp.. ix and 4.
17. James Bamberg, *The History of The British Petroleum Company: Vol. III, British Petroleum and Global Oil, 1950–1975*, Cambridge University Press, Cambridge, 2000, p. 357.
18. Glenhafod, Aberbaiden and Pentre collieries, all in the Port Talbot area, were closed by the National Coal Board in 1959.
19. See Chapter 2.
20. Coal consumption in the UK peaked in 1952 at 228 million tonnes, produced by roughly 1,334 deep mines.
21. Don Strawbridge and Peter Thomas, *Baglan Bay, Past: Present and Future*, BP Chemicals Ltd, Baglan Bay, 2001, p. 7.
22. James Bamberg, *The History of The British Petroleum Company: Vol. III, British Petroleum and Global Oil, 1950 -1975*, Cambridge University Press, Cambridge, 2000, p. 357.
23. Don Strawbridge and Peter Thomas, *Baglan Bay, Past: Present and Future*, BP Chemicals Ltd, Baglan Bay, 2001, p. 25.
24. Ibid., p. 10.
25. James Bamberg, *The History of The British Petroleum Company: Vol. 3, British Petroleum and Global Oil, 1950 -1975*, Cambridge University Press, Cambridge, 2000, p. 357.
26. The company overseeing Baglan was a joint venture of British Hydrocarbon Chemicals and Distillers. It later became absorbed into BP.
27. Don Strawbridge and Peter Thomas, *Baglan Bay, Past: Present and Future*, BP Chemicals Ltd, Baglan Bay, 2001, p. 11.
28. See *The Changing Skyline* film made by BP film unit about construction of Baglan Bay.
29. Don Strawbridge and Peter Thomas, *Baglan Bay, Past: Present and Future*, BP Chemicals Ltd, Baglan Bay, 2001, p. 25.
30. Ibid., p. 27.
31. The film *The Graduate*, released 22 December 1967, further cemented the idea that the future was in plastics. In it Mr McGuire advises the graduate Benjamin Braddock: 'There's a great future in plastics. Think about it. Will you think about it?'
32. Don Strawbridge and Peter Thomas, *Baglan Bay, Past: Present and Future*, BP Chemicals Ltd, Baglan Bay, 2001, p. 16.
33. Ibid., p. 244.
34. Ibid., p. 19.
35. Ibid., p. 15.

36. From 1 September 1964 EDC also exported by ship, some of it going to BP's Antwerp petrochemicals factory. See Ibid., p. 30.
37. Ibid., p. 33.
38. 'Derrick's' was established in 1956 in Port Talbot in a small unit in Talbot Street. It sold records, radios and various hi-fi equipment. A second shop followed a few years later on Church Street, Port Talbot, which was probably the first dedicated pop/rock shop in Wales. During the redevelopment of Port Talbot in 1966–7 the Talbot Street shop moved to Cwmavon Square and later to Station Road. This shop closed after the death of Derrick in 1985.
39. It is intriguing to reflect upon all the women and men involved in the production of this vinyl, from Dere to the tanker to Llandarcy, to Baglan, to Barry, to Abbey Road, to Hayes, to the truck drivers, to the shop workers – and then the Fab Four, including John Lennon in his Surrey mansion.
40. Possibly the most dense 'block' of plastics in the home was the collection of albums and singles in the corner.
41. The average number of direct employees rose from 450 in 1963 to 1,000 in 1970. See Don Strawbridge and Peter Thomas, *Baglan Bay, Past: Present and Future*, BP Chemicals Ltd, Baglan Bay, 2001, p. 337.
42. In the interwar years Port Talbot, and the towns of Aberavon and Taibach that merged with it, had held a fierce industrial workforce associated with Miners' Clubs, welfare halls and the Workers' Education Authority. A contingent of men from the area had joined the International Brigade in the late 1930s, travelling to Spain to fight Franco's fascists. The constituency of Aberavon had returned the MP W.G. Cove, a socialist, president of National Association of Labour Teachers and champion of comprehensive education. The Communist Party had strong support in the town before and after the war. However, by the 1960s this culture had been largely replaced by a new one based on drinking clubs, bingo halls and cheap holidays abroad. Spain was the most popular venue of these last, the Franco regime not withstanding.
43. The Afan Lido was also the location of BP Baglan Social and Recreational Club events, such as the 'Fashion Spectacular' when 350 people attended a catwalk show in 1967, and the BP Baglan staff and children's Christmas parties. The social world of the petrochemicals factory echoed the experience of staff at the Coryton and Shell Haven Refineries. See Don Strawbridge and Peter Thomas, *Baglan Bay, Past: Present and Future*, BP Chemicals Ltd, Baglan Bay, 2001, p. 46.
44. Angela V. John, *The Actors' Crucible: Port Talbot and the Making of Burton, Hopkins, Sheen and All the Others*, Parthian, Cardigan, 2015, p. 69.
45. Ibid., p. 70.
46. The attractions of oil-fired central heating in the UK preceded gas-fired heating by almost a decade, before natural gas became widely available following extraction from the North Sea fields.
47. See David Jeremiah, *Architecture and Design for the Family in Britain, 1900–1970*, Manchester University Press, Manchester, 2000, p. 179. 'Mrs 1970' was a huge advertising campaign designed by John May of S.H. Benson's advertising agency promoting oil-fired central heating. See www.telegraph.co.uk/news/obituaries/1388331/John-May.html.

48. This sense of being entirely cocooned in the world of plastics and oil came to be used to emphasise the 'common sense' of its indispensability and this in turn underpinned the industry's sense of invulnerability.

49. Insulation (propylene and PVC); ceiling tiles (polystyrene); double-glazing windows (PVC); emulsion paints (vinyl acetate from ethylene); electric switches and plugs (phenolics); cable and wire coatings (PVC); water pipes (PVC); home appliances such as refrigerators, washing machines, vacuum cleaners, electric irons (polystyrene); bowls and buckets (polyethylene); floor tiles (styrene monomer from ethylene); carpet underlay (styrene monomer from ethylene); rubber gloves and sponges (styrene monomer from ethylene); hoses (styrene monomer from ethylene); furniture foam seating (propylene); combined with melamine to make Formica (phenolics); TV and radio cabinets (styrene monomer from ethylene); food packaging (PVC); garden pots and trays (PVC); hoses (styrene monomer from ethylene); cosmetics (propylene); textiles (vinyl acetate from ethylene). All drawn from Don Strawbridge and Peter Thomas, *Baglan Bay, Past: Present and Future*, BP Chemicals Ltd, Baglan Bay, 2001.

50. Ibid., p. 43.

51. James Bamberg, *The History of The British Petroleum Company: Vol. 3, British Petroleum and Global Oil, 1950–1975*, Cambridge University Press, Cambridge, 2000, p. 367.

52. Don Strawbridge and Peter Thomas, *Baglan Bay, Past: Present and Future*, BP Chemicals Ltd, Baglan Bay, 2001, p. 16.

53. See BP Carshalton in Chapter 5.

54. Inua Ellams, *Three Sisters*, Oberon Books, London, 2019, p. 77.

55. Andy Rowell, James Marriott and Lorne Stockman, *The Next Gulf, London, Washington and the Oil Conflict in Nigeria*, Constable & Robinson, London, 2005, p. 73.

56. James Bamberg, *The History of The British Petroleum Company: Vol. 3, British Petroleum and Global Oil, 1950–1975*, Cambridge University Press, Cambridge, 2000, p. 472.

57. See Sarah Ahmad Khan, *Nigeria: The Political Economy of Oil*, Oxford University Press, Oxford, 1994.

58. James Bamberg, *The History of The British Petroleum Company: Vol. 3, British Petroleum and Global Oil, 1950–1975*, Cambridge University Press, Cambridge, 2000, p. 427.

59. Ibid., p. 451 and Ian Seymour, *OPEC: Instrument of Change*, London, 1981.

60. James Bamberg, *The History of The British Petroleum Company: Vol. 3, British Petroleum and Global Oil, 1950–1975*, Cambridge University Press, Cambridge, 2000, p. 454.

61. Ibid., pp. 473–4.

62. James Marriott and Mika Minio-Paluello, *The Oil Road: Journeys from the Caspian Sea to the City of London*, Verso, London, 2012, p. 317.

63. On 26 November 1973 an auction of Nigerian crude fetched $16.85 per barrel (it had been around $2 per barrel a year previously) and the total volume covered by the bids exceeded the whole of Nigeria's annual production. This was noted by Peter Walters, chairman of BP. See James Bamberg,

The History of The British Petroleum Company: Vol. 3, British Petroleum and Global Oil, 1950–1975, Cambridge University Press, Cambridge, 2000, p. 484.

64. Echoes the warnings of writer and thinker E.F. Schumacher in the 1960s and 1970s.

65. James Bamberg, *The History of The British Petroleum Company: Vol. 3, British Petroleum and Global Oil, 1950–1975*, Cambridge University Press, Cambridge, 2000, pp. 481–2.

66. Edward Heath, *The Course of My Life: The Autobiography of Edward Heath*, Hodder & Stoughton, London, 1998, p. 503.

67. Andy Beckett, *When the Lights Went Out: Britain in the Seventies*, Faber & Faber, London, 2010, p. 130.

68. There was no rationing of electricity supply to the construction yards of the North Sea oil platforms, which were exempt.

69. Henry Kissinger, *Years of Upheaval*, Little Brown US, Boston, 1982, p. 885.

70. See Chapter 5 and the impact of the Oil Crisis on the Canvey Refineries.

71. WTI Crude was trading around $21.68 per barrel in October 1970 when oil was first discovered at what became BP's Forties Field. See www.macrotrends. net/1369/crude-oil-price-history-chart.

72. WTI Crude was trading around $52.00 per barrel in September 1975 when oil was first came ashore from BP's Forties Field. See www.macrotrends. net/1369/crude-oil-price-history-chart.

73. See Donella H. Meadows, Dennis L. Meadows, Jorgen Randers and William Behrens III, *The Limits to Growth: A Report for the Club of Rome's Project on the Predicament of Mankind*, Potomac Associates, Virginia, 1972.

74. James Bamberg, *The History of The British Petroleum Company: Vol. III, British Petroleum and Global Oil, 1950–1975*, Cambridge University Press, Cambridge, 2000, p. 476.

75. Ibid., p. 476.

76. Timothy Mitchell, *Carbon Democracy: Political Power in the Age of Oil*, Verso, London, 2011, p. 186.

77. Ibid., p. 186.

78. John Browne, *Beyond Business: An Inspirational Memoir from a Visionary Leader*, Weidenfeld & Nicolson, London, 2010, pp. 34–9.

CHAPTER 4

1. This description is drawn closely from that given by offshore rig radio operator Robert Orrell in his memoir, *Blow Out*. See Robert Orrell, *Blow Out*, Seafarer Books, Woodbridge, 2000, pp. 23–4.

2. See *West Sole Story, 48/6, Celebrating 40 Years in the Southern North Sea*, BP Anniversary publication, 2005.

3. James Bamberg, *The History of The British Petroleum Company: Vol. 3, British Petroleum and Global Oil, 1950–1975*, Cambridge University Press, Cambridge, 2000, p. 199.

4. Angus Becket was serving the minister of power, the Conservative Fredrick Erroll MP, during the period of the licensing round.

5. James Bamberg, *The History of The British Petroleum Company: Vol. 3, British Petroleum and Global Oil, 1950–1975*, Cambridge University Press, Cambridge, 2000, p. 199.
6. Ibid., p. 199–200.
7. Five years earlier, in 1960, Smith's Yard had built its last oil tanker for Shell – the *Amoria*.
8. In the first few years of North Sea development BP supported British-owned yards but this shifted radically in the 1970s.
9. Not only was life in Aberdeen set apart from London, but in the capital there were stark contrasts of culture. Life in the City of London was different from life in London's West End in the mid-1960s. As David Steel was driven on his way from Chelsea to the sedate order of Britannic House he passed through the counter-culture world of mod boutiques on the Kings Road. The Rolling Stones were also living in Chelsea, channelling the music of the southern US states and laying down tracks such '(I Can't Get No) Satisfaction', released on 12 May 1965. The joy of cars was still there, but rebellion lay beneath the surface.
10. James Bamberg, *The History of The British Petroleum Company: Vol. 3, British Petroleum and Global Oil, 1950–1975*, Cambridge University Press, Cambridge, 2000, p. 203.
11. Ibid., p. 330.
12. Links to the life of the Coryton Refinery in the 1960s are described in Chapter 1.
13. William Fraser was the son of the previous chairman of AIOC/BP, Sir William Fraser, Lord Strathalmond.
14. Compared to the situation 35 years later, note that all these men were born in Britain and were British citizens.
15. The British government had owned 51 per cent of BP's shares since 1914, but this holding was increased to 68 per cent, partly on account of the crisis at Burmah Castrol.
16. In our interview with Lord Deben, John Selwyn Gummer, in February 2018, he remarked: 'Most politicians had a limited understanding of energy policy anyway, only the odd one – like Tony Benn – turned into an energy wonk. When you have a complex thing to master you can develop a different attitude. North Sea oil was seen as a contributor to the country's wealth to be used now when it was needed. Benn thought otherwise, but he was on his own.'
17. As a result of his full name and title, Anthony Wedgwood Benn Viscount Stansgate, and his education at the private Westminster School and Oxford University, Benn was often portrayed as an archetypal aristocrat. However, he famously won a legal battle to renounce his title, and it is more accurate to see him as coming from a left-wing family within the long tradition of English Dissenters.
18. In the wake of the *Sea Gem* disaster, as David Steel continued in his position as head of BP's Exploration and Production division, much of his work was devoted to developments in the UK North Sea. The *Sea Quest* drilling rig was launched as planned in January 1966 and four years later struck oil in block 21/10, which was then named the Forties Field.

19. James Bamberg, *The History of The British Petroleum Company: Vol. 3, British Petroleum and Global Oil, 1950–1975*, Cambridge University Press, Cambridge, 2000, p. 203.

20. Laurance Reed MP also wrote a book on the matter, *Political Consequences of North Sea Oil* (1973).

21. Andy Beckett, *When the Lights Went Out: Britain in the Seventies*, Faber & Faber, London, 2010, p. 189.

22. Ibid., p. 189.

23. In parallel with the exploration for oil in the North Sea came the construction of the web of gas mains across the country by the UK state to distribute natural gas initially from under the southern North Sea. This was a second bloodstream, an echo of the petrol pipelines laid down in 1941–3, and it bound the country deeper into the oil and gas industry.

24. The journey echoed that of the Queen Mother's boat from London to the Coryton Refinery in 1953, 22 years before. See Chapter 1.

25. Tony Benn, *Against the Tide: Diaries 1972–1976*, Arrow, London, 1999, p. 403, entry for Wednesday 18 June 1975.

26. Ibid., p. 454, entry for Monday 3 November 1975.

27. Benn was perhaps one of the first to identify how the North Sea industry echoes scenes in Iran and Nigeria. It was the model of empire brought home, a new offshore British colony.

28. Said to be the Bram Stoker's inspiration for Count Dracula's castle.

29. BNOC was the British National Oil Corporation, established under Benn's predecessor as energy minister, Eric Varley MP.

30. Tony Benn, *Against the Tide: Diaries 1972–1976*, Arrow, London, 1999, p. 482, entry for Friday 19 December 1975. The entry illustrates the process of Benn learning about the industry, as noted by Lord Deben.

31. The exploitation of gas in the southern North Sea, underway following the *Sea Gem*'s find in 1965, had begun to diversify the UK's sources of energy. For over a century Britain's industry and households had been supplied with 'town gas' manufactured from coal, but in 1971 it was ousted as the main resource by 'natural gas' from beneath the seabed off the east coast. Two years later the UK was self-sufficient in gas; however, every single barrel of the 117 million tonnes of oil that it consumed each year came in tankers from foreign states. Britain imported 80 per cent of its oil from just four states – Iraq, Iran, Libya and Kuwait – all of which where BP and Shell had until recently held dominant positions but who had been largely ousted by the wave of nationalisations.

32. Tony Benn, *Against the Tide: Diaries 1972–1976*, Arrow, London, 1999, p. 393, entry for Tuesday 10 June 1975.

33. Edward Heath entered office in 19 June 1970, the Forties discovery was on 7 October 1970.

34. Licensing took place under the secretary of state for trade and industry from 1970 to 1974: John Davies (15 October 1970–5 November 1972), Peter Walker (5 November 1972–4 March 1974). It was then transferred to the minister for energy: Peter Carrington (8 January 1974–4 March 1974), Eric

Varley (5 March 1974–10 June 1975), Tony Benn (10 June 1975–4 May 1979).

35. John Davies MP was avowedly monetarist. In BP and then Shell-Mex and BP, he'd worked alongside David Steel and Peter Walters.
36. John Browne, *Beyond Business: An Inspirational Memoir from a Visionary Leader*, Weidenfeld & Nicolson, London, 2010, pp. 43–4 and 258.
37. Interview, Terry Macalister with John Browne, London, 22 March 2018.
38. Among those assets that might have been at risk was the Dussek Campbell Ltd plant, a manufacturing specialist in waxes, oils, resins and polymers, at Crayford next to Dartford at the western edge of the Thames Estuary.
39. Observation on Eric Varley MP from interview by James Marriott with Paul Potts, 21 September 2017.
40. Benn's diaries record that BP said it was working in 80 countries on 15 July 1975, and said it was working in 70 countries on 19 December 1975. Doubtless this is Benn's inaccuracy, but it emphasizes the point that BP was trying to make it clear that it was an international corporation.
41. Tony Benn, *Against the Tide: Diaries 1972–1976*, Arrow, London, 1999, p. 419, entry for Tuesday 15 July 1975.
42. Ian Cummins and John Beasant, *Shell Shock: The Secrets and the Spin of an Oil Giant*, Mainstream, Edinburgh, 2005, pp. 159–60.
43. The BP CEO did not entertain self-doubt, www.independent.co.uk/news/people/obituary-sir-eric-drake-1350962.html.
44. Statoil was founded in 1972 to develop oil and gas resources in the Norwegian sector of the North Sea. It later moved into petrochemicals and established a retail arm with petrol stations etc. It was renamed Equinor in 2018. DONG (Dansk Olie og Naturgas A/S) was founded in 1972 to manage the oil and gas resources of the Danish sector of the North Sea. It was renamed Ørsted in 2017.
45. In an array of aspects, including labour relations, the role of corporations, the role of non-state actors and the role of US finance, the development of the North Sea in the late 1960s and 1970s diverged significantly from the social democratic settlement and anticipated Thatcherism.
46. Tony Benn, *Conflicts of Interest, Diaries 1977–1980*, Arrow, London, 1996, p. 175, entry for Friday 24 June 1977.
47. Interview, Terry Macalister with Nick Butler, 18 October 2017.
48. The distinct phases in the 'privatisation of BP' is described in the following table from Stephanie M. Hoopes, *The Privatization of UK Oil Assets 1977–87: Rational Policy-Making, International Changes and Domestic Constraints*, PhD dissertation, London School of Economics, June 1994:

Date	Proceeds (£mill)	Costs	Remaining government share (%)	Chair/chancellor
June 1977	535	20	51	Steel/Healey
November 1979	290	9.6	46	Steel/Howe
July 1981	15	7	46	Steel/Howe
September 1983	565	9.4	31.5	Walters/Lawson
October 1987	5,500	137.1	0	Walters/Lawson

49. John Browne, *Beyond Business: An Inspirational Memoir from a Visionary Leader*, Weidenfeld & Nicolson, London, 2010, p. 44.
50. Richard Milne and Thomas Hale, 'Norway's oil fund urged to invest billions more in equities', *Financial Times*, 18 October 2016; https://www.forbes.com/sites/davidnikel/2020/11/17/norway-oil-fund-hits-record-12-trillion-value-following-coronavirus-vaccine-boost/?sh=6f2c8fba694e.
51. Nick Butler made this point to us during an interview in 2017. John Browne likewise told us in an interview in 2019 that the failure to establish a sovereign wealth fund in the UK was an error.
52. Nicholas Faith, Obituary, 'Sir David Steel BP chairman who triumphed over Tony Benn', *Independent*, 20 August 2004.
53. Benn is a household name and he is seen as a man who shaped left politics of the late twentieth and early twenty-first century. Yet two men who, as we discover through this tale, were far more powerful than Benn, and more influential in the direction of Britain, are utterly unknown to the public. There is no sign of Steel or Walters in the biographies and autobiographies of Thatcher and Heath.

CHAPTER 5

1. As Mark Cocker describes, 'In half a century farmers ... made a psychological journey from viewing an entire chemical arsenal – herbicides, fungicides and pesticides – as an expensive and technical line of last resort, to a prophylactic stand-by.' See Mark Cocker, *Our Place: Can We Save Britain's Wildlife Before It Is Too Late?*, Vintage, London, 2018, p. 190.
2. The material coming from the process of refining crude oil into petrol, aviation fuel and other products, and also the base resource for petrochemical works such as Baglan.
3. Stephen Howarth and Joost Jonker, *Powering the Hydrocarbon Revolution, 1939–1973: A History of Royal Dutch Shell, Vol. 1*, Oxford University Press, Oxford, 2007, p. 350.
4. Ibid., p. 429.
5. Interview, Terry Macalister with Ben van Beurden, The Hague, 16 February 2018.
6. Stephen Howarth and Joost Jonker, *Powering the Hydrocarbon Revolution, 1939–1973: A History of Royal Dutch Shell, Vol. 1*, Oxford University Press, Oxford, 2007, p. 351.
7. ICI had a pesticides plant at Yalding nearby, producing agrochemicals in the Garden of England.
8. Kent History Forum website, www.kenthistoryforum.co.uk/index.php?topic=11021.0.
9. Stephen Howarth and Joost Jonker, *Powering the Hydrocarbon Revolution, 1939–1973: A History of Royal Dutch Shell, Vol. 1*, Oxford University Press, Oxford, 2007, p. 429.
10. Ibid., p. 429.
11. Rachel Carson, *Silent Spring*, Houghton Mifflin, Boston, 1962, p. 123.
12. Ibid., pp. 123–4.

13. See J.A. Baker, *The Peregrine,* HarperCollins, London, 1967, David Cobham, *The Sparrowhawk's Lament,* Princeton University Press, Princeton, NJ, 2014 and James Macdonald Lockhart, *Raptor,* Fourth Estate, London 2016.

14. J.A. Baker, *The Peregrine,* HarperCollins, London, 2010, pp. 21–2. Originally published 1967.

15. Stephen Howarth and Joost Jonker, *Powering the Hydrocarbon Revolution, 1939–1973: A History of Royal Dutch Shell, Vol. 1,* Oxford University Press, Oxford, 2007, pp. 427–8.

16. Different chemicals were banned in different years in the UK and EU. Heptachlor by 1981 in the UK. Aldrin by 1989 in the UK.

17. The amount of wastewater needing treatment at Beddington Sewage Treatment Works was so great by the late 1950s that plans were drawn up from 1962 to build a new works replacing the original that had opened in 1902.

18. Understanding of the origins of the Wandle Group came through James Marriott's conversations with the group's leading lights, Doug Cluett and Margaret Cunningham, in the mid-1990s.

19. See David Jeremiah, *Architecture and Design for the Family in Britain 1900–1970,* Manchester University Press, Manchester, 2000, p. 179. Also see Chapter 3: 'Mrs 1970' was a huge advertising campaign designed by John May of S.H. Benson's advertising agency promoting oil-fired central heating.

20. The experimentation with detergents by the oil companies echoed their impulse to increase the consumption of aviation fuel in the UK domestic market, both actions being in part driven by the need to utilise the capital that had been sunk into refining capacity during the World War II production ramp-up.

21. Jeffrey Harrison and Peter Grant, *The Thames Transformed: London's River and Its Waterfowl,* Andre Deutsche, London, 1976, p. 43.

22. Ibid., pp. 35–6.

23. The clean-up operation used BP 1002 detergent, with the company attempting to rescue the situation by increasing production of the detergent at BP's Hythe works, Southampton. See Crispin Gill, Frank Booker and Tony Soper, *The Wreck of the Torrey Canyon,* David Charles, Devon, 1967, p. 53. BP had previously used BP 1002 detergent to deal with an oil spill off the Kent Refinery in 1966. See Jeffrey Harrison and Peter Grant, *The Thames Transformed: London's River and Its Waterfowl,* Andre Deutsche, London, 1976, pp. 123–7 – it was clear that detergents had done little to improve the situation in the Medway, yet were used on *Torrey Canyon* the following year.

24. Both organisation chapters were founded in the UK in 1971.

25. Shell Film Production Unit, *The River Must Live,* 1966, www.shell.com/ inside-energy/shells-pioneering-films.html.

26. Jeffrey Harrison and Peter Grant, *The Thames Transformed: London's River and Its Waterfowl,* Andre Deutsche, London, 1976, p. 36.

27. John Davies, *Energy to Use or Abuse?,* Shell UK Ltd, London, 1976. John Davies had worked in Shell for 30 years, starting his career at Shell's Thornton Research Centre, Stanlow, Merseyside in the late 1940s.

28. James Piers Taylor, 'The Shadow of Progress', in *Shadows of Progress: Documentary Film in Post-War Britain 1951–1977*, BFI, London, 2010, p. 71.

29. The first international Liquid Natural Gas plant in the world was built on land owned by British Gas on Canvey Island, opening in 1959. Using two re-engineered ships, natural gas, frozen to a level at which it became liquid, was transported from Lake Charles in Louisiana, US. At the Canvey Methane Terminal it was turned back into gas and pumped through a pipeline to Romford Gas Works and used as town gas. The pioneering project was a success, so two new liquid natural gas (LNG) tankers were built, in Belfast and Barrow-in-Furness, to transport LNG from a new plant built at Arzew in Algeria, the first cargo arriving in October 1964.

30. Interview, Terry Macalister with Wilko Johnson, Cambridge, 14 October 2016, and all following quotes.

31. Wilko Johnson, *Don't You Leave Me Here*, Little Brown, London, p. 13.

32. Ibid., pp. 69–70.

33. Nick Logan and Bob Woffinden, *The Illustrated New Musical Express Encyclopedia of Rock*, Salamander Books, London, 1977, pp. 69–70.

34. Wilko Johnson, *Don't You Leave Me Here*, Little Brown, London, p. 67.

35. The photos show banners that denounced Rippon. The target of these protests was Geoffrey Rippon, the minister of the environment. Rippon, educated at Brasenose College, Oxford University, and with a constituency at Hexam in rural Northumberland, was clearly out of his depth with this Essex rebellion. Rippon was at the same time involved in the abortive plans to develop Maplin Airport, a few miles east of Canvey Island, which would have utterly transformed the Thames Estuary, creating what critics called 'Heathograd' next to Southend-on-Sea.

36. Daniel Yergin, *The Prize*, Simon & Schuster, New York, 1991, pp. 574–80.

37. Greater volumes than even Saudi Arabia and Kuwait.

38. James Bamberg, *The History of The British Petroleum Company, Vol. 3: British Petroleum and Global Oil, 1950–1975*, Cambridge University Press, Cambridge, 2000, p. 451.

CHAPTER 6

1. It is extremely difficult, perhaps impossible, to assess the impact on so many thousands of lives spread over four and half decades. How are we to understand the cancers this might have caused, or to enumerate how lives of short breath were further shortened? The land upon which the plant was built is there still, it can be tested and analysed, but the bodies that lived in the lee of the chimneys are scattered.

2. James Bamberg, *The History of The British Petroleum Company, Vol. 3: British Petroleum and Global Oil, 1950–1975*, Cambridge University Press, Cambridge, 2000, p. 388.

3. Don Strawbridge and Peter Thomas, *Baglan Bay: Past, Present and Future*, BP Chemicals Ltd, Baglan Bay, 2001, p. 48.

4. James Bamberg, *The History of The British Petroleum Company, Vol. 3: British Petroleum and Global Oil, 1950–1975*, Cambridge University Press, Cambridge, 2000, p. 389.

5. Ibid., p. 389.

6. Don Strawbridge and Peter Thomas, *Baglan Bay: Past, Present and Future*, BP Chemicals Ltd, Baglan Bay, 2001, p. 50.

7. Ibid., p. 44.

8. Ibid., p. 249.

9. Ibid., p. 59.

10. Ibid., p. 17.

11. Ibid., p. 17.

12. Ibid., p. 59.

13. Ibid., p. 248.

14. Lynne Rees, *Real Port Talbot*, Seren, Bridgend, 2013, p. 104.

15. The pipe that Rees recalled was completed in September 1962 and pushed a mile out into Swansea Bay. Land and Marine Contractors Ltd assured residents that the waste was 'far out to sea', but the bay is shallow and the pipe ended in water that at low tide was a mere 15 feet deep. Even if the pipe did not spill directly onto the beach, as Lynne remembered, when a storm brewed in the Severn Estuary it was impossible for the plastics pollution not to drift back onto Aberavon Beach.

16. Don Strawbridge and Peter Thomas, *Baglan Bay: Past, Present and Future*, BP Chemicals Ltd, Baglan Bay, 2001, p. 337. Calculated as 2,300 at Baglan, plus around 1,000 at Llandarcy, around 1,000 at Barry and an estimated 2,000 contractors.

17. The pollution in the Severn Estuary is evidenced in the level of heavy metals around the island of Lundy, shown in the 1973 survey carried out by the Lundy Field Society.

18. Don Strawbridge and Peter Thomas, *Baglan Bay: Past, Present and Future*, BP Chemicals Ltd, Baglan Bay, 2001, p. 63.

19. Ibid., p. 246.

20. 'Reengineer the Plant Not the Workers', www.marxists.org/history/etol/newspape/isj/1977/no098/briefing.html.

21. John S. Hunter, the managing director of BP Chemicals, visited Baglan and said: 'At the Baglan and Barry Factories particularly, in the past few months, a VCM health issue has arisen. Intense efforts were being made at both factories in collaboration with other producers, not only in this country, but in the USA and with authorities like Government and TUC medical people. I think all one can say is that since January when this particular situation came to light, a great deal of progress has been made and more will be done. The object is to make the plants making and handling VCM as safe as any other on the site. Enquiries are now turning to the PVC fabrication industry, a lot of people depend on it and one only hopes it will continue.' Don Strawbridge and Peter Thomas, *Baglan Bay: Past, Present and Future*, BP Chemicals Ltd, Baglan Bay, 2001, p. 71.

22. During this period of the early 1970s Charlie Clutterbuck of British Scientists for Social Responsibility conducted a close study both of workers at the

plant and residents in Sandfields. The TV current affairs programme *World in Action* got to work on a documentary, but it was never broadcast. Indeed Clutterbuck believed it was deliberately suppressed by the company and the trades unions.

23. Don Strawbridge and Peter Thomas, *Baglan Bay: Past, Present and Future*, BP Chemicals Ltd, Baglan Bay, 2001, p. 151.
24. Ibid., p. 152.
25. The first industrial action at Baglan took place on 8 May 1963 when 80 union members walked off site in a pay dispute. As with the protest by the women of Sandfields, this was indeed before the factory had begun to operate. Nine months later there was another walkout by the same union. Significantly, it was timed with a massive six-week strike by 1,300 union members at the steel works that dominated the other side of Port Talbot. Clearly BP was building its new factory in a town whose citizens were used to defending their rights. Relations between the labour force and the management settled in the mid-1960s, absorbed by a general sense of prosperity in Port Talbot and the new estate of Sandfields.
26. Don Strawbridge and Peter Thomas, *Baglan Bay: Past, Present and Future*, BP Chemicals Ltd, Baglan Bay, 2001, p. 57 – signed 1969.
27. Ibid., p. 330.
28. Ibid., p. 57.
29. Ibid., p. 53.
30. Ibid., p. 63.
31. On 29 September 1972 there was an arson attack by Alan Perkins on Baglan Bay plant causing £12,000 worth of damage. In October 1972 Alan Perkins and Malcolm Jones entered Baglan plant again and painted slogans on 16 railway wagons. Perkins and Jones were arrested and Perkins jailed. Ibid., p. 69.
32. Ibid., p. 63.
33. Ibid., pp. 63–4.
34. The 1983 Energy Act allowed local electricity boards – including the South West Electricity Board – to purchase electricity from industry not just from the CEGB. See ibid., p. 123.
35. The voting ran 93.12 per cent for (25,058) and 6.85 per cent (1,843) against.
36. Frank Ledger and Howard Sallis, *Crisis Management in the Power Industry: An Inside Story*, Routledge, London, 1995, p. 88.
37. Don Strawbridge and Peter Thomas, *Baglan Bay: Past, Present and Future*, BP Chemicals Ltd, Baglan Bay, 2001, p. 87.

CHAPTER 7

1. AIOC/BP was still dominated by Chairman Lord Strathalmond William Fraser until 1956.
2. James Bamberg, *The History of The British Petroleum Company, Vol. 3: British Petroleum and Global Oil, 1950–1975*, Cambridge University Press, Cambridge, 2000, p. 329.
3. Ibid., p. 324.

4. The combining of accounts was expressly designed to assist AIOC/BP in its negotiations with producer states such as Iran, for it enabled the company to hide just what profit it made on production in those states.

5. Daniel Yergin, *The Prize*, Simon & Schuster, London, pp. 722–3.

6. Ibid., p. 723.

7. Ibid., p. 723.

8. These events contributed to the dramatic development of the oil traders' market. See David Jamison's story of seeing jets attacking the Abadan refinery on the TV in Chapter 12.

9. The shifts of the culture within these two giant oil corporations both catalysed and echoed the shifts within UK politics, where the centrality of the 'market' came to dominate thinking. The battle of David Steel to effectively privatize BP had inspired thinkers and MPs on the right of the Conservative Party. On coming to power Thatcher and her chancellor, Geoffrey Howe, embraced the idea of further sales of the government shareholding in BP. By the time Walters became chairman in November 1981 the state shareholding was down to 46 per cent from the 68 per cent that the government had held four years earlier. Just as Benn had predicted, this shift highlighted the new reality that BP was now acting in the interests of capital and making little pretence that it was acting in the national interest.

10. See closure dates of key BP assets of the Kent Refinery on the Isle of Grain (August 1982), Sandbach (early 1980s), Stroud (mid-1982), Llandarcy (April 1985) and Carshalton (1991).

11. Malcolm Pitt, *The World On Our Backs: The Kent Miners and the 1972 Miners' Strike*, Lawrence and Wishart, London, 1979, p. 195.

12. Frank Ledger and Howard Sallis, *Crisis Management in the Power Industry: An Inside Story*, Routledge, London, 1995, p. 144.

13. Ibid., p. 250.

14. Frank Ledger and Howard Sallis, *Crisis Management in the Power Industry, An Inside Story*, Routledge, London, 1995, p. 197.

15. Ibid., p. 127.

16. On 18 December 1984 two tanker drivers were suspended at the Texaco oil terminal at Purfleet for refusing to cross picket lines and 'the strike spreads to Canvey Island refinery [*sic*]'. See:= Mike Simons, *Striking Back: Photographs of the Great Miners' Strike 1984–1985*, Bookmarks, London, 2004, p. 77. BP's Kent Refinery on Grain had already closed.

17. See http://news.bbc.co.uk/1/hi/wales/south_west/8418548.stm and www.walesonline.co.uk/news/wales-news/nooses-hampers-and-mr-scargill-2443811.

18. Lynne Rees, *Real Port Talbot*, Seren, Bridgend, 2013, p. 119. Stephen Kinnock was later MP for Aberavon.

19. Walters' Conservative Party affiliation was confirmed to us by former colleagues. Another former BP executive noted that BP supported the precarious, and politically significant, floatation of Mercury Telecoms PLC in 1981.

20. Frank Ledger and Howard Sallis, *Crisis Management in the Power Industry: An Inside Story*, Routledge, London, 1995, p. 194.

21. Ibid., p. 322.
22. Ibid., p. 227.
23. Indeed, on Tuesday 19 June 1984, 15 weeks into the strike, Sir Peter Walters visited Thatcher at Number 10 Downing Street. They met for an hour but unfortunately the transcript of their conversation remains hidden from public view. Immediately afterwards she had a half-hour Meeting of Ministers on Environmental Pollution, Acid Rain. Earlier that day she had hosted, at Number 10, a meeting of the Royal Academy Appeal. The seven people who attended included John Raisman, chairman of Shell UK and chairman of the RA Appeal. Margaret Thatcher Engagement Diary, Tuesday 19 June 1984, Margaret Thatcher Archive, document 147230.
24. Don Strawbridge and Peter Thomas, *Baglan Bay: Past, Present and Future*, BP Chemicals Ltd, Baglan, 2001, p. 118.
25. The average workforce in 1985 was 1,010. These were direct employees of BP, but contractors in that year are likely to have doubled that number. Ibid., p. 337.
26. The final part of Baglan, the isopropyl alcohol plant employing 55 staff, was closed on 31st March 2004, a year after Shell announced the closure of its IPA plant at Stanlow. Glenda Thisdell, 'BP to shut down IPA plant at Baglan Bay', Independent Commodity Intelligence Services, 18 February 2004.
27. Guy Standing, *The Plunder of the Commons: A Manifesto for Sharing Public Wealth*, Penguin, London, 2019, p. ix.
28. Daily Mail Reporter, '30 Second Guide: Tell Sid', 9 December 2011, hwww.thisismoney.co.uk/money/news/article-2072001/30-SECOND-GUIDE-Tell-Sid.html.
29. The sale of the final stake in British Gas was in December 1986. The sale of the final stake in BP was in October 1987. The sale of the final stake in Britoil Plc was in February 1988.
30. See *Gas Production from the UK Continental Shelf*, Oxford Institute for Energy Studies, July 2019.
31. The twelve regional electricity companies were sold off later along with the National Grid, which controlled the transmission pylons and pipelines. It was not long before the industry consolidated into the Big Six domestic power providers – the majority were then snapped up by foreign owners such as E.ON. Privatisation was sold by the Conservatives as bringing more efficiency to the industry and lower prices to consumers. But UK government statistics show electricity bills rising by a third between 1996 and 2018. Gas bills are up over half during the same period although overall power prices in Britain seem roughly average for continental Europe where state providers dominate. The Big Six have been repeatedly accused of 'profiteering' by consumer groups while executive pay in the sector has rocketed.
32. Analysis by Climate Change Capital, a private asset management group.
33. The break-up of the CEGB was announced in the Queens Speech 1987. It was published as a White Paper in 1988 and enacted 1990.
34. In the mid-1970s the British minister for energy, Tony Benn, battled with BP over the order books of John Brown Shipyards in Glasgow. Clearly part of the reasoning from BP's point of view against commissioning British yards

was the threat that industrial action posed to production. Labour disputes had significantly added to the delays and escalating costs in the expansion of the Baglan factory. BP and Shell were opposing trades unions in the North Sea offshore fields, in rig yards (Brown and Root) and in suppliers (John Brown and Cammell Laird). Shell continued to order tankers from the Harland & Wolff shipyard in Belfast, including eight VLCCs between 1971 and 1982. BP had two tankers built in the yard during this period, both in 1981.

35. Interview, Terry Macalister and James Marriott with Eddie Marnell, Liverpool, 4th April 2018.

36. Interview, Terry Macalister with Michael Heseltine, London, 19th December 2018.

37. A play about the sit-in, *Cammell Lairds 37: The Truth*, was written by local playwright Mike Howl and performed at the Casa Theatre, Liverpool, in October 2018. Eddy Marnell appeared on stage at the end, paying tribute to those who had passed away since, and he vowed to continue his fight for justice.

38. When Sir Brian Atkinson, chairman of British Shipbuilders (1980–4), presented the case for shipbuilding's future to the Conservative government, 'a very senior permanent government officer' told him, 'Brian … Margaret wants rid of shipbuilding'. The 'death warrant' for British Shipbuilders was the UK government acceptance of £140 million in EEC aid in 1985. Cammell Laird was designated as a warship yard and so denied access to the EEC Shipbuilding Intervention Fund. See David Roberts, *Cammell Lairds: Life at Lairds*, Avid, Flintshire, 2008, pp. 127–8.

39. The interweaving of the EEC/EU and senior executives of Shell and BP echoes that between the companies and the UK state. For example, in 1985 Peter Sutherland was appointed as European commissioner responsible for competition policy. He was chair of the Committee on Completion of Internal Market and issued the Sutherland Report, presented to European Council 1992. He served on the Delors Commission and was crucial in opening up competition, especially in the airlines, telecoms and energy sectors. In June 1997 he became chairman of BP, remaining in post for twelve years until June 2009. In 1997 he replaced David Simon as chairman of BP. Simon was made a life peer and took up a post in the Blair government as minister for trade and competitiveness in Europe.

40. The expanded Baglan plant was finally commissioned in January 1974. Eleven years later the first phases of closures were announced. Baglan continued to export chemicals to its sister plant, BP Chemicals Antwerp, at least until it was sold to Inspec in 1992, Jim Ratcliffe's company. In 1988 deliveries of chemicals from Baglan were at an all-time high – 640 ships per year (50 per month) from River Neath Dock or Queen's Dock facility and road and rail movements all across the UK. In August 1989 a new ethylene carrier, *Sunny Girl*, took its first cargo of 2,100 tonnes from Baglan to BP Antwerp.

41. In September 1992 David Wayne, MEP for South Wales Central, said: 'Negotiations were on-going between the European Commission and Middle East states to establish a Gulf Co-Operation Council Agreement. If this agreement was signed it would mean greatly increased competition in

the form of cheap imports. Both the 900 BP Chemicals employees and the 1,000 contractors had expressed fears about job losses at Baglan Bay.' Within the EU there was overcapacity of ethylene production, with 53 crackers producing 19.4 million tonnes against a demand of around 15.5 million tonnes. The trade agreement with the Gulf states exacerbated the problem. Baglan, which produced just under 10 per cent of EU ethylene, was one of the plants in the EU that was closed. The end of the Baglan Ethylene plant was announced in 12 January 1994. Don Strawbridge and Peter Thomas, *Baglan Bay: Past, Present and Future*, BP Chemicals Ltd, Baglan Bay, 2001, pp. 191–2.

42. Ibid., p. 197.
43. Ibid., p. 197.
44. Ibid., p. 197.
45. By March 2004 the closure of Baglan was complete. It had been preceded by the closure of Angle Bay, Llandarcy and Stroud plants and the sale of BP Barry to INEOS. INEOS Vinyls plant at Barry closed in 2010. The Severn Estuary oil complex which had evolved slowly over the previous 90 years had finally been destroyed, following coal into industrial history. However, the closure of oil was not fought over as the closure of coal had been by the NUM. And there was no mirror to the action of miners laid off at Tower Colliery, 15 miles north of Baglan, who organised to buy the mine from British Coal in January 1995.
46. The *Piper Alpha* tragedy on July 6, 1988, left 167 men dead, https://www.pressandjournal.co.uk/fp/news/aberdeen/1508442/piper-alpha-the-167-men-who-died-in-the-disaster/.

CHAPTER 8

1. The lyrics of 'Star of the Morning' in Ogoni run in translation: 'The Ogoni race, we greet you | Ogoni people, humble race | All men raise their hands in your honor | All women support your programme | Welcome, welcome, welcome | We thank God | Ken Saro-Wiwa, the star of the morning | The eye of the blind | The leg of the cripple | The bright day light of Ogoni | We welcome you | Welcome, welcome, welcome.'
2. Interview, Terry Macalister with Lazarus Tamana, London, 23 October 2019.
3. Andy Rowell, James Marriott and Lorne Stockman, *The Next Gulf: London, Washington and Oil Conflict in Nigeria*, Constable, London, 2005, p. 115.
4. Through MOSOP was published the Ogoni Bill of Rights which declared that some $30 billion worth of oil had been drilled from Ogoniland, but in return 'the Ogoni people had received NOTHING'. The Ogoni elders called for 'the right to the control and use of a fair proportion of Ogoni economic resources for Ogoni development'. See Ken Saro-Wiwa, *Genocide in Nigeria: The Ogoni Tragedy*, Saros International, Lagos, 1992, pp. 92–103.
5. Interview, Terry Macalister with Lazarus Tamana, 23 October 2019.
6. Lieutenant Colonel Okuntimo (1994) RSIS Operations: Law and Order in Ogoni Etc, Memo From the Chair of the Rivers State Internal Security

(RSIS) to His Excellency, The Military Administrator, Restricted, 12 May 1994.

7. Andy Rowell, James Marriott and Lorne Stockman, *The Next Gulf: London, Washington and Oil Conflict in Nigeria*, Constable, London, 2005, p. 72.

8. Artist and activist John Jordan remembers that his first use of the internet was to transmit information following a demonstration in London in support of the Ogoni in 1996.

9. Glenn Ellis and Kay Bishop, *The Drilling Fields*, Catma Films Production, London, broadcast on Channel 4, 23 May 1994.

10. Unrepresented Nations and Peoples Organization, https://unpo.org/.

11. Andy Rowell, James Marriott and Lorne Stockman, *The Next Gulf: London, Washington and Oil Conflict in Nigeria*, Constable, London, 2005, pp. 127–8.

12. Between 30 April and 23 May 1995 Greenpeace was engaged in the 'Battle over *Brent Spar*' in the UK North Sea and the two struggles became entwined.

13. Andy Rowell, James Marriott and Lorne Stockman, *The Next Gulf: London, Washington and Oil Conflict in Nigeria*, Constable, London, 2005, pp. 127–8.

14. The nine men executed on 10 November 1995 were: Baribor Bera, Saturday Dobee, Nordu Eawo, Daniel Gbokoo, Barinem Kiobel, John Kpuinen, Paul Levula, Felix Nuate and Ken Saro-Wiwa.

15. Nigeria was suspended for three and a half years until 29 May 1999 when the military handed rule back to a civilian government.

16. John Browne, *Beyond Business: An Inspirational Memoir from a Visionary Leader*, Weidenfeld & Nicolson, London, 2010, p. 104.

17. BP, ejected from Nigeria in 1979 via nationalisation of its holdings, had tentatively begun to return to the Delta. A joint venture with Statoil had been agreed and exploration work was underway. Irene Gerlach was part of the BP team in Nigeria working in community relations and later recalled: 'Philanthropy died and corporate social responsibility was born the day after Saro-Wiwa was murdered'.

18. How the *Brent Spar* controversy hurt Shell: http://priceofoil.org/2015/02/03/ghost-brent-spar-haunts-shell/.

19. *Brent Spar* had shown that the UK government had been unable to ensure the public acceptance of Shell's plans even though ministers had approved them.

20. Andy Rowell, James Marriott and Lorne Stockman, *The Next Gulf: London, Washington and Oil Conflict in Nigeria*, Constable, London, 2005, pp. 117–18.

21. This position effectively made John Jennings head of Shell in the UK.

22. John Browne, *Beyond Business: An Inspirational Memoir from a Visionary Leader*, Weidenfeld & Nicolson, London, 2010, p. 91.

23. The company utilized a careful strategy of engaging in civil society on the issue, some of it drawing from the analysis of John Elkington of Sustainability. In part this advised that NGOs should be seen as distinct, each playing different roles, each requiring different practices of engagement, characterized as four archetypes: sharks, orcas, sea lions and dolphins.

24. John Browne, *Beyond Business: An Inspirational Memoir from a Visionary Leader*, Weidenfeld & Nicolson, London, 2010, p. 104.

25. In 1896 Swedish mathematician Svante August Arrhenius developed the theory that the release of CO_2 into the Earth's atmosphere through the burning of fossil fuels would increase the Earth's temperature. Seventy years on there was, at least in the scientific community, growing acceptance of anthropogenic climate change. This term denotes that while the temperature of the Earth's atmosphere goes through cycles of its own, since the Industrial Revolution human activity had been warming the Earth and that this in turn is altering the climate across the world, leading to droughts and floods, and through the melting of polar ice to the rising of sea levels and the inundation of low-lying coastal land. From the late 1960s the issue was gaining not only scientific traction but some cultural impact too.

26. In 1986 the chairman of Shell Transport was Sir Peter Holmes, and chairman of the Royal Dutch/Shell Group was Lodewijk van Wachem.

27. Steven Mufson and Chris Mooney, 'Shell foresaw climate dangers in 1988 and understood Big Oil's big role', *Washington Post*, 5 April 2018.

28. Sir John Cadogan sadly passed away on 9 February 2020. He was 89.

29. Interview, Terry Macalister with Sir John Cadogan, Swansea, 7 June 2017.

30. Cadogan was appointed chief scientist by CEO Sir David Steel in 1979. In 1981 he was made BP's first worldwide director of research by CEO Peter Walters. He was also chair of BP Solar International, CEO of the BP Innovation Centre and CEO of BP Ventures.

31. Sunbury was the home of BP research since 1917. It had a key role in the company during World War II, positioned as it was across the Thames from the petrol pipeline hub built at Walton-on-Thames in 1941. It expanded rapidly in the 1950s and remains central to BP's global operations.

32. The BP archive, a possible treasure trove of information about the oil company's early research into climate and renewable energy, has gained almost mythical status. It may contain much – or even little – of value but BP seems determined that no one will find out. The archive is based in the UK's Modern Record Centre which itself is located in the middle of Warwick University's main campus. Terry's requests for access as energy editor of the *Guardian* were repeatedly rebuffed on the grounds that no material for the past 40 years was available to the public as it was commercially sensitive. This led to a campaign by students from Fossil Free Warwick University for BP to open up the archive and for the company to be removed from the campus. The university authorities subsequently announced they were pushing BP to see whether 'access can be widened'. Shareholders also raised the issue directly with the BP board at its AGM in April 2015. Carl-Henric Svanberg, then BP chairman, insisted that 'nothing is locked away. We share everything happily.' Despite this commitment, the archive remained closed when we asked for access. In frustration we visited the Modern Records Centre and tried to get in. On the right of an entry room was a grey door with reinforced glass. A white paper note was stuck there which said: 'Welcome to the BP Archive', and in smaller letters, 'Please ring the bell on the right for access'. We pressed the chrome intercom buzzer but there was no reply.

33. He had been a long-term friend of its founder Sir Ralph Harris and had overseen an annual donation of £10,000 from BP to the IEA during the 1980s when he was CEO of the company.

34. Brendan Montague Desmog, 'You will never guess who was really behind Britain's first climate denial propaganda', 7 February 2015, www.desmog. co.uk/2015/02/07/you-will-never-guess-who-was-really-behind-britain-first-climate-denial-propaganda.

35. Browne's announcement came on the back of sustained pressure from a number of institutional investors in BP.

36. The United Nations Framework Convention on Climate Change.

37. In the lead-up to his speech Browne spoke constructively with Eileen Clausen of the Pew Centre on Global Climate Change. See John Browne, *Beyond Business: An Inspirational Memoir from a Visionary Leader*, Weidenfeld & Nicolson, London, 2010, p. 79.

38. John Browne, *Beyond Business: An inspirational Memoir from a Visionary Leader*, Weidenfeld & Nicolson, London, 2010, pp. 76–7 and pp. 83–4.

39. Exxon under Raymond was sceptical around climate policies. See www. greenpeace.org/usa/global-warming/exxon-and-the-oil-industry-knew-about-climate-change/exxons-climate-denial-history-a-timeline/.

40. Jeremy Leggett, *The Carbon War: Global Warming and the End of the Oil Era*, Penguin, London, 1999, p. 252.

41. BP was followed by Shell in early 1998. See Mark Moody-Stuart, *Responsible Leadership: Lessons from the Front Line of Sustainability and Ethics*, Greenleaf, London, 2014, p. 142.

42. Jeremy Leggett, *The Carbon War: Global Warming and the End of the Oil Era*, Penguin, London, 1999.

43. Brendan Montague, 'When a BP boss finally called for action on climate change', *The Ecologist*, 28 September 2018, https://theecologist.org/2018/sep/28/what-happened-whenlord-browne-bp-boss-called-action-climate-change.

44. Don Strawbridge and Peter Thomas, *Baglan Bay: Past, Present and Future*, BP Chemicals Ltd, Baglan, 2001, pp. 239–40.

45. Environmental Defense Fund, 'British Petroleum announces plan to measure and report greenhouse gas emissions; to set targets, and to practice emissions trading', 30 September 1997, www.edf.org/news/british-petroleum-announces-plan-measure-and-report-greenhouse-gas-emissions-set-targets-and-pr. See Jeremy Leggett, *The Carbon War: Global Warming and the End of the Oil Era*, Penguin, London, 1999, p. 274.

46. By 2016, 20 or so years later, these limits had effectively begun to be imposed on exploration – Cairn had been blocked from drilling off Greenland in the north-west Atlantic, Shell in the US Alaskan Arctic Ocean and BP off south Australia – see Chapter 11.

47. Meanwhile, on 20 October 1997 Shell announced that it would invest heavily in renewable energy, and early the next year they followed BP and withdrew from the Global Climate Coalition.

48. Announced 14 March 2000.

49. Shell and Exxon tried to follow suit with similar mergers – Shell failed to undertake anything so dramatic, acquiring only the small UK company Enterprise Oil in 2002, but Exxon achieved the elusive merger with Mobil to become ExxonMobil.

50. The staff of the former Mobil refinery at Coryton, bought by BP, battled to retain the Mobil name for their social centre, the Pegasus Club, which then became a club for workers employed by BP. They kept the title built on the Mobil logo, Pegasus.

51. It was perhaps the zenith of the company's deep engagement with civil society, the pinnacle of the mountain of CSR that BP had climbed. Shell management could only look on in envy.

52. 'Friends of the Earth director Tony Juniper sensationally quits – and attacks hypocritical "green" celebrities', *Daily Mail*, 27 January 2008, www.dailymail. co.uk/tvshowbiz/article-510612/Friends-Earth-director-Tony-Juniper-sensationally-quits--attacks-hypocritical-green-celebrities.html.

53. Robert Horton was also a long-term Conservative supporter, and after his ejection from office as CEO of BP he returned to public life a year later as chairman of Railtrack, assisting the Major government with its flagship policy to privatise the UK's railways.

54. See John Browne, *Beyond Business: An Inspirational Memoir from a Visionary Leader*, Weidenfeld & Nicolson, London, 2010, p. 65. The date of this exchange is unclear in Browne's autobiography, but it must have been before he became CEO on 10 June 1995, and most likely after Blair became leader of the opposition on 21 July 1994.

55. By 1982 Butler had joined BP and was working on the 26th floor of Britannic Tower in Corporate Planning. As he explained, BP looked like a dying company with just 'two pipelines', one in the UK North Sea and the other in Alaska. The view was that BP would be finished by 1990. He experienced BP as very meritocratic, not public school, but preferring grammar school boys. In his first week he was invited to join the Christian Union and to come to a meeting of the Freemason's Lodge in the basement of Britannic House. Within a year he had set up a branch of a trades union at BP head office and was handing out leaflets to staff entering the building criticising BP's involvement in sanctions busting in support of Ian Smith's Southern Rhodesia regime and for involvement in apartheid South Africa. He and others were issued with disciplinary warnings, but these were subsequently lifted when the protestors got the issue raised in the House of Commons.

56. Nick Butler stood as the 1987 parliamentary election candidate for Labour in Lincoln (33.7 per cent of the vote) and in 1992 (42.8 per cent of the votes).

57. Interview, Nick Butler with Terry Macalister, London, 18 October 2017.

58. John Browne, *Beyond Business: An Inspirational Memoir from a Visionary Leader*, Weidenfeld & Nicolson, London, 2010, p. 301.

59. The removal of this spectre ties the company's fortune back to 1945–7 when the Labour government might have nationalized AIOC.

60. It is instructive to look at these appointments in relation to the fate of plants such as Baglan and Coryton. Both were impacted by 'competitiveness in Europe' and East Asia. At Baglan, September 1996 saw the demolition of

the ethylene plant begin. October 1996 saw the official announcement of the styrene plant closure. The long-term viability of Coryton was radically impacted by the announcement by Shell in June 1998 that they planned to close the next-door Shell Haven refinery and lay off 290 staff. A Shell executive at the time said: 'Our feeling was that Europe's surplus refining capacity, and tight refining margins, meant there was little prospect of getting a viable price for the plant; but we're still open to offers.'

61. There was a neat synergy between the rebranding of the Labour Party as 'New Labour' and the rebranding of BP PLC with the new Helios logo and the 'Beyond Petroleum' strapline. Both echo the work of Mark Leonard at the Labour think tank Demos, which talked of 'Branding Britain'. Presented at Tate Britain on 25 September 1997.

62. One of the problems with this scheme is that BP was not trading with a fixed set of assets. The company's assets, and their CO_2 emissions, were constantly shifting as a result of wider forces. Hence BP's company-wide emissions were reduced by the Baglan factory ethylene plant and styrene plant closing between 1996 and 1999, and increased by the acquisition of the Coryton Refinery and Birkenhead blending plant in 1996. To some degree the BP CO_2 emissions strategy was like banning smoking in a cigarette factory.

63. 'The EU Emissions Trading System is the first and largest international carbon-trading scheme in the world. Under the scheme, factories and power plants have a cap on their yearly emissions, but within this cap they are granted emissions allowances that can be traded.' John Browne, *Seven Elements That Have Changed the World*, Weidenfeld & Nicolson, London, 2013, p. 228.

64. Interviews, Terry Macalister with John Browne, London, 22 March, 25 April and 22 June 2018.

65. See John Browne, *Beyond Business: An Inspirational Memoir from a Visionary Leader*, Weidenfeld & Nicolson, London, 2010, p. 87. 'BP … carried out the first trades in the UK Emissions Trading Scheme (ETS) when it was launched in 2002, and helped customers trade in the market. The UK ETS went on to inform the European scheme (EU ETS). These will be important markets for the future as the world puts carbon-free and fossil-free energies on a level playing field.'

66. There is an additional challenge regarding how to account for the reduction of EU emissions by exporting them 'elsewhere'. For example, by the closing of the Coryton Refinery, or BP's Ingolstadt Refinery in Germany, and the reopening of such plants in India from whence fuel is imported to the EU with the emissions then being 'counted' as 'Indian'.

67. For further analysis of the failure of the emissions trading scheme and why it benefited BP and the wider industry, see Tom Bergin, *Spills and Spin: The Inside Story of BP*, Random House, London, 2011, pp. 55–8.

68. Thursday 18 April 2002.

69. The article was by Tobias Buck and David Buchan, 'Sun King of the oil industry', *Financial Times*, July 2002. 'The *Financial Times* had indicated that they wanted to run a substantial article on BP. I was pleased initially, but Nick Butler, my policy adviser, was concerned that this would be a personal profile

and would be ill advised ... The piece turned out to be a hagiography; I was embarrassed ... People thought I would be pleased but I felt very uncomfortable. Nick had been right. He usually was. No good would come of it.' See John Browne, *Beyond Business: An Inspirational Memoir from a Visionary Leader*, Weidenfeld & Nicolson, London, 2010, pp. 202–3.

70. In the general election on 7 June 2001 the Conservatives gained one seat while Labour lost five seats, gaining a very substantial Labour majority of over 200 seats in 'the quiet landslide'.

71. Interviews, Terry Macalister with John Browne, London, 22 March, 25 April and 22 June 2018.

72. Browne announced he was standing down from Huawei in July, 2020, https://news.sky.com/story/lord-browne-quits-as-huawei-uk-chairman-as-government-ban-looms-12028206.

CHAPTER 9

1. Bill Forsyth, *Local Hero*, filmscript, Goldcrest Films, 17 February 1983.

2. At Dire Straits' concert of 10 July 1985 at Wembley Arena they were accompanied by Hank Marvin at the end to play 'Going Home'. The concert was televised in the United Kingdom on *The Tube*, Channel 4, January 1986.

3. John McGrath, *The Cheviot, the Stag and the Black, Black Oil*, Methuen Modern Plays, London, 1981, p. 73.

4. By August 1974 there had been 15 oil fields found in the North Sea off Scotland: Montrose (September 1969), Forties (November 1970), Auk (February 1971), Brent (July 1971), Argyll (October 1971), Beryl (September 1972), Cormorant South (September 1972), Piper (January 1973), Thistle (July 1973), Dunlin (July 1973), Heather (December 1973), Ninian (January 1974), UK (April 1974), Claymore (May 1974) and Buchan (August 1974). The first to produce oil would be Argyll in June 1975, eleven months later.

5. Released May 1995.

6. Released August 1988.

7. Neal Ascherson, *Stone Voices: The Search for Scotland*, Granta, London, 2002, p. 188.

8. CEO of BP John Browne oversaw the sale of the Forties Field on 13 January 2003. The purchaser, the Apache Corporation, asserted that it would invest heavily in the declining oil wells and thereby extend the production life of the project. In the first decade Apache invested £2.8 billion in the Forties and proclaimed that that they had doubled production, found a further 80 million barrels of oil and would extend the life of the field by 20 years. When BP held the field its estimated lifespan ran to 2013, now Apache had extended that to 2033.

9. Terry's journey to the Forties Field left a powerful memory. After flying for a long time across the slate grey North Sea it seemed astonishing to see through a break in the clouds his destination, a flimsy island of black steel in the middle of the ocean. He had been thoroughly briefed on what to do in the event of a crash landing and was wrapped in layers of bright orange safety suits. The helicopter came down suddenly, its rota-blades clatter-

ing. The passenger door was pulled open, a strong wind blew in and he was ushered onto the helideck and through a steel door. It was warm and calm, only the gentle hum of diesel generators. Inside the belly of a beast it was part hotel, part drilling machine. It was clean and brightly lit. The buffet restaurant served sumptuous food, the pool tables and recreational area seemed comfortable but the sleeping quarters and their bunk beds were tiny. The work shifts were a brutal twelve hours on and the drill floor was dangerous. The offshore staff generally worked two weeks on and two weeks off. It was highly paid but highly disciplined and pretty disruptive to any domestic relationship. There was an air of camaraderie but also danger.

10. See Robert Orrell, *Blowout*, Seafarer Books, Woodbridge, 2000 and Paul Carter, *Don't Tell Mum I Work on the Rigs, She Thinks I'm a Piano Player in a Whorehouse*, Nicholas Brealey Publishing, London, 2006.

11. See BP Statistical Review 2015 and Dr Carole Nakhle, *Assessing the Future of North Sea Oil and Gas*, Crystol, 21 April 2016.

12. Esso was the brand name for the subsidiary of US oil company Exxon that operated in the UK.

13. Throughout the 1970s and 1980s there was a growing presence of the culture of the American oil world, and especially Texas, in British life. Foremost was the TV series *Dallas*, which drew audiences of over 20 million when broadcast on BBC 1 between 1978 and 1991. Several of the lead characters, such as J.R. Ewing, became popular icons, celebrated for their ruthlessness and the representative of the wedding of oil and money.

14. Government tax revenues from oil and gas production soared. https://assets.publishing.service.gov.uk/government/uploads/system/uploads/attachment_data/file/902798/Statistics_of_government_revenues_from_UK_oil_and_gas_production__July_2020_for_publication.pdf.

15. Steve Levine, *The Oil and the Glory: The Pursuit of Empire and Fortune on the Caspian Sea*, Random House, New York, 2007, p. 113.

16. Peter Maass, 'The triumph of the quiet tycoon', *New York Times*, 1 August 2004.

17. FCO, 'Minutes: Call on the Secretary of State by Chairman and Chief Executive of British Petroleum – HF/FOI '68, 2 December 1993, obtained through Freedom of Information Act.

18. John Browne, *Beyond Business: An Inspirational Memoir from a Visionary Leader*, Weidenfeld & Nicolson, London, 2010, p. 156.

19. Ibid., p. 156.

20. Greg Muttitt, *Fuel on the Fire: Oil and Politics in Occupied Iraq*, Random House, London, 2012, p. 43.

21. Under CEO Eric Drake, BP's assets were nationalised by Libya on 7 December 1971, and under Chairman and Managing Director David Barran, Shell had assets nationalised on 1 September 1973. One company successfully utilised this turbulence – Occidental – which under the guiding hand of its founder, Armand Hammer, retained its oil production. It was because of the need to process the immense amounts of Libyan crude that Occidental pushed to build a refinery on Canvey Island. This in turn led to the events in

which Hammer went head-to-head with Wilko Johnson in a Basildon hotel (see Chapter 5).

22. Richard Wray, 'BP marks return to Libya with $900m gas deal', *The Guardian*, 30 May 2007.

23. John Browne, *Beyond Business: An Inspirational Memoir from a Visionary Leader*, Weidenfeld & Nicolson, London, 2010, p. 129.

24. Terry Macalister, 'Secret documents uncover UK's interest in Libyan oil', *The Guardian*, 30 August 2009.

25. See Medard Gabel and Henry Bruner, *Globalinc: An Atlas of the Multinational Corporations*, The New Press, New York, 2003. In the Foreword BP CEO John Browne writes: 'There is other territory to be explored as well. In each of the corporate maps shown here there are large gaps. Multinational business, even after two decades of rapid growth, probably still reaches less than half the world's populations. Globalization is still a very partial phenomenon, and many still lack the means or the freedom to enjoy the choices we take for granted. For those of us who believe passionately that globalization is a good thing that represents the next frontier – the map is still to be drawn.'

26. Until just 13 years prior to 1989 the Chinese economy was strongly autarkic. For example the state had zero oil imports, and in fact in 1976 the People's Republic of China was exporting eight million tonnes of crude to Japan.

27. James Bamberg, *The History of The British Petroleum Company, Vol. 3: British Petroleum and Global Oil, 1950–1975*, Cambridge University Press, Cambridge, 2000, pp. 405–9.

28. Ibid., p. 417–18.

29. However, the progress of IT continued. Two years after Walters expressed his concern the young John Browne was creating computer models of the oil fields while working for BP in Alaska. Although these techniques were novel in the British company, Browne was essentially learning what had become standard practice among BP's American corporate peers. As he described, 'I soon began to feel our [computing] technology was lagging behind that of the American companies, particularly ARCO and Humble. I knew BP had to get up the curve and quickly to solve problems ahead of others. I was experiencing something very different from my peers back home in the government-protected BP. Working in America, and collaborating and competing with great American companies, taught me what competition really meant.' See John Browne, *Beyond Business: An Inspirational Memoir from a Visionary Leader*, Weidenfeld & Nicolson, London, 2010, p. 31.

30. Ibid., p. 31.

31. Brian Pietrzyba was employed for 15 years in the post room of Britannic Tower, delivering to all floors of the office block, from 1975 until he was laid off under the Robert Horton restructuring of BP head offices.

32. Don Strawbridge and Peter Thomas, *Baglan Bay: Past, Present and Future*, BP Chemicals Ltd, Baglan, 2001, p. 141.

33. The intimacy of the link between BP and WPP is illustrated by a story from Tom Bergin. He details the struggle between Peter Sutherland, chair of BP, and John Browne over the latter's retirement date. Sutherland even-

tually won, but realised that he'd been battling against Martin Sorrell, head of WPP, who had assisted Browne's side. 'When the dust settled, Sutherland called WPP's Sorrell, furious that one of BP's largest suppliers had engaged in what he saw as a putsch against the board. He got straight to the point. "I understand", he said "you've been calling newspapers in relation to corporate governance matters here at BP. I'm just wondering if that was in a personal capacity or as CEO of a company which provides BP with $300 million worth of marketing services each year?"' Tom Bergin, *Spills and Spin: The Inside Story of BP*, Random House, London, 2011, pp. 105–6.

34. Technological revolutions such as 3-D seismic testing which enabled the companies to use IT to model far more accurately the structure of oil- and gas-bearing rocks deep beneath the Earth's surface and better assess what the level of resource there was in an exploration block, and essentially to estimate at what oil price it could be profitably extracted. The process of finding profitable oil, which in its heroic heyday from the 1900s to the 1960s had required teams of geologists working in places such as Alaska and the Niger Delta, assisted by aerial surveying, now increasingly became the realm of the computer modelling team.

This shift was manifest in the layout of the companies. The IT hubs of Shell and BP grew radically in size and influence in the 1990s. In 2001 and 2002 BP was able to move into a set of smaller head offices on St James' Square, Piccadilly, despite having dramatically grown in staff and capitalization over the previous five years, because more and more of its functions took place at Sunbury-on-Thames next to Heathrow and in Houston, Texas. The same morphing could be seen in Shell.

The revolutions in the application of Big Data to oil and gas exploration also altered its geopolitics. When the Soviet sphere opened to Western oil corporations in the early 1990s, the repeated line was that the communist industry was light years behind the technological advances of the likes of Shell, BP and Exxon. In part this was commercial hype as the corporations attempted to overawe the energy ministries of Russia, Azerbaijan and other states. However, the technological disparity between the two sides did accentuate the disparity of political power.

The Azeri-Chirag-Guneshli oil field in the Caspian Sea was developed by BP from 1994 onwards. The seismic analysis of what lay beneath the seabed, the design for the offshore platforms, the data flows from the systems that monitored its daily function were all run through BP-owned and -operated computers. In the 1990s much of this IT hardware was based in the BP offices in Baku. However, by the mid-2000s these computers and the staff that operated them were based in Sunbury-on-Thames.

Gradually the daily operation of Azerbaijan's most valuable oil field, and the terminals and pipelines that allowed the Azeris to export crude, came to be run out of Sunbury. The minute-by-minute functioning of this massive machine was undertaken by men and women who constantly studied screens on the BP campus and at the end of each shift drove to homes in London or the Home Counties. This was a revolution in industrial practice enabled by Big Data, and like the political impact that had arisen from the closure of plants like Baglan and the concentration of company activity in London, this shift also had a geopolitical impact.

By the mid 2000s what Big Data was enabling the companies to enact was a mirror of drone warfare. The use of drones armed with laser- and computer-guided missiles had transformed military operations. No longer did the US need to put 'boots on the ground' in Afghanistan or Somalia as it had in Iraq, but now it could attack its 'targets' and 'win' conflicts without invading countries, without toppling regimes and without attempting to build states. It altered both the style of warfare and the politics of the US around it. Obama had come to office in 2009 partly on the back of his pledge to end the US occupation of Iraq, and indeed from 2010 there was a steady wind-down of the number of troops stationed in the country. However, he also oversaw, in both of his presidential terms, a massive increase in drone strikes. The legitimacy of Obama's actions was constantly questioned, but in effect the shift in the technology of warfare had bypassed the US constitution that placed limits on the rights of the president to declare and wage war on other states. The drone strikes were a form of warfare without war. And the practice of the oil companies came to mirror that of the US military, of oil development without development.

From the first days of exploring for crude in Persia in the 1900s, the Anglo-Persian Oil Company built up a massive infrastructure of extraction in the country. This encompassed drilling rigs, pipelines, the world's largest refinery, export docks, R&D facilities, airfields and harbours, petrol stations and shops across Iran, compounds for workers and towns for staff and their families. All the expertise to extract, refine and export oil was locked within this infrastructure, which was effectively a 'state-within-a-state' located in Persia, which later became Iran. This is the world that John Browne had lived in as a teenager with his parents.

The life of this 'state-within-a-state' was assured by the security umbrella provided by the Royal Navy and the British Foreign and Colonial Office. The ability of the UK to defend the oil infrastructure – impossible without the added assistance of the US military and Washington – was demonstrated by the overthrow of the government of Mossadegh.

Just over 50 years later, the ability of the US to assure the security of an equivalent infrastructure across the border in Iraq looked highly questionable. Furthermore, the very limited capacity of British forces and diplomacy to provide any such security was obvious to all. Much was made of how valuable it was to the US that the UK provided a small element of the occupying troops, but nobody presumed to suggest that the British would be able to exert control over Iraq alone.

BP and Shell learned rapidly how to adapt to the inability of the Western powers to secure Iraq in the manner that had been envisioned prior to the invasion. There were to be no towns for staff families as in 1950s Iran, no extensive company offices as in 2000s Azerbaijan, and no oil law with the state which could give a long-term guarantee to an investment of billions made in the oil fields. But for both BP and Shell the prospects that they obtained a share were too great to be forgone despite these highly unfamiliar circumstances. The Rumaila field constituted a substantial part of BP's oil production in 2020. New strategies would need to be applied.

Significantly, BP's joint venture partner for this giant oil field was not ExxonMobil, Total or Statoil but CNPC, a corporation that was wholly owned by the Chinese state. Once again Big Data assisted BP in the operation of this field and Shell in the exploitation of the Iraqi Majnoon gas field. As had been shown in the development of the platforms and pipelines in Azerbaijan, all the data and much of the day-to-day operation of this vast machine could be run from Sunbury-on-Thames or Aberdeen. This remote-working technique was clearly vital in Iraq where anything held 'in country' was liable to sudden attack or sabotage, and at the very least placing staff 'in country' required hugely expensive insurance and an extensive web of private security.

After BP and Shell signed the exploration agreements with Qaddafi in 2007 they set to work developing the offshore prospects. The stability of the Qaddafi regime, and especially the question of succession, loomed over the long term, but the companies pursued their projects regardless. However, the end came sooner than expected. In 2010 and 2011 the Arab Spring spread across the Middle East and North Africa. Libya was, after some delay, not immune. In February 2011 there was a substantial uprising against Qaddafi. The regime was brutal in its suppression but the dam could not hold and soon the country was in a full-scale civil war. The UK and France – with US logistical support – weighed in on the side of the anti-Qaddafi uprising with aerial bombing raids. By October 2011 Qaddafi had been driven from power, hunted down and shot; just like Saddam Hussein before him.

However, if the West or the oil companies had believed that military and diplomatic support for the uprising would produce a stable state that was compliant to foreign powers and corporations, they were soon disappointed. Libya collapsed into warfare on a scale arguably more intense than in Iraq. Onshore the oil companies discovered that no extraction or export infra-structure could be secured.

Yet despite this turmoil, work on the development of the offshore oil fields by BP and Shell largely continued unabated. The BP Libya business unit and Shell Libya business unit were all underpinned by the extraordinary power of Big Data. These operations took place far from the scrutiny of the media in the UK. The Foreign Office and the Ministry of Defence kept a watchful eye, but there was little the UK could do to try to assist the establishment of a more stable state in Libya.

BP joined the hunt for hydrocarbons off the Libyan coast. By utilising the technology of floating production vessels refined in the North Sea, it may well be that – should exploration be successful there – a stable Libyan state is not necessary in the production phase either. For oil extracted from Libya's offshore fields could be loaded directly onto tankers and exported without ever touching Libyan soil. Big Data has enabled remote working to operate at an entirely new level, or to become what we call 'drone drilling'.

Drone drilling has helped drive a new set of geopolitical realities in which resources can be extracted from countries that are, or become, 'failed states' successfully enough for there no longer to be a necessity for the West to establish a level of security. In turn this means that BP's and Shell's need for UK military and diplomatic power is becoming ever slighter, and it does so when Britain's power beyond its borders is becoming ever weaker.

CHAPTER 10

1. Interview, Terry Macalister with Andy McClusky, London, January 2018.
2. Interview, Terry Macalister and James Marriott with Kenny Cunningham. Chester, 4th April 2018.
3. David Holmes, 'Shell to close lab at Thornton Research Centre near Chester', *Chester and Cheshire News*, 23 July 2013.
4. Andrew Miller MP did talk to the press in defence of the 100 jobs at the linked Shell Stanlow lubricants plant. 'Ellesmere Port and Neston MP Andrew Miller vows to fight for jobs at Shell', *Cheshire Live*, 23 July 2013.
5. By contrast, four months after Terry's article the newspapers reported that BP's Coryton Refinery was to close with up to 850 jobs to go. This was met by howls of rage by MPs, unions and the media. The local MP, Stephen Metcalfe, was dogged in his attacks on the chancellor despite being in the same party. He had little effect.
6. See Chapter 8.
7. See Lynne Rees, *Real Port Talbot*, Seren, Newport, 2013, p. 125: 'the local authority wards of Sandfields West and East, along with Aberafan, are identified as among the most deprived in the Port Talbot area, according to Welsh Index of Multiple Deprivation statistics, with parts of the wards in the top 10 per cent of the most deprived wards in Wales' (calculated on income, employment, health, education, housing and geographical access to services).
8. 'BP-led joint venture proposes Chinese acetate esters plant', *Oil and Gas Journal*, 26 September 2000.
9. Don Strawbridge and Peter Thomas, *Baglan Bay: Past, Present and Future*, BP Chemicals Ltd, Baglan, 2001, p. 239.
10. The 24th G8 Summit was held in Birmingham on 15–17 May 1998.
11. KPMG opens private members club in Mayfair: www.ft.com/content/d03822co–3b6o-11e5-bbd1-b37bco6f59oc.
12. Interview, Terry Macalister with Vivienne Cox, London, 27th February 2019.
13. BP's Gas, Power and Renewables business which, under her direction, was rebranded BP Alternative Energy.
14. Cox made business woman of the year: www.telegraph.co.uk/finance/newsbysector/2937655/Business-profile-Business-woman-of-the-year.html.
15. Cox takes up government role: www.gov.uk/government/people/vivienne-cox#biography.
16. Cox chair of the newly created Rosalind Frankling Institute: www.rfi.ac.uk/about/governance/.
17. Ex-BP board member John Manzoni held a key role as chief executive of the civil service but stood down in 2020: www.gov.uk/government/people/john-manzoni.

18. Ex-BP executive Neil Morris made chief executive of the Faraday Institution: https://faraday.ac.uk/author/neil-morris/.

19. John Manzoni knighted before stepping down in 2020: www.thegazette.co.uk/notice/3454532.

20. Others moving from BP or Shell into the civil service include: Lord Browne, ex-government lead non-exec director (BP), Tony Meggs, chair of Crossrail and former chief executive of the Infrastructure and Projects Authority (BP) and Simon Henry, chair of the Defence Audit Committee (Shell).

21. John Browne, *Beyond Business: An Inspirational Memoir from a Visionary Leader*, Weidenfeld & Nicholson, 2010, pp. 24–5.

22. Oil Change International campaigns to 'Separate Oil and State': http://priceofoil.org/campaign/oilandstate/.

23. I. Bickerton et al., 'Marriage after a century of cohabitation: Shell prepares for the next merger round', *Financial Times*, 28 June 2005.

24. Cable refused to condemn Shell as he writes in his autobiography. Vince Cable, *Free Radical*, Allen and Unwin, London, 2010, pp. 180–200.

25. Others active in Parliament in 2019 included the following. House of Commons: Alan Duncan, minister of state for Europe (Shell); Greg Barker, minister of state for climate change (Sibneft of Russia); Andrew Robathan, minister of state for Northern Ireland (BP); Maria Miller, secretary of state for culture (Texaco). House of Lords: Lord John Browne of Madingley (BP); Lord David Simon of Highbury (BP); Lord Ronald Oxburgh (Shell); Lord John Kerr of Kinlochard (Shell).

26. Former BP employee David Lidington in government: www.gov.uk/government/people/david-lidington.

27. Former Shell employee Liz Truss in government: www.gov.uk/government/people/elizabeth-truss.

28. Former Shell employee Alan Duncan in government: www.gov.uk/government/people/alan-duncan.

29. Former BP employee Lord Robathan in government www.gov.uk/government/people/andrew-robathan.

30. Shell sharpened its target in April 2020, promising to become 'net zero' by 2050: www.theguardian.com/business/2020/apr/16/shell-unveils-plans-to-become-net-zero-carbon-company-by-2050.

31. Interview, Terry Macalister with Greg Muttitt, London, 23rd October 2019

32. Donna J. Haraway, *Staying with the Trouble: Making Kin in the Chthulucene*, Duke University Press, Durham, NC, 2016, pp. 34–5.

33. Tom Bergin, *Spills and Spin: The Inside story of BP*, Random House, London, 2011, pp. 105–6.

34. John Browne, *Beyond Business: An Inspirational Memoir from a Visionary Leader*, Weidenfeld & Nicolson, London, 2010, p. 221.

35. Browne quickly recovered his public standing. Two months after being sacked he was appointed chair of trustees at Tate. In 2010 he was given the government post of lead non-executive director. Soon after he published his autobiography, *Beyond Business: An Inspirational Memoir from a Remarkable Leader*.

36. Interview, Terry Macalister with John Browne, London, 25 April 2018.

37. Interview, Terry Macalister with John Browne, London, 22 March 2018.
38. In this deal BP also acquired Mobil's Birkenhead lubricants plant. See 'BP, Mobil to merge European refining, sales', *Oil and Gas Journal*, 4 March 1996.
39. In July 2005, John Browne had attended the G8 Summit at Gleneagles and announced an innovative project, just 70 miles north, close to Aberdeen. BP, with UK state financial assistance, would convert an existing gas-fired electricity power station into a mechanism to capture carbon dioxide emissions and assist in addressing climate change. The Peterhead power station had for 13 years been supplied with gas from the Miller oil and gas field 170 miles offshore. As the output of the field began to decline, the system would now be run in reverse. The CO_2 emissions from the power station would be pumped back down the gas pipeline, thrust back beneath the rocks of the North Sea and sealed off. A massive conveyor belt had been shifting carbon from under that cold ocean, up through drills and engines, through pipes and furnaces and into the atmosphere above. Now the machine was to be run in reverse, capturing the exhaust from the power station chimneys before it escaped into the heavens, and driving it back underground to be imprisoned in perpetuity. In principle the idea of CCS is a bold weapon in the struggle against climate change. But it is barely tested. There are a handful of CCS systems actually operating in the world, and many challenges remain, such as just how sure can we be that any CO_2 stored underground in former gas fields will not, at some point in the future, leak back into the atmosphere. Furthermore, the few projects in existence are largely double-edged swords. The CO_2 is pumped back into the rocks to drive out the last, hard-to-reach cubic metres of gas or oil from the field in a technique known as 'enhanced oil recovery'. Thus the overall benefit to the atmosphere is reduced. Nevertheless, CCS has been, and continues to be, presented by the oil industry as the panacea for climate change, in which the oil wells will be reused as stores for captured carbon and the problem of CO_2 emissions solved by the technological prowess of the corporations. At the same time profit can be generated from the provision of that storage, especially from public subsidies if the state pays a private company to undertake this civic role.

CCS offers a solution that perfectly combines the twin drivers of engineering and capital. It was a key tool of the oil industry as they lobbied in the UNFCCC process against the terms of any agreement, especially in Paris, that would limit their ability to extract oil and gas. Ben van Beurden had firmly placed it among the armoury that Shell would use in its battle against climate change, for some climate scientists believe CCS will have to play a part in the energy transition.

However, these schemes need commitment: they are long-term ventures and very expensive. Trevor Garlick, during his term as head of BP North Sea, saw the Peterhead scheme overridden by short-term corporate priorities. Less than two years after announcing the CCS project at Gleneagles, John Browne had been ousted from his post, and his replacement, Tony Hayward, wielding a new broom, ordered that the scheme be shut down. By the time that Garlick took up his position two years later in October 2009, the Miller Field had begun its decade-long phase of decommissioning, which would make any future use of the old reservoir for a CCS scheme potentially more complex and costly. Garlick's role was to keep a distant eye on this orderly process of decommissioning Miller. Soon after Shell explored the possibility of its own CCS scheme at Peterhead linked to the Goldeneye gas field, but in 2015 this too was shelved as a £1 billion government funding commitment to CCS was axed.

In Autumn 2018 we visited Peterhead. We pulled into the empty car park at the perimeter fence for the power station, then owned by the corporation SSE. It seemed there was no one around. We knocked on the door of the Portakabin where a security guard was housed. Out came a guy, black zip top, badge on his lapel for SIA Security. He was friendly and helpful. 'Yes, two years ago there was a big presentation to the local community by Shell as to what would happen. But then it stopped. They've finished here now. If you want information you need to contact Shell in Aberdeen.' We thanked him and walked back to the car. So this was the outcome of a decade of projects trying to get CCS off the ground.

40. Remit of Forum for the Future and Craig Bennett working at Judge Business School, Cambridge University in the Sustainable Business unit.

41. Forum for the Future continued to work closely with BP and it was not until after Browne's fall that the ties began to fray. It was Jonathon Porritt who highlighted this with a striking reversal of views. In an article in 2015 he declared that 'hydrocarbon supremacists' at the oil companies had successfully ousted reformers wanting to diversify into green energy. Speaking on behalf of Forum for the Future he said: 'We came to the conclusion that it was impossible for today's oil and gas majors to adapt in a timely and intelligent way to the imperative of radical decarbonisation. We felt we had no option but to end our longstanding partnerships with both Shell and BP.'

Clearly, many of the hopes that the British NGOs had held since the later 1990s were dashed. In the decade between 2005 and 2015 they slowly withdrew their support for the oil industry, eroding a relationship that had existed since the early 1960s. A parallel process was also taking place, as BP and Shell incrementally forsook their strategy of helping to form the opinion of UK NGOs and thus the opinion of a section of British society. Just as surely as capital was reallocated from plants such as Coryton and Baglan, so too it was reallocated away from a number of cost centres at head office such as CSR. With the reduction in staff numbers came a diminution in the calibre of those staff. Alongside this came a steep reduction in the amount of public information that the companies produced about themselves. BPs monthly magazine *Horizon*, for many years printed in hard copy and distributed throughout the staff and beyond, was moved online and then ceased entirely a few years later.

Shell's and BP's AGMs had been held as public affairs since the late 1970s. The companies saw that they should display an openness to their shareholders and the wider public. This was especially the case for BP as it was steadily privatized in the 1980s and endeavoured to embrace a form of 'shareholder democracy'. The AGMs under Browne were great events of state and, as if to emphasise the role of this private company as a 'national champion', the

AGMs were held at the Royal Festival Hall, the People's Palace, built at the dawn of Britain's era of social democracy. Here was the corporation apparently laying itself bare in the heart of London, in a building which symbolises the cultural life of the nation, open to all citizens, funded by the public purse and in which BP had only relatively limited and temporary powers of control.

After Browne's fall in 2007, the AGMs were relocated to the more humble, anodyne and relatively far-flung corner of London: the Excel Centre in Docklands. This venue, owned by the Abu Dhabi Sovereign Wealth Fund, was clearly a private location available for private hire by corporations, where those who wished to criticise any company could only do so at that company's behest. The AGM was being literally withdrawn from public view and the attention of all journalists, except those whose work required them to attend. This retreat from London-centred media scrutiny was further emphasised a decade later when the 2018 BP AGM was held at a conference centre in Manchester. In 2019 it was held in Aberdeen and in 2020 in Glasgow. The same slipping out of the limelight was manifest in Shell's withdrawal from holding two great AGMs each year in London and The Hague, to limiting the main event to the Dutch venue. The company was subtly leaving Britain. The window of corporate openness was being closed.

This change involved reducing CSR staff, winding up contracts with designers who worked on the in-house magazine and reducing the budgets of the AGM. Relative to the whole of BP the shift in capital allocation was minute, and these cuts had fractional social impacts compared to, say, the shutting down of Baglan. Partly as a consequence of this they were entirely hidden from public view and considered to be matters internal to the companies. However, with hindsight they provide key indicators of the company's change of direction.

42. See James Marriott and Mika Minio Paluello, *The Oil Road: Journeys from the Caspian Sea to the City of London*, Verso, London, 2012.

43. In October 1980 crude from Shell's Brent Field was pumped through a subsea pipeline to the Sullom Voe terminal in Shetland. Here some of it could be loaded onto a tanker and shipped to Bromborough Oil Terminal. From the terminal it was piped fifteen miles to Stanlow, refined into petrol and transported by road tanker to nearby petrol stations, where it filled the likes of Andy McCluskey's car. Some of the petrol was moved by river tanker to the Salford Docks and unloaded by Rebecca Long-Bailey's father.

44. The relocation of BP's blending capacity to China continued to grow in the coming two decades. One plant was built in Shenzhen in 1998 and the second in Taicang in 2005. In December 2017 the company announced the construction of a third plant in China, and its largest in the world, at Tianjin, covering 150,000 square metres with an annual production capacity of 200,000 tones.

45. 'Shell to shut Carrington ethoxylates, polyols plants', *ICIS News*, 1 December 2005, www.icis.com/resources/news/2005/12/01/1025408/shell-to-shut-carrington-ethoxylates-polyols-plants/.

46. Over in Wilton the plant that had taken on the ethoxylates production closed in 2010.

47. Interview, Terry Macalister and James Marriott with Colin McMullen, Carrington, 6 April 2018.
48. Interview, Terry Macalister with Ben van Beurden, The Hague, 16 February 2018.
49. John Browne, *Beyond Business: An Inspirational Memoir from a Visionary Leader*, Weidenfeld & Nicolson, London, 2010, p. 223.

CHAPTER 11

1. In contrast the threat of BP collapsing in 2010 did not induce panic that the oil and gas system would implode, that cars would stop dead on the motorways, but rather that pensions would take a hit. Would the UK government have stepped in to bail out BP?
2. The Occupy camp did not take place in Canary Wharf as this seemingly public space is in fact corporate property that is heavily policed by private security forces.
3. Prime Minister Margaret Thatcher's famous phrase stated on a number of occasions during her premiership.
4. Crude oil prices – 70 year historical chart by Macrotrends: www.macrotrends.net/1369/crude-oil-price-history-chart (30 March 2020). The idea of a global oil price, which today we take as an essential indicator of the world economy, is in fact a relatively novel concept. Before the first commercial exploitation of crude oil in Pennsylvania in 1859, all the oil that has since been extracted around the world, and all the oil that today remains to extract, still lay within the Earth and had apparently no value. Subsequent to John Rockefeller's aggressive seizure of a monopoly on US oil production in the 1870s, the price of oil was effectively set by the handful of corporations who extracted it. Sometimes in the first century of oil they operated as a formal cartel, sometimes their collusion was more subtle. By the 1960s 'The Seven Sisters', as the Western corporations had come to be called, still dictated the global oil price.

   However there was no 'price' in the way that we'd normally understand it, for there was no 'open market' on which oil was traded – none of the daily, hourly, volatility in the cost of a barrel of crude in the manner that we now take as a fact of life. The 'Oil Crisis' of 1973 changed this structure. As national oil companies took control of the sovereign geology of producer states, the Western corporations lost control of the oil fields that had guaranteed their supply of crude. Now the oil companies had to buy crude from each other and the national companies in order to meet the demands of their own refineries and other customers. As they bought and sold oil, a marketplace was developed and with the market came a price, or set of prices, for world oil. Different qualities of crude, which would require different techniques to refine, were established – Saudi Light, West Texas Intermediate and Brent, whose name derived from the oil being exacted from the UK's new North Sea fields.
5. It is not just today's earnings that determine the value of shares in a corporation. For the price of a share derives not only from a share or portion of the

profits today, out of which the dividend is paid, but also from the fact that it gives the holder a right to part of the profits of the future. Therefore the value of a share in BP or Shell is linked to the prospects of each company's future profits and this is linked to the future price of oil. To buy a share today is effectively to bet on future profits, on the future price of oil.

The ability of Shell or BP to sell oil, in say 10 or 20 years, depends upon each company having stocks to extract and sell at that point. These future stocks, or reserves as they are called, need to be constantly replenished. This is an energy source that is inherently non-renewable. If Shell is extracting oil faster than it is finding new reserves, then its ability to sell oil in the future will be curtailed.

6. John Browne wrote in his autobiography that the size of BP, by capitalisation, increased nearly five times during his period as CEO between 1995 and 2007 and the value of dividends paid out had tripled. The rise of BP's profits during Browne's rule was partly on account of the high oil price, partly on account of BP buying up competitors such as Amoco, but also due to a business strategy which included significant cost-cutting. This drove the sale or closure of plants in the UK such as Coryton and Baglan, while obtaining petrol and chemicals from lower-cost locations in Asia. These shifts created, and depended upon, the infrastructure of globalization.

The use of the capital generated was unevenly spread. Profits enabled the company to sustain the 'Beyond Petroleum' initiative with its £4.6 billion investments in alternative energy. Some oil analysts described the clean power strategy as 'a bull market concern', meaning it was a product of the largesse that usually accompanies periods when corporations are making high profits. Implicit in this comment seemed to be a belief that the green portfolio could vanish when the boom years were inevitably followed by a period of bust as is the pattern in the ever cyclical commodity markets.

7. BP Chairman Carl-Henric Svanberg caused further offence on 16 June 2010 by stating: 'I hear comments sometimes that large oil companies are greedy companies who don't care. But that is not the case in BP. We care about the small people.'

8. Ben Feller, 'BP CEO Tony Hayward meets with Obama at the White House', *Christian Science Monitor*, 16 June 2010.

9. Jeff Mason, 'BP agrees to a $20 billion spill fund, cuts dividend', *Reuters*, 16 June 2010.

10. Nick Serota, interviewed by the *Jewish Chronicle*: www.thejc.com/culture/interviews/interview-sir-nicholas-serota-1.16648.

11. Further emphasized by BP's decision to sell on the lease of their head offices and go more online in August 2020.

12. Tate's decision was not to renew a sponsorship contract with BP. That deal had run from 2011 to 2016. There had been expectations among environmental campaigners that BP itself would not renew as it needed to save money in the shadow of the *Deepwater Horizon* accident in the Gulf of Mexico. But the company went further, celebrating a five-year contract not only with Tate, but also with the Royal Opera House, the National Portrait Gallery and the British Museum. Would the withdrawal of Tate lead to the

other three dominos falling? Pressure was built up, especially on the British Museum. There were flash mobs coordinated by BP or Not BP?, banner drops by Greenpeace, an unlicensed concert by a quartet and lobbying of the incoming director of the British Museum, Hartwig Fischer. Campaigns sprang to life in Oslo, Paris, New York and Amsterdam targeting a number of corporations including Total, ExxonMobil, Statoil and the Koch Brothers. At the same time a new initiative, Oil Sponsorship Free, was launched to encourage cultural institutions and individual artists to sign a pledge never to take funding from the fossil fuel companies. And a number of UK professional bodies such as the Museums Association, recognising the need to address climate change at a systematic level, began to support the movement to drive oil out of culture.

13. In response to the judicial murder of Ken Saro-Wiwa in 1995, the arts and activism group Platform, in alliance with others such as Helix Arts, began to campaign against Shell's and BP's use of cultural institutions and the sponsorship of environmental organisations as a means of improving their brands and building their 'social licence to operate'. The vigour of the campaign had died away by the late 1990s, but came back to life in 2002 around the issue of BP's construction of the Baku–Tbilisi–Ceyhan oil pipeline through Azerbaijan, Georgia and Turkey. To battle against the environmental and human rights impacts of this infrastructure project a new coalition, Art Not Oil, was born.

Art Not Oil brought together London Rising Tide, Platform and several other organisations with a particular focus on BP's sponsorship of the National Portrait Gallery. For seven years demonstrations were mounted outside the gallery's keynote event, the opening night of the BP Portrait Award, as well as against Shell's sponsorship of the National Gallery, the Science Museum and the South Bank. Pressure was sustained and powerful. It established an understanding in the UK environmental community that corporate sponsorship was a key part of the oil companies' machinery of fossil fuel extraction, and to campaigners a part of the system driving forward climate change.

Just as Art Not Oil was beginning to fade, yet another wave of activism began, catalysed by Liberate Tate. This had ironically been born out of a workshop commissioned by Tate from the Laboratory of the Insurrectionary Imagination in February 2010. John Jordan had been engaged to teach students as part of Disobedience Makes History, but had been specifically warned by Tate staff not to mention BP's sponsorship of the galleries. It was a red rag to a bull. Jordan asked Platform to present analysis of how BP used Tate to manufacture its social licence to operate and from those two days the group Liberate Tate was born. See Evans, Mel *Artwash, Big Oil and the Arts*, Pluto Books, 2015.

14. Financial watchdog fines Shell for 'unprecedented misconduct': www.investegate.co.uk/ArticlePrint.aspx?id=200408241529542697C.

15. St Michael's Church: www.warfield.org.uk/Groups/260805/Warfield_Church/Locations/St_Michaels/St_Michaels.aspx.

396 · CRUDE BRITANNIA

16. Keetie Sluyterman, *Keeping Competitive in Turbulent Markets, 1973–2007: A History of Royal Dutch Shell, Vol. 3*, Oxford University Press, Oxford, 2007, p. 396.

17. Terry witnessed the deep concern of the Inupiat peoples during a journey to Barrow, the largest community that stood to be hit by any oil spill from Shell's drilling campaign off Alaska.

18. Shell understood this and Jeremy Bentham's Shell Scenarios Team produced a special report entitled 'After Macondo'.

19. Meanwhile the global price of oil remained unpredictable, averaging $90 during 2012 but plunging to below $40 by the end of 2015.

20. Simon Henry left in 2017 and in 2019 was made head of the UK government's Defence Audit Committee and put on the wider Defence Board alongside another ex-Shell senior executive, Paul Skinner. www.gov.uk/government/people/simon-henry.

21. 'This Bitter Earth' was originally written and produced by Clyde Otis and performed by Dinah Washington in 1960. Robbie Robertson laid Washington's vocals over Max Richter's string piece 'On the Nature of Daylight' for the soundtrack of Martin Scorsese's 2010 film *Shutter Island*. When Mel Evans approached Charlotte Church about performing as part of *Requiem for Arctic Ice*, Church had recently seen *Shutter Island* and suggested she perform Washington's vocal over the quartet's rendition of Richter's string piece.

22. So large and so up to date that it had a swimming pool and rifle range for the Shell staff in the basement.

23. Shell scraps Arctic drilling: www.theguardian.com/business/2015/sep/28/shell-ceases-alaska-arctic-drilling-exploratory-well-oil-gas-disappoints.

24. The new Cold War in the Arctic: www.amazon.co.uk/Polar-Opposites-Opportunities-threats-Guardian-ebook/dp/B0076KWQ7M and www.theguardian.com/environment/ng-interactive/2015/jun/16/drilling-oil-gas-arctic-alaska.

25. France bans new oil exploration: www.independent.co.uk/news/world/europe/france-ban-new-oil-gas-exploration-stop-granting-licences-macron-hulot-renewable-energy-drive-a7806161.html.

26. BP had sponsored an exhibition of Australian aboriginal art at the British Museum during this period.

27. Shell sells out of tar sands: www.theguardian.com/world/2017/mar/10/shell-sells-canadian-oil-sands-as-boss-warns-of-losing-public-support.

28. Three years on, in August 2019, BP too announced a withdrawal from Alaska after 60 years. It sold the Trans Alaskan pipeline and its stake in the once mighty Prudhoe Bay oil field – where John Browne cut his professional teeth as a young man – to a privately owned US company, Hilcorp. Jamison's trouble-hit company, NordAq, now rebranded as Borealis Alaska, would be flying the faded British flag.

29. By February 2020, 189 of the 196 states that negotiated the agreement had become parties to it.

30. Interview, Terry Macalister with Jeremy Leggett, London, 19th December 2019.

31. Professor Buchanan: www.imperial.ac.uk/people/d.buchanan.

32. Leggett had a played a lead role in the oil debates of the mid-2000s, arguing the case that the coming energy crisis was one in which peak oil would play a key role. Peak oil proposed that the world was facing geological and financial limits to the extraction of oil: that resources which could be extracted at profit were becoming ever more inaccessible, and that this in turn would force up the global oil price, driving the replacement of fossil fuels with renewable energy systems. The oil companies, including BP led by Hayward, found themselves under pressure to rebuff these concerns.

33. See Jeremy Leggett, *Global Warming: The Greenpeace Report*, Oxford University Press, Oxford, 1990.

34. See Jeremy Leggett, *The Carbon War*, Penguin Books, London 1999 – the first of five books by Leggett on these issues.

35. Browne's *chef de cabinet*, Matthew Powell, disputed this version of events saying in an email to us that Browne's piece in the *Financial Times* was about the failures of the oil and gas sector not a defence of it. He, Powell, wrote to Leggett only to 'correct factual errors' and not because anyone was 'upset'. Powell added: 'There was no reason for Lord Browne to contact Jeremy himself, not least because he did not and has not read Jeremy's blog.'

36. As there are almost no opportunities for the public to challenge the heads of this industry, they do so at these annual events, which, although they are controlled by the companies, the board are required to attend. Business leaders are known to hate these AGMs. The media also knows this and enjoys the spectacle.

37. A study published by Columbia University Earth Institute explains that these emissions could remain for 1,000 years and their impacts felt for 10,000 years.

38. A campaign to highlight what Exxon knew about climate change: https://exxonknew.org.

39. Polly Higgins and her work to make destruction of the planet a crime: www.stopecocide.earth/polly-higgins.

40. At times pensions were held by companies in their industrial sector, but directly in rival companies. Thus in the 1990s and 2000s the largest part of the BP company pension scheme was held in Shell shares, and the largest part of the Shell company pension schemes in BP shares.

41. Bill McKibben launches his Do the Math campaign on divestment: https://math.350.org.

42. New York city declares plans to divest from oil: www.theguardian.com/commentisfree/2018/jan/11/new-york-city-oil-industry-war-divestment.

43. More investors flee fossil fuels: www.prnewswire.com/news-releases/global-fossil-fuel-divestment-movement-reaches-6–24-trillion-in-assets-under-management-120x-increase-from-four-years-ago-report-says-300710204.html.

44. Norway sells off fossil fuel investments: www.bloomberg.com/news/articles/2019-06–11/norway-s-1-trillion-fund-set-to-win-go-ahead-for-oil-divestment.

45. Big union pulls out of fossil fuel investment: www.unison.org.uk/content/uploads/2018/01/Divest-from-carbon-campaign.pdf.

46. But not all are sympathetic to the divestment campaign. Helen Thompson criticised the analysis of this movement as too simplistic and argued that we will need to finance the extraction of oil in order to avoid the dysfunctionalities that come from future oil price shocks. Vivienne Cox seemed shocked by the divest demand.

CHAPTER 12

1. INEOS reopens pipeline: www.ineos.com/businesses/ineos-fps/news/ineos-forties-pipeline-system-media-update-0900–28122017-fps/.
2. Vitol is said to have had its best year ever financially in 2009 after the oil price collapse, and in 2014 it distributed over $1 billion in a special dividend to its 350 employee shareholder partners.
3. Now belonging to Essar.
4. This is a century-old pattern, but what is novel is the level of information that Big Data has enabled traders to access at any one time about this exchange. Sensors can give an exact flow rate at the terminal in Nigeria, satellite-based global positioning systems can give an exact location of a tanker in the mid-Atlantic and the oil demands of the refinery at Stanlow are available.
5. Vitol 2019 Volumes and Review: www.vitol.com/vitol-2019-volumes-and-review/.
6. BP Annual Report: www.bp.com/content/dam/bp/business-sites/en/global/corporate/pdfs/investors/bp-annual-report-and-form-2of-2018.pdf.
7. World's biggest oil consumers: www.eia.gov/tools/faqs/faq.php?id=709and t=6.
8. Shell sells businesses in Africa and Australia to Vitol: www.ft.com/content/be31fe62–3d99-11e0-ae2a-00144feabdcoandwww.ft.com/content/ca22ca16–9a8a-11e3-8e06-00144feab7de.
9. The oil corporations built their businesses along the lines developed by early oil tycoon John D. Rockefeller in the 1870s, controlling oil from the well to the pump. These 'integrated' businesses were significantly hit by nationalisations, the 'Oil Crisis' of 1973 and the subsequent birth of a global oil market. The integrated model remains in place but the large Western-owned independents hold a much smaller share of the total market. As Jamison told us, 'BP are now following the Vitol model which is more trading, less physical assets. ... but Vitol are now buying more physical assets. They've bought up all Shell's downstream in Africa except for South Africa. They bought a lot of Shell stuff in Australia as well. In a way they're going the opposite of the way the majors are going.'
10. BP's trading arm made profits when other divisions made losses: www.ft.com/content/340784a0-9d15-11e6-a6e4-8b8e77ddo83a.
11. Vivienne Cox described to us her role in the oil-trading division of BP: 'I set up our commodity derivatives business from scratch in 1991. At the time a few jet swaps had been done and they said "go and see if you can set up a new business". So we bought a few screens, got a few traders and it is now a very big part of BP.' Cox explained how she had effectively created the IST division at BP. 'I put all the trading businesses together. I ran Benelux

Refining and Supply, which was the Rotterdam refinery and all the supply trading around the Rotterdam refinery. Buying the crude oil to go into the refinery and then selling the product in Europe mainly. On the back of this, I put all the trading activities together including BP Finance Trading who were buying and selling currencies and debt. It was pulled all together as IST.'

After John Browne became head of BP, the IST arm was moved from the City to 20 Canada Square in Canary Wharf. By the mid-2000s the division was growing fast. After the oil price crash of 2008, BP profited handsomely on the back of oil market volatility. Traders told us that the regulatory authorities repeatedly investigate transactions for potential wrongdoing. Within BP, IST was beginning to have a pivotal position. As an illustration of this shift, when in 2012 BP's chief financial officer, Byron Grote, retired after ten years in post, his replacement was Dr Brian Gilvary, who had earlier been chief executive of the trading division for five years. Cox confirmed that experience of IST was of vital importance: 'All the senior figures in BP had been through the trading side of the business at some point in their career.'

12. It would be perfectly possible for the UK to cease producing any oil and gas, or cease consuming oil and gas, and to be entirely powered on renewable electricity, and yet this trade in these commodities at Canary Wharf to continue. The UK could finally declare itself zero carbon and yet the British economy could still be underpinned by the oil and gas of elsewhere. It is worth remembering that in the seventeenth and eighteenth centuries, while the British economy was partly driven by the wealth accrued through the trade in enslaved Africans, slavery was illegal within Britain herself, that bastion of English liberty.

13. Interview, Terry Macalister with David Jamison, West Sussex, 29th May 2018.

14. Jamison told us how he learned much of the business from Richard Oaten, who had himself been taught by 'two of the finest oilmen of that period'.

15. World class oil find in Alaska: www.adn.com/business-economy/energy/2016/10/04/caelus-claims-world-class-offshore-arctic-oil-discovery-that-could-among-alaskas-biggest/.

16. On 27 August 2019, less than three years later, BP finally sold all its Alaskan assets to Hillcorp, ending its 60-year presence in the state: www.reuters.com/article/us-bp-divestiture/bp-to-quit-alaska-after-60-years-with-5-6-billion-sale-to-hilcorp-idUSKCN1VH21N. This was part of a continuum with Shell's failed attempt to drill in the Alaskan Arctic Ocean.

17. Alaskan oil scandal: www.adn.com/business-economy/energy/2017/02/27/claiming-millions-of-dollars-went-missing-an-alaska-exploration-company-sues-2-former-executives/.

18. Interview, Terry Macalister and James Marriott with Trevor Garlick, Aberdeen, 22nd November 2019.

19. Garlick is keen to tell us about how his organisation is encouraging diversification by the oil and gas supply sector, particularly the underwater engineering sector that has traditionally looked after seabed production systems, rather than say large steel platforms. 'Nearly all those subsea technologies and the skills to, for example, run robots and monitor things, could

be applied to anything that is in the physical space between the sea level and seabed ... they [companies] could work across wind, wave and tidal [power]. They could work in marine agriculture, they could work in submarine repair. They could work in seabed mining', he tells us.

20. Prime Minister David Cameron talks to oil workers: www.facebook.com/ DavidCameronOfficial/photos/a.658575084166813/736793963011591/?ty pe=1andtheater.

21. The Clair Ridge field is in West of Shetland with BP holding 28.6 per cent, the operator's share, and Shell 28 per cent.

22. In defiance of the narrative of North Sea decline, the £4.5 billion Clair Ridge scheme was the biggest UK project for decades. When the very first oil field in West of Shetland was brought on-stream in 1998 in the Schiehallion field, a special FPSO, also called *Schiehallion*, was constructed by Harland & Wolff's workforce in Belfast. It was anchored in 1,400 feet of water and oil pumped out from rocks a mile and a half below sea level. Some of the production from West of Shetland was shipped to Bromborough Oil Terminal on the Mersey where it supplied the Stanlow Refinery.

After a decade the *Schiehallion* workhorse was exhausted. However, its replacement, the *Glen Lyon* FPSO, was constructed in the Hyundai Heavy Industries yard, South Korea. Garlick is clearly uneasy that the project should have largely bypassed a British shipbuilding industry that desperately needed work. 'We tried very hard to use local content across the board – but you can't buy local if it's not there', he says.

23. With thanks to Professor Gavin Bridge, who found figures for UK oil production for Shell and BP for 2019 in their annual reports and/or US Securities and Exchange Commission filings: for Shell, 517,000 barrels per day; and for BP, 534,000 barrels per day. The BP Statistical Review of World Energy gave the oil production figure for the UK for 2001 as 2,667,000. BP and Shell combined accounted for about 39 per cent of production that year. So, in round numbers, the two companies represented two-fifths of UK production in 2000 but in 2019 were around only one-fifth (18.7 per cent).

24. The question 'when will the oil run out?' has plagued the UK North Sea since its earliest days in the 1960s. Time and again the resources of this colony have lasted longer than many predicted. It seems that searching for a year in which it 'runs out' is like searching for a phantom ship, it slips away into the sea fog just as you approach. 'Running out' depends upon technology, on the global price of oil, on the relative difficulty of accessing oil in other basins such as the Alaskan Arctic Ocean, or South Australian offshore, or the decline of global oil demand driven by the rise of technologies such as electric vehicles or events such as the Covid-19 pandemic. However, regardless of these multiple unknowns and the elusive nature of the end date, UK North Sea oil is entering a decline, as a simple graph of production output shows. How quick that decline will be is heavily dependent on UK government policies on tax, offshore regulations and how to address climate change.

25. Interview, Terry Macalister and James Marriott with Professor Alex Kemp, Aberdeen, 22nd November 2018.

26. See Alex Kemp, *The Official History of North Sea Oil and Gas, Vol. 1: The Growing Dominance of the State*, Routledge, London, 2012 and Alex Kemp, *The Official History of North Sea Oil and Gas, Vol. 2: Moderating the State's Role*, Routledge, London, 2012.

27. The Chrysoar deal included a percentage of Shell's holding in the BP-operated Schiehallion Field in West of Shetland.

28. Interview, Terry Macalister and James Marriott with Jake Molloy, Aberdeen, 22nd November 2018.

29. DONG (now Ørsted) sells its oil business to INEOS: www.reuters.com/article/us-dong-energy-m-a/dong-energy-to-sell-oil-gas-business-to-ineos-for-1-3-billion-idUSKBN18K0OB.

30. See *Offshore, Oil and Gas Workers' Views on Industry Conditions and the Energy Transition*, Platform, Friends of the Earth Scotland and Greenpeace UK, London, 2020.

31. Saipem crane barge: www.saipem.com/en/identity-and-vision/assets/saipem-7000.

32. Kvaerner dismantling Miller: www.offshore-mag.com/field-development/article/16755814/s7000-removes-first-structures-from-decommissioned-miller-platform.

33. Sixty-seven per cent owned by NSMP and 18 per cent by INEOS.

34. Interview, Terry Macalister with Mike Wagstaff, Guildford, 13th February 2019.

35. PX Group manages operations and maintenance at St Fergus.

36. Sparkling future for oil man's grapes: https://localfoodbritain.com/surrey/articles/sparkling-future-for-greyfriars-vineyard/

37. Later we asked Wagstaff if he thought there was a climate emergency. He replied, 'Absolutely. I think there is an environmental catastrophe and I think that democracy has absolutely no chance in solving it because everyone's only interested in keeping their job rather than solving the problem … The question is, is it going to hit the Chinese sufficiently badly? They can do something because they don't have democracy, because they have a political system that can take 20- to 50-year decisions. We can't take a 20- or a 50-minute decision.'

Did he think the oil industry, and people who've profited from it like himself, bore any kind of responsibility for the climate situation, we asked bluntly? He replied: 'No, I think the oil industry in terms of the environmental impact has been actually way ahead, if you look at of the amount of pollution that is created by producing oil and gas, it is tiny compared to what it was. Burning fossil fuels, you know, is bad for the environment above a certain amount. This is an extremely complex problem and everybody's got to grasp the bullet. It's not just the oil companies, it's the people that burn it. So you know, if you're a power station, the answer is you put a massive great scrubber on the top or change fuel source.'

The idea that carbon dioxide can be removed by putting a 'massive scrubber' on a power station seems wilfully naive for a man so quick-witted. A scrubber is for removing the sulphur from power station smoke, not CO_2, and as Peterhead shows, to imagine we can rely on CCS schemes to deal with emissions from oil and gas requires massive optimism. We can hear the words of Greta Thunberg: 'You don't listen to the science because you are only interested in solutions that will enable you to carry on like before.' For 35 years the oil industry has been a casino in which to make cash. The activity remains separate from the environmental catastrophe he sees.

38. The Labour Party has a policy of bringing the energy networks of the UK into public ownership. Perhaps this in part explains Wagstaff's antipathy: 'I'm a thoughtful Conservative simply because the alternative is much worse. If there was one thing that might make me leave this country that would be the current policies of the Labour Party because they are hell-bent on bankrupting the country.'

39. The UK's Pipelines and Storage System built during World War II was sold in 2015 by the Ministry of Defence to the Spanish combine, Compania Logistica de Hidrocarburos, for £82 million.

40. *Sunday Times* Rich List 2018.

41. Ratcliffe was 13 when, on Boxing Day 1965, the *Sea Gem* gas rig went down forty miles off the East Yorkshire coast and 18 of those who survived were helicoptered to Hull Hospital.

42. Jim Ratcliffe and Ursula Heath, *The Alchemists: The INEOS Story, an Industrial Giant Comes of Age*, Biteback, London, 2018, p. 57.

43. Ibid., p. 63.

44. Ibid., p. 64.

45. Ibid., p. 68.

46. Ibid., p. 80.

47. Ibid., p. 82.

48. Bryan Sanderson, head of BP Chemicals, 1990–2001.

49. It had held a stake in it since the 1950s, but the company was rapidly pulling out of all chemicals investments in Europe. In 1994 it put the nail in the coffin of Baglan in south Wales, which had a ship dedicated to exporting raw product from the Severn Estuary to the Antwerp plant. Not long after closing Baglan's ethylene works, BP invested heavily in a chemicals plant in China, taking a 51 per cent stake in the Yangtze River Acetyls Co. Ltd. The corporation was shifting capital from Western Europe to the Far East and leaving in its wake areas such as south Wales blighted by deindustrialisation. BP had tried to sell its Baglan plant to an American and then an Indian buyer but failed. It looked as though the same fate might befall Antwerp.

50. Jim Ratcliffe and Ursula Heath, *The Alchemists: The INEOS Story, an Industrial Giant Comes of Age*, Biteback, London, 2018, p. 90.

51. Ibid., p. 88.

52. See LSE Alumni Energy Group, 'Fireside chat with Sir Jim Ratcliffe', 2019.

53. How INEOS raised cash: www.ft.com/content/68360108–928a-11da-977b-0000779e2340.

54. Morgan Stanley, Barclays and Merrill Lynch.

55. Jim Ratcliffe and Ursula Heath, *The Alchemists: The INEOS Story, an Industrial Giant Comes of Age*, Biteback, London, 2018, p. 108.

56. Ibid., p. 113.

57. Interview, Terry Macalister and James Marriott with Mark Lyon, Grange-mouth, 24th November 2018.

58. Six years later INEOS entered into a 50:50 joint venture with Petrochina, the state oil company of China, and formed PetroINEOS which took control of Grangemouth. A long cycle of change was being completed. In 1994 BP had closed its chemicals plants in Europe, such as Baglan, and invested in plants in China. Now, in 2011, the Chinese state was purchasing ex-BP works in the UK. It was a perfect symbol of the shifting geopolitics of the age.

59. Details of the battles between 2005 and 2013 are described in Mark Lyons' own book, *The Battle of Grangemouth*, Lawrence and Wishart, London, 2017. Lyon told us about the struggle with INEOS: 'Here's one for you, even among all this craziness in 2013 and the threats to shut the plants and investment for the government and handouts, millions of pounds of handouts to Ratcliffe. Millions of pounds. Do what you want. Among all that, there wasn't even a single guarantee sought on anything. Even in that model, there was nothing asked for [by government from INEOS].'

60. See LSE Alumni Energy Group, 'Fireside chat with Sir Jim Ratcliffe', 2019.

61. NSMP gained control of the St Fergus terminal and gas pipeline systems in 2015 and then sold it on to Kuwait Overseas Investments in 2018.

62. There was particular concern that the chemicals used to break up oil-bearing rocks would seep into the water table.

63. Neither BP nor Shell chose to pursue fracking for gas in the UK, but they did so abroad.

64. Owned 45 per cent by US private equity group Riverstone Holdings where John Browne had briefly worked.

65. INEOS lost a court action trying to curb protests: https://friendsoftheearth. uk/climate-change/ineos-defeated-campaigners-celebrate-right-free-speech-and-peaceful-protest-upheld.

66. Ratcliffe hits out at government: www.ft.com/content/75b99568-2888-11e9-a5ab-ff8ef2b976c7.

67. INEOS wins government support: www.ineos.com/inch-magazine/articles/issue-11/government-support-was-crucial-says-ineos/. This illustrates the long decline of shipbuilding for the oil and gas industry in the UK.

68. Part of the armoury that the oil industry has used in its battle to prevent a ban on the extraction of fossil fuels is to argue that gas is a transition fuel, that it has substantially lower carbon emissions per unit of energy than either coal or oil, and thus that it should be supported as part of the technology required by the low-carbon economy.

   This questionable calculation is especially flawed in relation to LNG. The sheer scale of energy required to build and maintain a system that pumps gas from western Pennsylvania, freezes it into a liquid, ships it across the Atlantic and then turns it back into gas, means that LNG as a process is extremely carbon heavy. It is profoundly more carbon intense than gas pumped from the offshore fields of the UK North Sea down the pipeline from St Fergus to Grangemouth. The system INEOS created was fundamentally to generate a higher profit on the output of Grangemouth and other plants, not to save on CO_2 emissions. Rarely in the 300 pages of Ratcliffe's autobiography, *The*

Alchemists, or in interviews has he even mentioned climate, unless he is criticising policies around it. This position is the polar opposite to that of van Beurden, who barely lets an interview pass without some comment on the climate.

69. Ratcliffe's autobiography leaves no doubt about his energy, determination, grit, deal-making flair, numbers brilliance, boyish camaraderie and loyalty to those to whom he's closest. And all of this despite his alleged personal shyness. The very title of the book, *The Alchemists*, bestows on 'Jim' almost magical powers of transformation in its portrayal of his epic struggles to exert control through the tools of accountancy and the power of money. It describes his battles to overcome those forces that threaten to take things out of his control – such as BP, financial crises or the trades unions. And how he has overcome nature – by walking to the North Pole and South Pole, by mountain biking and by off-road motor biking in Africa. Ratcliffe has focused his or INEOS's wealth on financing football clubs such as Nice in France, luxury yachts and an extensive array of homes, one on the river near Hythe in Hampshire. He is presented as the rugged plutocrat, utterly different from van Beurden, Browne, Watts, or Walters, indeed all those who have previously held sway over Crude Britannia. Yet is also possible that without Ratcliffe, Grangemouth could have closed like Coryton, Carrington or Baglan, or at least be 100 per cent foreign owned like Stanlow. Doubtless there are many in Grangemouth who are grateful to INEOS in that they have a job at all.

70. Ratcliffe leaves UK: www.telegraph.co.uk/business/2018/08/08/britains-wealthiest-man-sir-jim-ratcliffe-leaves-uk-move-monaco/.

71. Meanwhile the 'aromatics and acetyls' operation bought in June 2020 by INEOS is also a good indicator of the territories where BP currently concentrates activities. Of the 15 manufacturing sites, eight are in Asia, five in America, one in Belgium and one in Britain, the latter at Saltend near Hull. The new acquisition was described by Longden as 'very good for us', not least because it consolidated its hold on the Saltend site where INEOS already made chemicals and which is fed by pipeline from its Grangemouth plant.

We asked who was providing the cash for the latest purchase in such troubled Covid times. 'We are not disclosing where the financing is coming from but we will in time,' said Longden, sidestepping the issue. He denied the company was heavily indebted. 'If you look at 2019 group sales of $60 billion and EBIDA [profits] of just shy of $6 billion then our leverage [borrowings to profits] is only two times.' And Longden insisted that INEOS was doing fine in turbulent times of low oil and gas prices. 'We have a gas business so we have felt the effect of low prices but we are also a chemical producer and that is a natural hedge', meaning cheap gas 'feedstock' prices help the chemicals arm to make more money. INEOS, probably one of the top three largest chemical companies in the world, says it is also benefitting from the fact that as many as 300 of its products are used in medical applications. Covid has provided opportunities as well as threats.

72. INEOS accounts: www.ineos.com/globalassets/investor-relations/public/annual-reports/annual-report-blocks/2018-igh-sa-annual-report_final.pdf.

CHAPTER 13

1. Exxon only just within the top ten companies: www.investopedia.com/markets/quote?tvwidgetsymbol=xom.
2. Handful of staff share $1 billion: www.dailymail.co.uk/news/article-2127343/Facebook-buys-Instagram-13-employees-share-100m-CEO-Kevin-Systrom-set-make-400m.html.
3. Alibaba is a retailer without an inventory: www.investopedia.com/articles/investing/062315/understanding-alibabas-business-model.asp.
4. We were keen to discover whether Shell perceived these changes as threats to their future. Jeremy Bentham had said little on the matter, only that yes, 'we are aware of it, and for reasons of my pension, if nothing else, I want us to get it right'.
5. Interview Terry Macalister and James Marriott with Jeremy Bentham, Den Haag, 16th February 2018.
6. Shell Scenarios website, 'Meet the Shell Scenarios Team page', www.shell.com/energy-and-innovation/the-energy-future/scenarios/meet-the-shell-scenarios-team.html.
7. Bentham is also a visiting professor at Oxford University's School of Business.
8. Bentham's comment seems remarkably insightful, given that this interview was made nearly two years prior to Johnson winning the December 2019 election and cementing his position as prime minister.
9. In 2019 Shell was still the second largest producer of oil and gas in the UK North Sea, although only producing around 9 per cent of the UK's oil. In terms of the UK, Shell is a key player in the province. In terms of Shell, its oil production in the UK is just a tiny fraction of its global production.
10. Energy futures by Shell: www.shell.com/energy-and-innovation/the-energy-future/scenarios/a-better-life-with-a-healthy-planet.html.
11. 'Power trader' in the US economy denotes a company in the business of buying and selling electricity.
12. Eighteen months later, by July 2019, Ben van Beurden of Shell was saying he supported the EU's target of net zero by 2050: www.shell.com/media/speeches-and-articles/2019/getting-to-net-zero-emissions.html.
13. This forecast is set to be overtaken by events. In September 2020, DNV of Norway produced a report asserting that global oil demand had peaked in 2019, partly as a consequence of Covid-19 in 2020.
14. Jeremy Ball, 'Inside oil giant Shell's race to remake itself for a low-price world', *Fortune Magazine*, 24 January 2018.
15. Shell even owns its own staff hotel: https://shell-la-residence-the-hague.hotelmix.co.uk/.
16. Interview, Terry Macalister with Ben van Beurden, Den Haag, 17th February 2018.
17. Shell's £5 million plan to aid the reforesting of Scotland caused political upset in October 2019: www.forestryjournal.co.uk/news/17972293.scotgov-criticised-5m-shell-funded-tree-planting-scheme/.
18. Twenty-seven million UK households: www.ons.gov.uk/peoplepopulationandcommunity/birthsdeathsandmarriages/families.

19. Trump takes Scottish government to court: www.bbc.co.uk/news/uk-scotland-north-east-orkney-shetland-34471449.

20. Interview, Terry Macalister and James Marriott with Natalie Ghazi and Kevin Jones, Aberdeen 22nd November 2018.

21. Vattenfall is a Swedish wind developer: https://group.vattenfall.com/who-we-are/contact-us.

22. One of Nordic region's largest groups: www.statista.com/statistics/555103/sweden-20-largest-companies-by-turnover/.

23. Farmer takes power company to court over climate: www.theguardian.com/environment/2017/nov/30/german-court-to-hear-peruvian-farmers-climate-case-against-rwe.

24. Who causes carbon pollution? https://b8f65cb373b1b7b15feb-c70d8ead6ced550b4d987d7c03fcdd1d.ssl.cf3.rackcdn.com/cms/reports/documents/000/002/327/original/Carbon-Majors-Report-2017.pdf?1499691240.

25. The worst polluters: www.theguardian.com/environment/2019/oct/09/revealed-20-firms-third-carbon-emissions.

26. Vattenfall, *Energy Solutions, Cleaner Smarter Possibilities*, brochure, 2017.

27. Wind farms did play a role in the TV series *Wild Bill*, BBC, June 2019.

28. Lack of local employment in building British wind farms: www.dailyrecord.co.uk/news/scottish-news/fears-scots-workers-miss-out-16180370.

29. Foreign firms dominate some elements of the wind industry: www.offshorewind.biz/2018/05/02/aegir-takes-on-eowdc-jackets-at-port-of-tyne/.

30. Interview, Terry Macalister with Danielle Lane, London, 24th April 2019.

31. Russian workers paid £5 per hour on UK wind farm: www.theguardian.com/uk-news/2018/oct/21/migrants-building-beatrice-windfarm-paid-fraction-of-minimum-wage.

32. It is impossible not to notice that most of the oil executives we interviewed were men but renewables has a much more gender-balanced workforce. Oil and gas was framed largely by a male macho culture from the southern US, the new wind world may be framed by Nordic states (gender equal, socially liberal, feminized, etc.). And many of the key climate campaigners are women – often young ones such as Greta. Vivienne Cox at BP was a notable exception as a woman in the world of oil and she said – largely – it was just a matter of doing the job well: 'I never positioned myself as a woman. I positioned myself in the job that I was doing. I had the belief that if I did a good job and delivered on what I was supposed to do I would be recognized for it and I would say for almost all my career it was true.'

33. After meeting Lane we talked to a pro-renewable power government official who did not want to be named. We asked him about why foreign and not British operators and investors dominate the British offshore. 'It's an old British malaise. We often fail to invest in our own academic and technological successes. We invented onshore wind in this country but have not invested in it. The Norwegians, Germans and Danes have come in and stolen our picnic. The reason why CCGT [combined cycle gas turbine] gas-fired power stations were so successful is because Thatcher invested billions in making it so. The state has to invest in the early stages of a new technology,

you cannot just leave it to the free market. One reason Thatcher was prepared to pour money into CCGT was because she was trying to close down the coalmines and the left-led miners' union. But we in this country have a long-standing commitment to free market economics and an aversion to taxes so we fail to raise public funds that are needed to invest in things like education and infrastructure.'

34. Labour Energy Forum, 'Who owns the wind, owns the future', September 2017.
35. Even the industry lobby group, Renewable UK, has a list of 13 'sponsoring' member businesses on its website of which over half are owned abroad.
36. Around 90 per cent of London Array contracts went to foreign firms: www.theguardian.com/environment/2012/oct/02/uk-windfarm-little-british-involvement.
37. As part of the reshuffle of Shell assets instituted by Jeroen van der Veer following the 'Reserves Crisis'.
38. Why Shell said it needed to sell off a giant UK wind farm: http://news.bbc.co.uk/1/hi/business/7377164.stm.
39. Shell may start buying UK wind farms: www.shell.com/energy-and-innovation/new-energies/wind.html.
40. Shell has a 'difficult' relationship with UK wind farms: www.energyvoice.com/opinion/213939/shells-difficult-relationship-with-uk-wind-energy/.
41. The government initially pushed for the process to slow down as the Norwegian state had, while the corporations pushed for the process to speed up. Later the Thatcher government was happy to see the oil brought out as fast as possible to use the tax revenues to feed its planned neoliberal economic and social revolution.

CHAPTER 14

1. Interview, Terry Macalister with Gail Bradbrook, Stroud, 5th June 2019.
2. Hallam took a lower profile in Extinction Rebellion in 2019 after intense disagreements within the movement.
3. This action was stopped by the courts in February 2020.
4. Bradbrook explains it is not XR's job to take oil companies to court. 'If you look at the Shell action, there was some research done to make sure you got your messaging right, but we are not there to be a think tank. That's not our role. Our job is to get people into action, civil disobedience. Honestly it's been an uphill struggle.'
5. Interview, Terry Macalister with Elsie Luna, London, 15th April 2019.
6. Young girl stirs strikes: www.thetimes.co.uk/article/girl-10-takes-fossil-fuel-protest-to-the-top-rjvbw7wr9.
7. Climate hero: www.grimsbytelegraph.co.uk/news/local-news/climate-coalition-hero-awards-2019-2612153.
8. See Greta Thunberg, *No One Is Too Small to Make a Difference*, Penguin, London 2019.

9. Elsie Luna Podcast, 'Hear, hear!', https://twitter.com/hearhearpodcast; award given in parliament: www.grimsbytelegraph.co.uk/news/local-news/elsie-luna-climate-coalition-awards-2640670 of award ceremony.

10. Quotes from *Can You Hear Me?* speech, London, 23 April 2019, in Greta Thunberg, *No One Is Too Small to Make a Difference*, Penguin, London 2019.

11. British minister accepts failure to act on climate: www.theguardian.com/environment/2019/apr/23/greta-thunberg.

12. OPEC rails at climate protestors: www.theguardian.com/environment/2019/jul/05/biggest-compliment-yet-greta-thunberg-welcomes-oil-chiefs-greatest-threat-label.

13. BP pledges to cut carbon emissions by 2050: www.ft.com/content/e1ee8ab4-4d89-11ea-95a0-43d18ec715f5.

14. DONG has exited oil and morphed into Ørsted. Vattenfall – though never an oil company – will do something similar as it goes 'fossil free in a generation'.

15. See John Browne, *Make, Think, Imagine: Engineering and the Future of Civilisation*, Bloomsbury, London, 2019.

16. Browne says no silver bullet: www.theguardian.com/environment/2019/jun/08/john-browne-engineering-fracking-greta-thunberg-huawei.

17. Business needs the right cost incentives to tackle climate: www.theguardian.com/environment/2019/jun/08/john-browne-engineering-fracking-greta-thunberg-huawei.

18. We are heading for death if we don't do something: www.theguardian.com/environment/2019/jun/08/john-browne-engineering-fracking-greta-thunberg-huawei.

19. Speech by Cllr Lena Šimić, Liverpool, 22nd May 2019.

20. See Donna J. Haraway, *Staying with the Trouble: Making Kin in the Chthulucene*, Duke University Press, Durham, NC, 2016.

21. On 29 June 2019, Merseyside Labour for a Green New Deal hosted the 'Developing a Zero Carbon Manifesto for Merseyside' event, which encouraged participants to contribute ideas on how to radically cut carbon emissions and start developing a manifesto for the region. The event was chaired by Lena Šimić, with a panel of speakers that included Mika Minio-Paluello (Labour Energy Forum), Abhijith Subramanian (Youth Climate Strikes), Dan Carden MP (at the time acting shadow secretary for international development) and Clara Paillard, PCS union. The event was instrumental to producing the 'Local Manifesto 2021: Merseyside Labour for a Green New Deal', with a number of suggested policies for the region.

22. https://liverpoolexpress.co.uk/liverpool-declares-climate-change-emergency/.

23. Ørsted group: https://orsted.com/en.

24. Battery prices generally are rapidly falling in price, down by almost 90 per cent since 2010. They fell from $1,100 per kilowatt hour at the turn of this century to $150 in 2019. They are expected to have fallen close to $100 by 2023. It has been the same with the plunging cost of solar arrays and offshore wind farms. They can both be built commercially with a small, or even no,

subsidy. This is a massive change from the past when consumers had to help finance renewable power projects through higher energy bills.

25. Liverpool wants a green new deal: www.liverpoolecho.co.uk/news/liverpool-news/huge-230m-new-deal-liverpool-16664348.
26. www.bbc.co.uk/news/uk-england-merseyside-51677360.
27. Interview, Terry Macalister with Mark Shorrock, via Zoom, 19th December 2019.
28. When we met Jeremy Bentham he talked of Steven Fries, Shell's chief economist, who now works in Bentham's Scenarios Team. Bentham told us, with unabashed pride, 'I'm his line manager. I was the one who initially recruited Steven from the EBRD [European Bank of Reconstruction and Development]. Then he worked for four or five years in Shell. Then he was headhunted from Shell by the Department for Energy and Climate Change, where he spent three or foiur years redesigning the structure of the power markets in the UK. Then I headhunted him back to Shell. He's probably the macroeconomist with the deepest understanding of energy policy and the energy industry of any economist in the world, because he's now had ten years in industry and government.' It was a precise description of the interlocking of the state and private capital corporations and it echoed the words of his CEO, Ben van Beurden, who explained part of Shell's competitive advantage over Big Data by saying, 'And they might not be the right people to sit down with governments to talk about market design to enable growth in this sector.'
29. John Manzoni disputed these interpretations of events. Manzoni said he supported clean technology but '[I] remain sceptical about the lagoon as the most promising or economic of renewable technologies'. He added: 'I have always felt renewables to be an important part of the future energy mix and am still involved and interested in their development.'
30. EDF – French state-backed giant, one of the Big Six utilities in Britain – is building the new Hinkley Point C nuclear plant elsewhere in the Severn Estuary.
31. Shorrock, who supports climate movements such as XR, has been targeted by the tabloid newspapers for making money out of renewable power. He is married to another renewable power executive: Juliet Davenport, the chief executive of Good Energy. Davenport was covered by the *Daily Mail* for earning £290,000 and she and her husband were accused of being 'wealthy hippies' and 'fat cat green entrepreneurs'. Shorrock shrugs this off and wants to talk about music. '"Heaven" by Emile Sande is an important song to me because it reminds me of the time I made a presentation to Iberdrola, the Spanish owner of Big Six UK utility SSE. I was elated because a room full of executives who totally understood wind power took my hydro scheme completely seriously.'
32. Interview, Terry Macalister with Rebecca Long-Bailey, Salford, 1st November 2019.
33. Friends forever with Princess Di: www.standard.co.uk/news/politics/rebecca-longbailey-interview-labour-leadership-race-a4382071.html.

34. Skater girl: www.standard.co.uk/news/politics/rebecca-longbailey-interview-labour-leadership-race-a4382071.html.
35. By 2020 failed Labour leadership contender and then shadow education secretary.
36. See Kate Raworth, *Doughnut Economics*, Random House, London, 2017.
37. Burbo Bank wind farm: https://orsted.co.uk/energy-solutions/Offshore-wind/Our-wind-farms.

EPILOGUE

1. Oil prices fall below zero: https://ftalphaville.ft.com/2020/04/20/1587407982000/Oil-goes-sub-zero/.
2. www.dailypost.co.uk/news/north-wales-news/giant-tanker-anchored-anglesey-days-18063639.
3. Tanker boom: www.ft.com/content/32563eae-fae5-4f24-b3ad-459f6ebca495.
4. Ratcliffe's wealth plummets: www.thetimes.co.uk/sunday-times-rich-list.
5. Shell writes off $22 billion: https://uk.reuters.com/article/uk-shell-outlook/shell-to-cut-asset-values-by-up-to-22-billion-after-coronavirus-hit-idUKKBN2410SJ.
6. Industry on its way down: www.forbes.com/sites/davidblackmon/2020/06/16/bps-big-writedown-a-harbinger-for-a-declining-industry-or-of-a-struggling-company/#5d233952d465.
7. BP retreats from oil and gas: www.reuters.com/article/us-bp-outlook/bp-to-cut-fossil-fuels-output-by-40-by-2030-idUSKCN2500NH.
8. While colleges at Helen Thomson's Cambridge University surrendered to student demands for divesting pension funds from BP and Shell, there was still little focus on the private equity-backed and state oil companies buying up discarded North Sea fields.
9. Exxon removed from leading share index: www.ft.com/content/76ecd406-b08e-4c2c-8643-cd23acb7cf2c.
10. Thurrock invests millions in solar power: www.ft.com/content/7a01b39d-a5df-4e3f-b9a6-dfcf9339368f.
11. www.newpower.info/2019/08/welsh-development-framework-sets-out-15-priority-areas-with-presumption-in-favour-of-large-scale-onshore-wind-and-solar/.
12. https://renews.biz/63286/wood-sgn-draw-up-scottish-decarbonisation-roadmap/.
13. Interview, Terry Macalister with Dave Randall, London, 1st February 2019.
14. Lyrics from the song 'Deliver Us' by Slovo, written by Dave Randall, Camberwell Recordings, 2020.

Selected Bibliography

Abercrombie, Patrick, *Greater London Plan*, HMSO, London, 1944.

Ackroyd, Peter, *London: The Biography*, Vintage, London, 2001.

Ascherson, Neal, *Stone Voices: The Search for Scotland*, Granta, London, 2002.

Baker, J.A., *The Peregrine*, Harper Collins, London, 1967.

Bamberg, James, *The History of the British Petroleum Company, Vol. 2: The Anglo-Iranian Years, 1928–1954*, Cambridge University Press, Cambridge, 1994.

Bamberg, James, *The History of The British Petroleum Company, Vol. 3: British Petroleum and Global Oil, 1950–1975*, Cambridge University Press, Cambridge, 2000.

Becket, Andy, *When the Lights Went Out: Britain in the Seventies*, Faber & Faber, London, 2010.

Benn, Tony, *Against the Tide: Diaries 1972–1976*, Arrow, London, 1999.

Benn, Tony, *Conflicts of Interest, Diaries 1977–1980*, Arrow, London 1996.

Bergin, Tom, *Spills and Spin: The Inside Story of BP*, Random House, London, 2011.

Browne, John, *Beyond Business: An Inspirational Memoir from a Visionary Leader*, Weidenfeld & Nicolson, London, 2010.

Browne, John, *Seven Elements That Have Changed the World*, Weidenfeld & Nicolson, London, 2013.

Cable, Vince, *Free Radical*, Allen and Unwin, London, 2010.

Carson, Rachel, *Silent Spring*, Houghton Mifflin, Boston, 1962.

Carter, Paul, *Don't Tell Mum I Work On the Rigs, She Thinks I'm a Piano Player in a Whorehouse*, Nicholas Brealey Publishing, London, 2006.

Cocker, Mark, *Our Place, Can We Save Britain's Wildlife Before It Is Too Late?*, Vintage, London, 2018.

Crane, Nicholas, *The Making of the British Landscape*, Weidenfeld & Nicolson, London, 2016.

Cummins, Ian and John Beasant, *Shell Shock: The Secrets and the Spin of an Oil Giant*, Mainstream, Edinburgh, 2005.

Curtis, Mark, *Web of Deceit*, Vintage, London, 2003.

Davies, R.E.G. and Philip J. Birtles, *Comet: The World's First Jet Airliner*, Paladwr Press, Virginia, 1999.

Davies, John, *Energy to Use or Abuse?*, Shell UK Ltd, London, 1976.

Ellams, Inua, *Three Sisters*, Oberon Books, London, 2019.

Evans, Kathy and Douglas Marsh, *Who's Hoo: A Century of Memories*, Running Dog Press, UK, 2008.

Evans, Mel, *Artwash, Big Oil and the Arts*, Pluto Press, London, 2015.

Gabel, Medard and Henry Bruner, *Globalinc: An Atlas of the Multinational Corporations*, The New Press, New York, 2003.

Gill, Crispin, Frank Booker and Tony Soper, *The Wreck of the Torrey Canyon*, David Charles, Devon, 1967.

Gretton, Dan, *I, You, We, Them, Vol. 1*, William Heinemann, London, 2019.

Haraway, Donna J., *Staying with the Trouble: Making Kin in the Chthulucene*, Duke University Press, Durham, NC, 2016.

Harrison, Jeffrey and Peter Grant, *The Thames Transformed: London's River and Its Waterfowl*, Andre Deutsche, London, 1976.

Harvey, W.J. and R.J. Solly, *BP Tankers: A Group Fleet History*, Chatham Publishing, London, 2006.

Heath, Edward, *The Course of My Life: The Autobiography of Edward Heath*, Hodder & Stoughton, London, 1998.

Howarth, Stephen, *Sea Shell: The Story of Shell's British Tanker Fleets 1892–1992*, Thomas Reed Publications, London, 1992.

Howarth, Stephen, *A Century in Oil: The 'Shell' Transport and Trading Company 1897–1997*, Weidenfeld & Nicolson, London, 1997.

Howarth, Stephen and Joost Jonker, *Powering the Hydrocarbon Revolution 1939–1973: A History of Royal Dutch Shell, Vol. 2*, Oxford University Press, Oxford, 2007.

John, Angela V., *The Actors' Crucible: Port Talbot and the Making of Burton, Hopkins, Sheen and All the Others*, Parthian, Cardigan, 2015.

Johnson, Wilko, *Don't You Leave Me Here*, Little Brown, London.

Jonker, Joost and Jan Luiten van Zanden, *From Challenger to Joint Industry Leader, 1890–1939: A History of Royal Dutch Shell, Vol. 1*, Oxford University Press, Oxford, 2007.

Kemp, Alex, *The Official History of North Sea Oil and Gas, Vol. 1: The Growing Dominance of the State*, Routledge, London, 2014 .

Kemp, Alex, *The Official History of North Sea Oil and Gas, Vol. 2: Moderating the State's Role*, Routledge, London, 2011.

Kissinger, Henry, *Years of Upheaval*, Little Brown US, Boston, 1982.

Ledger, Frank and Howard Sallis, *Crisis Management in the Power Industry: An Inside Story*, Routledge, London, 1995.

Leggett, Jeremy, *The Carbon War: Global Warming and the End of the Oil Era*, Penguin, London, 1999.

Levine, Steve, *The Oil and the Glory: The Pursuit of Empire and Fortune on the Caspian Sea*, Random House, New York, 2007.

Logan, Nick Logan and Bob Woffinden, *The Illustrated New Musical Express Encyclopedia of Rock*, Salamander Books, London, 1977.

Lyons, Mark, *The Battle of Grangemouth*, Lawrence and Wishart, London, 2017.

Marriott, James and Mika Minio-Paluello, *The Oil Road: Journeys from the Caspian Sea to the City of London*, Verso, London, 2012.

McGrath, John, *The Cheviot, the Stag and the Black, Black Oil*, Methuen Modern Plays, London, 1981.

Meadows, Donella H., Dennis L. Meadows, Jorgen Randers and William Behrens III, *The Limits to Growth: A Report for the Club of Rome's Project on the Predicament of Mankind*, Potomac Associates, Virginia, 1972.

Mitchell, Timothy, *Carbon Democracy: Political Power in the Age of Oil*, Verso, London, 2011.

Moody-Stuart, Mark, *Responsible Leadership: Lessons from the Front Line of Sustainability and Ethics*, Greenleaf, London, 2014.

Mumford, Lewis, *The City in History*, Penguin, London, 1973.

Muttitt, Greg, *Fuel on the Fire: Oil and Politics in Occupied Iraq*, Random House, London, 2012.

Nockolds, Harold, *Shell War Achievements, Vol. 4: The Engineers*, The Shell Petroleum Company, London, 1949.

Orrell, Robert, *Blow Out*, Seafarer Books, Woodbridge, 2000.

Pirani, Simon, *Burning Up, A Global History of Fossil Fuel Consumption*, Pluto Press, London, 2018.

Pitt, Malcolm, *The World on Our Backs: The Kent Miners and the 1972 Miners' Strike*, Lawrence & Wishart, London, 1979.

Platt, Edward, *Leadville: A Biography of the A40*, Picador, London 2001.

Randall, Dave, *Sound System: The Political Power of Music*, Pluto Press, London, 2017.

Ratcliffe, Jim and Ursula Heath, *The Alchemists: The INEOS Story, an Industrial Giant Comes of Age*, Biteback, London, 2018.

Rees, Lynne, *Real Port Talbot*, Seren, Bridgend, 2013.

Roberts, David, *Cammell Laird: Life at Lairds, Memories of Working Shipyard Men*, Avid Publication, Gwespyr, 2008.

Rowell, Andy James Marriott and Lorne Stockman, *The Next Gulf: London, Washington and the Oil Conflict in Nigeria*, Constable Robinson, London, 2005.

Saro-Wiwa, Ken, *Genocide in Nigeria: The Ogoni Tragedy*, Saros International, Lagos, 1992.

Sebald, W.G., *On the Natural History of Destruction*, Hamish Hamilton, London, 2003.

Simons, Mike, *Striking Back: Photographs of the Great Miners' Strike 1984–1985*, Bookmarks, London, 2004.

Sluyterman, Keetie, *Keeping Competitive in Turbulent Markets, 1973–2007: A History of Royal Dutch Shell, Vol. 3*, Oxford University Press, Oxford, 2007.

Stanton Hope, W.E., *Tanker Fleet: Shell War Achievements, Vol. 1*, The Shell Petroleum Company, London, 1948.

Strawbridge, Don and Peter Thomas, *Baglan Bay: Past, Present and Future*, BP Chemicals Ltd, Baglan Bay, 2001.

Taylor, James Piers, *Shadows of Progress: Documentary Film in Post-War Britain 1951–1977*, BFI, London, 2010.

Thunberg, Greta, *No One Is Too Small to Make a Difference*, Penguin, London, 2019.

Yergin, Daniel, *The Prize*, Simon & Schuster, London, 1991.

Yergin, Daniel and Joseph Stanislaw, *The Commanding Heights: The Battle for the World Economy*, Simon & Schuster, New York, 2002.

Ziegler, Philip, *London at War*, Sinclair-Stevenson, London, 1995.

Acknowledgements

With thanks to . . .

. . . all those who agreed to be interviewed. Especially those who work at the heart of the oil world but are rarely described in the history of the industry such as Tony Wade, Stefan Rogocki, Dave Musson, Eddie Horrigan, Colin McMullen, John Pickering, Kenny Cunningham, Ron Wood, Trevor Garlick, Sir John Cadogan and Eddie Marnell. A key impulse for this book came from a meeting with former refinery workers on the Thames Estuary who made us feel this story had never been – but must be – told.

To all those men and women in the industry who generally prefer to keep out of the limelight such as David Jamison, Vivienne Cox and Mike Wagstaff. This story could not have been told properly without their involvement and that of better-known industry figures such as Lord John Browne, Ben van Beurden, Jake Molloy, Prof Alex Kemp, Jeremy Bentham, Mark Lyon and Nick Butler who gave generously of their time. All offered invaluable insights and personal testimony of life at the centre of a business that remains famously opaque.

The romance of this industry was lit up by meetings with songwriters such as Wilko Johnson, Andy McCluskey, Jon King, Dave Randall and film maker Peter Pickering, all of whom had themselves been inspired by oil's strange mechanical and political processes.

Thanks also to those we interviewed who have been involved as critics of the industry and as ecological campaigners, including Lazarus Tamana, Suzanne Dhaliwal, Gail Bradbrook, Elsie Luna, Greg Muttitt, Louise Rouse and Jeremy Leggett.

Thanks to those academics who agreed to be interviewed including Prof Paul Stevens, Prof Helen Thompson, James Bamberg and Tim Mitchell. All those who are politicians and in the realm of politics including Lord Michael Heseltine, John Selwyn Gummer Lord Deben, Clive Lewis MP, Rebecca Long-Bailey MP, Adam Vaughan and Paul Potts. Those in the world of finance and Big Data including Mark Campanale, Nick Robins and David Hockin. And those in the coming world of renewables, including Mark Shorrock, Bridgit Hartland-Johnson, Danielle Lane, Kevin Jones and Natalie Ghazi.

We would also like to thank all those who kindly agreed to be interviewed and yet we could not find a way to wind their words into the story. Their thinking helped shape the text. These include David Miller, Pauline Murray, Joe Corré, Grace Petrie, Billy Bragg, Boff Whalley, Jonty Colchester, Brian Madderson, Matthew Wright, Justin Monaghan, Clementine Cowton, Owen Thorn, Paul

Stott, Tony Carty and Demi Donnelly. And those industry insiders who gave so much assistance but prefer to remain anonymous.

We'd like to thank all those who helped us so much with our journeys around Britain and who provided places to write including Mike Innes, Carolyn Scott, Jane Marriott, Charlie Marriott, Llewela Gibbons, Rory Gibbons, Lena Šimić, Gary Anderson, Tim Jeeves, Zoë Svendsen, Ken Dunlop, Fran Crowe and Bill Parker.

Enormous thanks to all with whom we have discussed the ideas around the book over the last five years and gave us so much, including Greg Muttitt, Kolya Abramsky, Prof Gavin Bridge, Dr Nana de Graaff, Dr Gisa Weszkalnys, Alex Dodge, Andy Rowell, Dan Gretton, Nick Robins, Simon Pirani, Gareth Evans, John Jordan, Stuart Weir, Adam Ramsey, Simon Armstrong, James Meek, Farzana Khan, Ha Joon Chang, Aditya Chakrabortty, Luke Harding, Anne Robbins and Soft Wax.

All those we drew inspiration and advice from including Mel Evans, Charlie Kronick, Chris Garrard, Mae Hank, Mika Minio-Paluello, Anna Galkina, Jess Worth, Emma Hughes, Suzanne Dhaliwal, Paul Horsman, Mark Brown, Jess Worth, Danny Chivers and Jim Footner.

All those who assisted us in making links including Sally Donaldson, Andy Norman, Alice Nutter, Jo Pickering, Zoë Howe and Chris Mitchell.

Especial thanks to those who have been assisting us in expanding the ideas of Crude Britannia into other forms beyond the book, including Emma Davie, Sonja Henrici, Courtney Mulvay, Laura Wadha, Julian Schwanitz, Lindsay Poulton, Sue Jones and Colette Bailey.

Thanks to several guiding writers and thinkers whose work greatly assisted us including Prof Doreen Massey, Tim Mitchell, Naomi Klein, Paul Mason, James Bamberg, Prof Andrew Barry, Greg Muttitt and Simon Pirani.

We are thankful to all the songwriters and their lyric license holders, notably Leah Mack at Sony/ATV, who allowed us to publish songs free or at relatively minimal cost.

The British oil industry of the past century has been almost entirely the domain of white men. However, the communities impacted have often been those of people of colour, and those who have led the way in questioning the plans of the industry have often been women. We have tried to represent that reality. The paths out of Crude Britannia, as we try to tackle climate change, have often been made by women, and Part III of the book reflects this. We have tried to bear witness to the brave pioneering work of several of them including Greta Thunberg, Gail Bradbrook, Rebecca Long-Bailey, Helen Thompson, Danielle Lane, Lena Šimić, and ten-year-old Elsie Luna.

Crude Britannia would never have got off the drawing board without the initial and ongoing enthusiasm and support from other writers, particularly Nick Robins, Greg Muttitt, Stuart Weir, Aditya Chakrabortty and Luke Harding.

Especial thanks to David Shulman, our commissioning editor, who never erred from his valuable belief we had something important to say. We thank him and all the other Pluto Press editors and production staff who fed the project but kept our feet to the fire on deadlines. We are very grateful to all those at other publishing houses who helped us hone the book in outline such as George Owers and Leo Hollis, and especially to Tom Penn for all his patient advice. And likewise to Matthew Hamilton, who acted as agent to the project for several years.

Great gratitude is due to Madeleine Bunting who gave so generously with her time assisting us in editing.

And our gratitude also to Sally Fenn, Simone Brenneis, Kolya Abramsky, Andy Rowell, Prof Gavin Bridge, Farzana Khan, Jane Trowell, Chris Mitchell, Aditya Chakrabortty, Richard Fredman, Nick Robins, Robert Noyes, Rhianon Trowell, Emma Hughes and others who read drafts of the text and outlines along the way and gave us invaluable feedback.

There are numerous others who contributed with either advice, help, accommodation and crucially friendship, around the book.

For Terry these include: John Naughton, Meg Westbury, Senan Clifford, Helen Alder, Richard Brown, Richard Wachman, Phillip Inman, Rupert Neate, and John O' Sullivan plus Liz Macalister, the CCC and Bookishboys, Also, special thanks to Winnie Brenneis MD, plus NHS staff at Addenbrookes for putting Terry back together following a serious bike crash. Closer to home it was Callum and his brother Rory Fenn Macalister who provided the love, laughs and invaluable support for *Crude Britannia*. And Sally Fenn who acted as inspiration, multi-task editor and all-round cheerleader on a project that threatened on occasions to overwhelm us.

For James these include: the support of Platform companions who made the book possible including Emma Hughes, Gaby Jeliazkov, Farzana Khan, Sarah Legge, Ben Lennon, Adam Ma'anit, Rowan Mataram, Mika Minio-Paluello, Jo Ram, Laurie Mompelat, Robert Noyes, Mark Roberts, Sakina Sheikh, Sarah Shoraka, Jane Trowell, Kennedy Walker and Rose Ziaei. James's work comes out of the collective understanding of all at Platform and the legacy of those who built it including Greg Muttitt, Ben Amunwa, Mel Evans, Kevin Smith, Ben Diss, John Jordan, Suzanne Dhaliwal, Charlotte Leonard, Ewa Jasiewicz, Lorne Stockman, Emma Sangster, Emma McFarland, Nick McCarthy and Dan Gretton, and those who helped from outside especially Irene Gerlach, Tim Fairs, Andy Rowell, Cindy Baxter, Charlie Kronick, Nick Hildyard, Sokari Douglas-Camp, Ken Wiwa Jr, Celestine Akpobori, and Steve Kretzman. And finally, James is grateful for the immense support and care of beloved who have listened with such patience and given so much encouragement including Greg Muttitt, Nick Robins, Dan Gretton, John Jordan, Lena Šimić, Mark Brown, Ginny Farman, John de Falbe, Dany Steinert and Fi Spirals. And most especially Jane Trowell who has carried and encouraged this work on a daily basis for so many years.

COPYRIGHT ACKNOWLEDGEMENTS

Index

'A compelling read of post-war Britain's inconvenient histories – the manipulation, corporate smash and grab, and the boom and bust of an oil economy. Picking over the bitter aftermath, Marriott and Macalister take on the roles of sleuth, archaeologist, and witness to tell a story of oil, money and politics which changed millions of people's lives.'

—Madeleine Bunting, writer

'A marvellously rich account of how the oil industry has come to shape contemporary Britain, and how, as now an arm of international finance, it continues to influence government policy and policy makers even as its future looks increasingly uncertain.'

—David Beetham, Emeritus Professor of Politics, Leeds University

'A poignant and wonderfully crafted journey that connects the oil industries and global capitalism with local stories, carrying the reader gently towards holding the context of the world that we live in and the many hidden stories and gems that have brought us to this moment. The authors are thoughtful and conscientious storytellers guiding us through this journey.'

—Farzana Khan, writer and Executive Director of Healing Justice London

'A vivid and revealing portrait of how oil has shaped British society. Through a kaleidoscope of events and places, Marriott and Macalister introduce us to some of the landscapes and companies that created British "petro-culture" and the people now challenging it. Told with passion and wit, *Crude Britannia* is a brilliantly original account of oil's lasting national imprint. At a moment of transformation, with the oil sector arguably at its most precarious for some time, it shows how we got here and how change may come.'

—Professor Gavin Bridge, Durham University

'As a former oil geologist who worked offshore during the heyday of the North Sea oil boom, I was transported back to those times. This book beautifully captures the mood and spirit of the time, and with a forensic approach it unravels the various political and financial events that took place between the UK government and the oil companies.'

—Tim Fairs (former oil geologist), BSc MSc CGeol FGS

Crude Britannia

'A truly remarkable book of deep scholarship and great eloquence. The authors offer a unique insight into Britain's role and experiences in an oil-addicted world. Their book is both very human and almost brutally clinical in its portrayal of the toxic legacy of "carboniferous capitalism" to future generations.'
                    —Herbert Girardet, executive committee member, Club of Rome

'A vivid, compelling and very human account of how big oil has infiltrated our lives, the people it's enriched and those it's abandoned, and the commitment of those determined to see its end.'
                                                              —Caroline Lucas MP

'Superbly illustrates how the UK's toxic relationship with oil has defined our politics, our lives, and our culture. An engrossing read.'
                                                        —Jon King, Gang of Four

'The authors have taken a subject often given arid treatment and turned it into the stuff of our lives. In a book dripping with delicious detail, they show how oil affects so much of today's Britain.'
                    —Aditya Chakrabortty, Senior Economics Commentator, *Guardian*

'Tells you all you need to know about oil's part in the industrialisation and deindustrialisation of Britain – how lives were built, how they were destroyed and how we now need to urgently build a green, just and sustainable economy.'
                                                        —Rebecca Long-Bailey MP